NATIONAL ACADEMIES Sciences Engineering Medicine

NATIONAL ACADEMIES PRESS
Washington, DC

Municipal Solid Waste Recycling in the United States

Analysis of Current and Alternative Approaches

Committee on Costs and Approaches for Municipal Solid Waste Recycling Programs

Board on Chemical Sciences and Technology

Board on Environmental Studies and Toxicology

Board on Science, Technology, and Economic Policy

Board on Life Sciences

Division on Earth and Life Studies

Policy and Global Affairs Division

Consensus Study Report

NATIONAL ACADEMIES PRESS 500 Fifth Street, NW Washington, DC 20001

This project was supported by a contract between the National Academy of Sciences and U.S. Environmental Protection Agency (Contract No. AWD-002044). Any opinions, findings, conclusions, or recommendations expressed in this publication do not necessarily reflect the views of any organization or agency that provided support for the project.

International Standard Book Number-13: 978-0-309-72701-3
Digital Object Identifier: https://doi.org/10.17226/27978
Library of Congress Control Number: 2025941949

This publication is available from the National Academies Press, 500 Fifth Street, NW, Keck 360, Washington, DC 20001; (800) 624-6242; https://nap.nationalacademies.org.

The manufacturer's authorized representative in the European Union for product safety is Authorised Rep Compliance Ltd., Ground Floor, 71 Lower Baggot Street, Dublin D02 P593 Ireland; www.arccompliance.com.

Copyright 2025 by the National Academy of Sciences. National Academies of Sciences, Engineering, and Medicine and National Academies Press and the graphical logos for each are all trademarks of the National Academy of Sciences. All rights reserved.

Printed in the United States of America.

Suggested citation: National Academies of Sciences, Engineering, and Medicine. 2025. *Municipal Solid Waste Recycling in the United States: Analysis of Current and Alternative Approaches*. Washington, DC: National Academies Press. https://doi.org/10.17226/27978.

The **National Academy of Sciences** was established in 1863 by an Act of Congress, signed by President Lincoln, as a private, nongovernmental institution to advise the nation on issues related to science and technology. Members are elected by their peers for outstanding contributions to research. Dr. Marcia McNutt is president.

The **National Academy of Engineering** was established in 1964 under the charter of the National Academy of Sciences to bring the practices of engineering to advising the nation. Members are elected by their peers for extraordinary contributions to engineering. Dr. John L. Anderson is president.

The **National Academy of Medicine** (formerly the Institute of Medicine) was established in 1970 under the charter of the National Academy of Sciences to advise the nation on medical and health issues. Members are elected by their peers for distinguished contributions to medicine and health. Dr. Victor J. Dzau is president.

The three Academies work together as the **National Academies of Sciences, Engineering, and Medicine** to provide independent, objective analysis and advice to the nation and conduct other activities to solve complex problems and inform public policy decisions. The National Academies also encourage education and research, recognize outstanding contributions to knowledge, and increase public understanding in matters of science, engineering, and medicine.

Learn more about the National Academies of Sciences, Engineering, and Medicine at **www.nationalacademies.org**.

Consensus Study Reports published by the National Academies of Sciences, Engineering, and Medicine document the evidence-based consensus on the study's statement of task by an authoring committee of experts. Reports typically include findings, conclusions, and recommendations based on information gathered by the committee and the committee's deliberations. Each report has been subjected to a rigorous and independent peer-review process and it represents the position of the National Academies on the statement of task.

Proceedings published by the National Academies of Sciences, Engineering, and Medicine chronicle the presentations and discussions at a workshop, symposium, or other event convened by the National Academies. The statements and opinions contained in proceedings are those of the participants and are not endorsed by other participants, the planning committee, or the National Academies.

Rapid Expert Consultations published by the National Academies of Sciences, Engineering, and Medicine are authored by subject-matter experts on narrowly focused topics that can be supported by a body of evidence. The discussions contained in rapid expert consultations are considered those of the authors and do not contain policy recommendations. Rapid expert consultations are reviewed by the institution before release.

For information about other products and activities of the National Academies, please visit www.nationalacademies.org/about/whatwedo.

COMMITTEE ON COSTS AND APPROACHES FOR MUNICIPAL SOLID WASTE RECYCLING PROGRAMS

DON FULLERTON (*Co-Chair*), University of California, Santa Barbara; University of Illinois Urbana-Champaign (*Emeritus*)
DEBRA REINHART (*Co-Chair*), University of Central Florida
MALAK ANSHASSI, Florida Polytechnic University
FATIMA HAFSA, World Bank
TRACY HORST, Choctaw Nation of Oklahoma
DEREK KELLENBERG, University of Montana
MARQUISE McGRAW, American University (*through June 2024*)
JEREMY O'BRIEN, Solid Waste Association of North America (*retired*)
SUSAN ROBINSON, Waste Management (*retired*)
HILARY SIGMAN, Rutgers University
MITCHELL J. SMALL, Carnegie Mellon University (*Emeritus*)
REBECCA TAYLOR, University of Illinois Urbana-Champaign
SOFIA B. VILLAS-BOAS, University of California, Berkeley

Study Staff

ANDREW BREMER, Study Director (*through March 2025*)
LYLY LUHACHACK, Study Director (*from March 2025*)
NATALIE ARMSTRONG, Program Officer
KAVITA BERGER, Board Director, Board on Life Sciences
GAIL COHEN, Senior Board Director, Board on Science, Technology, and Economic Policy
ANTHONY DePINTO, Program Officer
CLIFFORD S. DUKE, Board Director, Board on Environmental Studies and Toxicology (*through May 2025*)
CHARLES FERGUSON, Senior Board Director, Board on Chemical Sciences and Technology
THOMASINA LYLES, Senior Program Assistant
CHRISTL SAUNDERS, Program Coordinator
LIANA VACCARI, Program Officer
RAY WASSEL, Scholar (*through June 2024*)

Sponsor

U.S. ENVIRONMENTAL PROTECTION AGENCY

NOTE: See Appendix B, Disclosure of Unavoidable Conflict of Interest.

Reviewers

This Consensus Study Report was reviewed in draft form by individuals chosen for their diverse perspectives and technical expertise. The purpose of this independent review is to provide candid and critical comments that will assist the National Academies of Sciences, Engineering, and Medicine in making each published report as sound as possible and to ensure that it meets the institutional standards for quality, objectivity, evidence, and responsiveness to the study charge. The review comments and draft manuscript remain confidential to protect the integrity of the deliberative process.

We thank the following individuals for their review of this report:

BOBBI ANNE BARNOWSKY, Tribal Solid Waste Advisory Network
NATALIE BETTS, Recycled Materials Association
JANIE CHERMAK, University of New Mexico
ANNE GERMAIN, National Waste & Recycling Association
JOEL HUBER, Duke University
RICHARD McHALE, Austin Resource Recovery
CHAZ MILLER, Miller Associates
SCOTT MOUW, The Recycling Partnership
TIMOTHEE ROUX, ExxonMobil
KATHLEEN SEGERSON (NAS), University of Connecticut
MICHAEL TIMPANE, Resource Recycling Systems

Although the reviewers listed above provided many constructive comments and suggestions, they were not asked to endorse the conclusions or recommendations of this report nor did they see the final draft before its release. The review of this report was overseen by **DANNY REIBLE (NAE),** Texas Tech University, and **DAVID DZOMBAK (NAE),** Carnegie Mellon University. They were responsible for making certain that an independent examination of this report was carried out in accordance with the standards of the National Academies and that all review comments were carefully considered. Responsibility for the final content rests entirely with the authoring committee and the National Academies.

Acronyms and Abbreviations

AI	artificial intelligence
ARL	Australasian Recycling Label
C&D	construction and demolition
CAA	Circular Action Alliance
CO_2	carbon dioxide
CVM	contingent valuation method
DRS	deposit-return system
EIO-LCA	economic input-output life cycle assessment
EPA	U.S. Environmental Protection Agency
EPR	extended producer responsibility
EPSR	European Pillar of Social Rights
GDP	gross domestic product
HDPE	high-density polyethylene
ISO	International Organization for Standardization
LCA	life cycle assessment
LDPE	low-density polyethylene
MRF	materials recovery facility
MSW	municipal solid waste
OECD	Organisation for Economic Co-operation and Development
PCR	postconsumer recycling
PET	polyethylene terephthalate
PFAS	perfluoroalkyl and polyfluoroalkyl substances
PMC	private marginal cost
PPW	plastic packaging waste
PRO	producer responsibility organization
PRO Europe	Packaging Recovery Organization Europe
SDG	Sustainable Development Goals
USD	U.S. dollars
WTE	waste-to-energy

Contents

SUMMARY 1

1 INTRODUCTION 12
    1.1 The Composition and Regulation of Solid Waste, 12
    1.2 The Importance of Recycling, 13
    1.3 MSW Recycling Costs and the Role of Public Policy, 14
    1.4 Study Scope and Approach, 15
    1.5 Organization of the Report, 17
    References, 17

2 MUNICIPAL SOLID WASTE MANAGEMENT AND RECYCLING SYSTEMS 18
    2.1 Today's Recycling Systems, 18
    2.2 MSW Management, 29
    2.3 Technological Advances in MSW Recycling, 37
    2.4 Key Policy Options and Recommendations, 40
    References, 44

**3 BENEFITS AND CHALLENGES OF RECYCLING PROGRAMS AND THE
ROLE OF POLICY** 49
    3.1 What Problems Should Recycling Policies Fix?, 49
    3.2 The Heterogeneity of Recycling, 54
    3.3 Policy Approaches That Respond to Heterogeneity, 56
    3.4 Conclusions for Recycling Programs and Policy Choices, 61
    References, 62

4 DIRECT COSTS AND FINANCING OF RECYCLING PROGRAMS 64
    4.1 Curbside Recycling Programs, 64
    4.2 Traditional Financing Approaches, 71
    4.3 Evaluating Traditional Financing for Recycling, 73
    4.4 Alternative Financing: Recycling Programs, 75
    4.5 Key Policy Options and Recommendations, 88
    References, 94

5 MATERIALS AND MARKETS 99
    5.1 Recycling Rates, 99
    5.2 Demand for Recyclable Materials: End Markets, 109
    5.3 Global, Regional and Local Recycling Markets, 116
    5.4 Public Policies for End Markets, 125
    5.5 Key Policy Options, 128
    References, 133

6 BEHAVIORAL CONSIDERATIONS AND SOCIAL IMPACTS OF RECYCLING PROGRAMS 138
    6.1 Program Availability, 138
    6.2 Attitudes Toward and Barriers to Recycling, 142
    6.3 Strategies for Promoting Behavior Changes, 153
    6.4 Overview of Social Impacts, 162
    6.5 Community Impacts, 165
    6.6 Pursuing Fairness in Access to Benefits of Recycling, 168
    6.7 Key Policy Options, 171
    References, 175

**7   BENEFITS AND MEASURING ENVIRONMENTAL IMPACTS AND
     EXTERNALITIES OF RECYCLING PROGRAMS** 183
     7.1 Key Recycling Benefits, 183
     7.2 Recycling Rates: Current Sustainability Metrics, 185
     7.3 Life Cycle Assessment, 186
     7.4 Reported Environmental Footprints for Recycling Materials, 192
     7.5 Sustainability Materials Management, 193
     7.6 Reducing Environmental Footprints, 194
     References, 196

APPENDIX A: COMMITTEE MEMBER BIOGRAPHICAL SKETCHES 201

APPENDIX B: DISCLOSURE OF UNAVOIDABLE CONFLICT OF INTEREST 205

APPENDIX C: PUBLIC MEETING AGENDAS 206

Boxes, Figures, and Tables

BOXES

1-1   Statement of Task, 15

2-1   Lithium-Ion Battery Fires, 27
2-2   Case Study: Evaluating Selected Waste Systems in New York State, 32
2-3   Case Study: Assessing Recycling Programs in Florida, 33
2-4   ReMADE Institute, 37
2-5   Case Study: Rumpke Recycling & Resource Center, 39

3-1   Is Recycling Worth It?: A Formula for Assessing Net Social Costs, 57
3-2   Case Study: Recycling Glass Jars or Bottles in Fargo, 57
3-3   Optimal Incentives via Taxes or Deposit Return System, 59

4-1   Case Study: Incorporating Social and Environmental Costs in Oregon, 70
4-2   Waste and Recycling Streams from Renewable Energy Technologies, 71
4-3   U.S. Environmental Protection Agency Grant Programs, 72
4-4   EPA National Strategy to Prevent Plastic Pollution: EPR Framework, 78
4-5   Case Studies: State-Level Packaging and Recycled Content Legislation, 78
4-6   Case Study: Colorado's Extended Producer Responsibility Policy, 80
4-7   Case Study: Oregon's EPR Policy, 81
4-8   Case Studies: State-Specific Needs Assessment Laws, 81
4-9   California's DRS Programs: A Redemption Value Case Study, 83
4-10 States with Landfill Tipping Surcharges, 91

5-1   Recycling and Composting Accountability Act, 100
5-2   Case Study: Recycling Excellence in the United States, 103
5-3   Case Study: Navigating Recycling Challenges for Toothpaste Containers, 107
5-4   Closed- and Open-Loop Recycling, 114
5-5   Case Studies: State Recycling Market Development Programs, 124
5-6   Case Studies: Washington's NextCycle Program, 125
5-7   Case Study: UK Plastic Packaging Tax, 128
5-8   Superfund Excise Tax on Hazardous Substances, 129

6-1   Opposition and Support for Contingent Valuation and Willingness to Pay, 152
6-2   Social Feedback Case Study, 157
6-3   Case Study: Bin-Tagging as Feedback in Seattle, 157
6-4   Case Study: Seattle's Online Recycling Search Tools, 160
6-5   Case Study: Quality Disconnect Between MRFs and Wisconsin End Users, 167
6-6   Case Study: The Minnesota-Based Wood From the Hood, 167
6-7   Case Study: Community-Centered Recycling in Phoenix, 168

FIGURES

S-1   A simplified system diagram for municipal solid waste management systems primarily highlighting processes and materials discussed in this report, 2
S-2   Major technological or behavioral decision stages and influences affecting recyclability of solid waste, 11

2-1   A simplified system diagram for municipal solid waste management systems primarily highlighting processes and materials discussed in this report, 19
2-2   Key stages and actors in the U.S. recycling system, 21

xiii

Boxes, Figures, and Tables

2-3	Major technological or behavioral decision stages and influences affecting recyclability of solid waste, 22
2-4	Hub-and-spoke recycling model, 28
2-5	Environmental return on investment associated with the effects of keeping a recycling program (Contract A and B), 33
2-6	State-by-state residential recycling rates, 34
2-7	State-by-state residential recyclable material lost in tons per year, 35
2-8	Relationship of state-by-state annual recyclable materials lost per capita versus residential recycling rate, 36
2-9	Examples of automation sorting technologies, 38
4-1	Processing costs per ton for materials recovery facilities in the Northeast, 2019–2024, 66
4-2	Revenues per ton recyclables for materials recovery facilities in the Northeast, 2019–2024, 67
4-3	Estimated annual waste management costs for 2021, 2020, and 2011 by region, 68
4-4	Bottle bill fact sheet, 84
4-5	Comparison of redemption rates for deposit-return systems by deposit level, 87
5-1	Recycling tonnages in the United States (1960–2018), 101
5-2	Recycling rates for five categories of materials in the United States, 1970–2015, 102
5-3	Trends of U.S. municipal solid waste (MSW) expenditures by gross domestic product (GDP), 1970–2015, 103
5-4	Plastics waste composition in the United States (percent by weight), for each material type and resin code, 106
5-5	Virgin and recycled plastic prices in 2023 and 2024, 111
5-6	Historical recycled commodity prices, 112
5-7	Patents for recycling by material in countries in the Organisation for Economic Cooperation and Development, 2000–2021, 115
5-8	Legislation related to chemical recycling, 116
5-9	Recyclable waste and scrap materials (four types) traded worldwide, in thousands of metric tons, 2012–2023, 120
5-10	U.S. waste and scrap paperboard imports and exports, in thousands of metric tons, 2012–2023, 120
5-11	U.S. waste and scrap aluminum imports and exports, in thousands of metric tons, 2012–2023, 121
5-12	U.S. waste and scrap plastics imports and exports, in thousands of metric tons, 2012–2023, 121
5-13	U.S. exports of plastic waste and scrap to OECD and non-OECD countries, in thousands of metric tons, 2012–2023, 122
5-14	U.S. imports of plastic waste and scrap from OECD and non-OECD countries, in thousands of metric tons, 2012–2023, 122
5-15	Conceptual map of the proposed fee-and-reward program, 131
6-1	Recyclable materials disposal channels, 139
6-2	State-by-state levels of recycling access and participation, 141
6-3	Fate of materials in residential recycling, tons per year, 141
6-4	Number of U.S. households with food waste collection, 2005–2023, 142
6-5	Food waste collection methods, 142
6-6	Residential food waste collection programs by state, 143
6-7	Importance of recycling, 144
6-8	Barriers to recycling more household plastic waste, 145
6-9	Support for regulations, 147
6-10	Local officials' agreement with statements regarding recycling benefits, 150
6-11	Sample intervention graphic: Comparative social feedback versus non-comparative feedback, 157
6-12	U.S. recycling laws, 158
6-13	Screenshot of the online tool offered by Seattle Public Utilities, 160
7-1	Life stages included in the sustainable materials management framework, 194

Boxes, Figures, and Tables

TABLES

S-1		Summarized List of Data Needs and Their Uses for Recycling Approaches, 5
2-1		MSW Recycling in Residential and Commercial/Multifamily Sectors, 30
2-2		Estimated Increases in Greenhouse Gas Emissions Based on Discontinuing Recycling Programs in Florida, 33
2-3		Summarized List of Data Needs and Their Uses for Recycling Approaches, 43
3-1		Examples of the Wide Variation Across Components of Cost for Waste and Recycling, 55
3-2		National Average Prices per Ton, 56
3-3		Comparing Pigovian Tax Rates with Deposit-Return System Costs, 59
4-1		Costs of Curbside Recycling and Residential Refuse Collection in North Carolina, 2023, 69
4-2		Colorado Needs Assessment—Estimated Recycling Outcomes per Scenario, 80
4-3		Colorado Needs Assessment—Estimated Costs per Scenario, 80
4-4		Deposit-Return Programs in the United States, 82
5-1		Recycling Tonnages in the United States, 1960–2018, 102
5-2		Examples of Recycling Excellence in the United States, 103
5-3		Resin Identification Codes, 105
5-4		End Uses and Prices for Recycled Commodities, 110
5-5		Key Definitions of Recycling Technologies, 115
6-1		Time Costs of Recycling Under Different Scenarios, 145
6-2		Five Types of Recyclers, 148
6-3		Literature on Willingness-to-Pay Estimates, 151
6-4		Social Dimensions in Waste Management, 163
6-5		Potential Research Questions for Assessing the Recycling Value and Its Social Impact, 166
6-6		How2Recycle Labels, 173
7-1		Produce and Waste Life Cycle Assessment (LCA) Models, 187
7-2		Key Input Assumptions and Their Defaults for Each Assessment Model, 190
7-3		Environmental Footprints for Recycling Materials, 192
7-4		Political Feasibility of Each Scenario, Defined by Evaluating Similar Programs and Policy Challenges, 195

Preface

Municipal solid waste (MSW) recycling programs play a crucial role in the U.S. waste management system, aiding in the reduction of waste disposed and the conservation of resources. These programs also drive economic activity and provide other social and environmental benefits. While most contemporary MSW recycling activities are local programs by nature, they are also impacted by policies and economic forces at the state, national, and international levels.

These programs face a multitude of challenges today that complicate their stability, efficacy, and economic efficiency. Some of these are historic challenges for the recycling industry—for example, contamination in the recycling material stream and securing suitable end markets for recycled materials. Meanwhile, new challenges have arisen over the past few years—for example, adapting to the impacts of the COVID-19 pandemic on the recycling industry and a changing international policy landscape that have placed additional limits on end markets for recycled materials.

Considering these challenges, the task of this committee was to assess the costs of MSW recycling programs in various contexts and to identify policy options to facilitate the effective implementation of these programs. As this report details, public policy is an invaluable tool in addressing these challenges and in shepherding solutions to the various challenges faced by MSW recycling programs. Well-designed policy at different levels of government can ultimately support the communities that administer recycling programs and can unlock the benefits they provide.

This report explains several major findings. First, many of the environmental benefits of local recycling spill over jurisdictional boundaries to the rest of the state and nation. No local government can be expected to shoulder the entire cost of providing these diffuse benefits, so they need additional financial help. Second, this financial help can be in the form of targeted incentives such as grant-making or policies that shift costs to producers to do the right kinds of recycling. However, beyond financial challenges, consumer behavior must also be modified to overcome recycling challenges, in some cases simply by providing more and better information to consumers and businesses. And third, a significant challenge in overcoming obstacles to effective MSW recycling programs is the heterogeneity across the industry. These differences include local material streams, local economies, proximity to end markets, existing infrastructure, geography, cultures and norms, and more. These forms of heterogeneity all impact outcomes of decision-making about options related to administering effective MSW recycling programs. Still, the committee has identified policy options and several recommendations to support these programs at all levels moving forward.

This report is the result of the committee's deliberations. It describes the committee's consensus conclusions and recommendations, and it identifies key policy options based on extensive information gathering, committee discussions, and input from a variety of experts who work in the MSW recycling industry. The committee met twice in-person over the course of its study and held several additional virtual information gathering meetings in 2024.

We thank the U.S. Congress for requesting this report, because it recognizes both the importance of recycling programs in meeting many of the waste management goals our nation has set as well as the need to identify and deploy novel policy solutions to support them. We also thank the U.S. Environmental Protection Agency, which sponsored the study.

The work of this committee was supported by several National Academies staff members. We are particularly grateful to Lyly Luhachack, Andrew Bremer, Thomasina Lyles, Natalie Armstrong, Anthony DePinto, Liana Vaccari, and Ray Wassel for their efforts.

Finally, we thank our fellow committee members for their hard work and contributions to the development of this report, and for offering their insight and expertise in our many discussions. Especially important is the cross-disciplinary nature of the committee—which was composed of engineers, economists, and industry experts in MSW recycling programs. We enjoyed the many fruitful discussions, and we are grateful for the opportunity to engage with and learn from our colleagues. They all enhanced the outcomes of this study.

Don Fullerton, *Co-chair*
Debra Reinhart, *Co-chair*
Committee on Costs and Approaches for
Municipal Solid Waste Recycling Programs
June 2025

Summary

The municipal solid waste (MSW) management system in the United States is a complex and distributed system that has evolved over time in response to policies, materials managed, infrastructure and technology, and actors at local, regional, state, national, and international levels. MSW is generally defined as the non-hazardous solid waste generated by the residents, commercial businesses, and institutions of a community. It excludes industrial and construction waste. In 2018, the United States generated approximately 292 million tons of MSW annually, most of which (about 68 percent) were not recycled or composted.[1] Other studies have found that as much as 79 percent of recyclable material in the MSW stream is not actually recycled, because of factors such as the material not being targeted for recycling, not being economically viable to recycle, limited access to recycling programs, and low participation rates.

Figure S-1 depicts a generalized process for a MSW management system with consideration of the actors and materials involved. This includes subsystems for (1) production and manufacturing, (2) waste generation, (3) waste collection, and (4) sorting and processing. The collection, processing, and marketing of recyclables are interdependent components, and each must be considered when designing and operating a recycling system since they affect one another. For example, changes in collection methods and materials collected will impact the design and operation of the materials recovery facility (MRF); how the MRF is designed and operated will determine whether materials will be produced that meet market specifications; and changes in market requirements may lead to changes in how materials are collected and processed.

In support of improving recycling outcomes, this study, authorized by the Consolidated Appropriations Act of 2022, reviews available information on programmatic and economic costs of MSW recycling programs in municipal, county, state, and tribal governments and provides advice on potential policy options for effective implementation. The focus of this work was on publicly accessible data on the policies and systems in place that relate to the collection, sorting, processing, transport, and sale of recyclable materials, especially those that are traditionally processed in MRFs. Additionally, the report presents several case studies to illustrate how local conditions impact the design, cost, and effectiveness of MSW recycling programs and provide context to inform policy.

Benefits and Challenges of Recycling Programs and the Role of Policy

Recycling involves choices by households, businesses, and many levels of government. The costs associated with managing MSW recycling may require public policy at local, state, and national levels to address financing of these systems effectively. While some recycling programs can be sustained through local policy and resources, others face difficulties such as high infrastructure costs and fluctuating commodity values for recyclables. Smaller municipalities, especially in rural areas, may struggle to achieve economies of scale, making recycling programs financially unsustainable without external support. Markets alone do not provide the necessary incentives for households, businesses, or local governments to engage in effective recycling practices. Heterogeneity—including a broad range of differences across costs, benefits, existing capabilities, material volumes, transportation distances, access to end markets, and cultural norms across regions—is a significant factor in recycling programs, complicating the policy needs.

---

[1] This Summary does not include references. Citations for the information presented herein are provided in the main text.

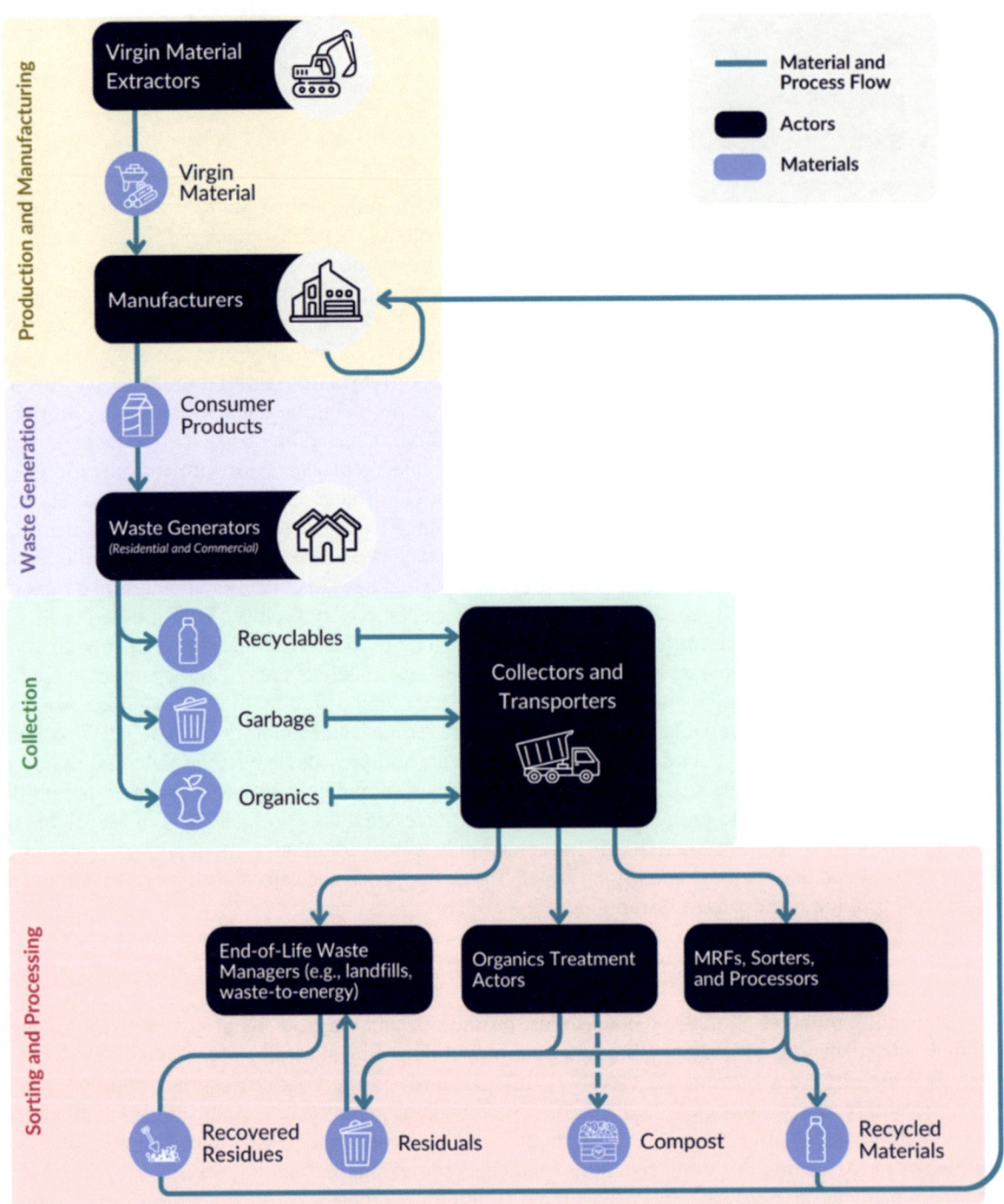

FIGURE S-1 A simplified system diagram for municipal solid waste management systems primarily highlighting processes and materials discussed in this report.
NOTES: Other non-residual organic products (e.g., animal feed, energy) may result from treatment of organics but are not discussed in this report. MRF = materials recovery facilities.

Despite these challenges, well-designed and supported MSW recycling programs hold many benefits. These programs can lead to measurable economic gains and the circular economies they enable create jobs, promote business development, and provide further positive social impacts for communities across the country. In addition to economic gains, associated environmental benefits include reducing use of non-renewable virgin materials, reducing use and extending the service life of landfills, and reducing greenhouse gas emissions.

Local, state, or federal government intervention through public policy can significantly improve the efficiency, affordability, and accessibility of recycling initiatives. Tailored national policies that address regional and local constraints and provide targeted support can enhance the effectiveness and sustainability of recycling programs across the United States. After consideration of available information on MSW recycling programs, the committee determined that it is helpful to identify and articulate the objective(s) that effective recycling policy is designed to achieve.

Conclusion 3-3[2]: Effective recycling policy targets some or all of the following objectives:
1. *Enhance end markets for recyclable materials*
2. *Provide stable financing of recycling systems*
3. *Clarify information for consumers, including what is recyclable, how to recycle, and which products best support recycling goals*
4. *Track and evaluate recycling activities through improved data collection and distribution*
5. *Increase the cost-competitiveness of recycled materials (relative to virgin material inputs) and of recycling (relative to landfilling)*
6. *Improve access to recycling collection and processing*
7. *Increase the cost effectiveness of recycling collection and processing*
8. *Decrease contamination of postconsumer recycling streams*
9. *Enhance social and environmental benefits associated with recycling*
10. *Maintain affordability, without undue burdens on low-income households*

Evaluation of the Effectiveness of Policies and Programs

Recycling goals may be used to identify benchmarks, measure progress, evaluate success, and simplify the communication of a policy or program's purpose to important stakeholders (e.g., constituents and citizens of a community, businesses, company shareholders). The most widely used metric for evaluating recycling progress remains the U.S. Environmental Protection Agency's 2020 "MSW recycling rate," calculated as the total weight of recycled MSW divided by the total weight of generated MSW. The popularity of this metric stems from its simplicity and applicability across states and regions, making it accessible to a range of stakeholders. In general, however, weight-based recycling rates and material-specific rates are incomplete metrics for recycling efficiency because they do not adequately account for changes in packaging material composition, waste reduction efforts, and all costs and benefits of using and reusing materials over their life cycle (e.g., economic, social, environmental costs and benefits). Compared with using only weight-based metrics, a sustainable materials management approach, which includes consideration of weight-based recycling, provides a more complete picture of the costs and benefits of using and reusing materials across their life cycles.

Recommendation 2-1: Goals for recycling policy should expand beyond weight-based recycling rates to include informative metrics for sustainable materials management. To support these efforts, the U.S. Environmental Protection Agency should study how to combine multifaceted sustainability goals into an overall policy framework, provide guidance for state and local governments to set and measure progress toward those goals, and use this information to evaluate progress. National recycling goals should be material specific but flexible to account for heterogeneity across regions and municipalities. These goals should include environmental, social, and economic targets, including cost-effectiveness. Goal-setting should be

---

[2] The conclusions, recommendations, and policy options in this Summary are numbered according to the chapter of the main text in which they appear.

leveraged to design a policy framework and set new national recycling goals using best practices such as life cycle assessment and SMART (specific, measurable, accessible, relevant, and time-bound) metrics.

Recommendation 2-2: The U.S. Environmental Protection Agency (EPA) should enhance data collection and reporting efforts related to municipal solid waste (MSW) and MSW recycling programs to fill significant data gaps, to ensure sufficient and contemporary data are available to inform policy decisions, and to aid in developing and evaluating recycling goals based on sustainable materials management. These efforts should include appropriate input from stakeholders including other federal partners; state, local, and tribal governments; and industry partners. Additionally, EPA's efforts should include:

- **Updating its publicly available website on at least a biennial basis with national-level facts and figures about materials, waste, and recycling. Where possible, this information should expand from input-output modeling figures to include direct observational data. Where necessary, EPA should continue to provide sufficient funding for collecting and reporting these data.**
- **Developing standard definitions of recycling and methodologies on data collection and reporting for recycling and MSW generation. These definitions and collection methodologies should distinguish between pre- and postconsumer recycling and differentiate between open- and closed-loop recycling. This public information should include, at a minimum, material-specific data on MSW generation, recycling, composting and other food and yard waste management, combustion with energy recovery, and landfilling. To the extent possible, these data should also be reported at regional, state, and local levels.**

A summarized list of data needs and their uses are provided in Table S-1. This list is not exhaustive but is representative of the need to improve data availability for decision making.

Key Policy Option 2-1: The U.S. Environmental Protection Agency (EPA) could support studies to update or otherwise fill missing data gaps to ensure sufficient data are available to inform policy decisions on recycling. These include:

- Tracking household time spent by single and multifamily households on recycling to support more complete and accurate estimates of the economic and social costs of recycling and ensure that life cycle assessment models are as current and accurate as possible.
- Regularly collecting and reporting direct observations of household and commercial behavior related to recycling. In addition to filling knowledge gaps, these data would complement top-down modeling in the recycling system and enable empirical study of the impact of public policy. As part of these efforts, EPA could consider a periodic household and commercial survey for waste and recycling akin to the Energy Information Administration's Residential Energy Consumption Survey.

Financing of Recycling Programs

Financing of the recycling system in the United States comes from both private and public sources. Typically, local governments, households, and commercial establishments pay for recycling collection and processing with limited funding from state or federal sources. Some local governments use a "general fund" approach in which recycling does not have a dedicated revenue source and is funded along with other categories of expenditure. Other municipalities rely on an "enterprise fund" approach by collecting fees for recycling (or for recycling and garbage collection together), sometimes as an item on property tax bills or utility bills, or as an explicit charge for businesses. By contrast, businesses typically hire and directly pay for private waste management companies to provide their recycling services rather than relying on government systems. This traditional financial approach includes four considerations: incentives for recycling, cost control, risk management, and distribution of financial burdens.

TABLE S-1 Summarized List of Data Needs and Their Uses for Recycling Approaches

Domain	Data/Units (where applicable)	Purpose and Use of Data	Primary Actors to Collect and Report Data
Product characteristics	• Product recyclability, composition, recycled content • Aggregate producer sales records by NAICS code and region	Ensure that related policies (e.g., EPR/PRO, interventions, product bans) are working, support recyclable labeling, project future material flows	EPA, manufacturers, U.S. Census Bureau, Federal Reserve Board, Bureau of Labor Statistics
Waste generation and composition	• Timely solid waste generation estimates, bin survey results for major composition categories (tons/year)	Help advance the understanding of how the recycling system is performing, estimate level of contamination in recycling streams, complete LCA and LCI, evaluate recycling goals	EPA, states, local governments
MSW recycling systems	• Number and capacity of recycling programs • Capital costs, operating costs, revenue	Evaluate the availability of programs, recycling capacity, level of consumer access and participation, and fiscal stability	States, municipalities, local MRF owners
System costs	• Distribution of system capital costs and operating costs • Performance versus cost histories for reporting MRFs • Consolidated fossil fuel and other virgin material taxes and subsidies	Estimate consumer cost of recycling more completely and accurately	Local government, compiled by each state
State policies and rules	• MSW facility operating rules and reporting requirements • Economic Incentives (taxes/subsidies) • Listing and brief description of state recycling targets, tipping fee surcharges, recyclable content goals, recycling rates, public participation rates • EPR data	Identify the objectives, economic impact, and constraints of government policy	EPA, state, nongovernmental organization, and industry experts on policy and regulation
Technological innovation	Descriptions and inventories of new MRF technologies, sorting technologies for consumers, and new patents	Understand how recycling performance could improve in the future	MSW research and development experts from industry
Environmental impacts and improvements	Input data for LCA and LCI; greenhouse gas emissions and air, water, land pollutants from waste and recycled material transport and processing; exposures and health impact estimates (environmental, economic, and social metric units)	Ensure LCA and LCI are updated and accurate	EPA

continued

TABLE S-1 *continued*

Domain	Data/Units (where applicable)	Purpose and Use of Data	Primary Actors to Collect and Report Data
Inventory of recyclable and recycled materials	Performance metrics, including fraction recycled (tons/day) and other impact-based indices	Improve markets and enable potential buyers and sellers of materials to be matched more easily	Local governments, MRF owners and operators, states
Consumer knowledge and behavior	• Household and establishment survey on waste and recycling • Summaries of survey studies in literature (links to key studies and papers)	Provide regular direct observations of household and commercial behavior; improve the ability to evaluate the empirical impacts of public policies; measure social impacts of recycling, including health, distribution of programs; evaluate true cost and benefits of recycling	EPA - Surveys and bin audits measuring behavior and contamination; local governments
Macroeconomic impacts	• Recycling process data describing inputs to recycling supply chains • Jobs associated with recycling and composting • Commodity values over time (i.e., price of scrap and recycled materials per ton)	Survey recyclers to enable estimates of material flowrates, enabling estimations of material availability; evaluate recycling impacts on economy	States, manufacturers

NOTE: EPA = U.S. Environmental Protection Agency; EPR = extended producer responsibility; NAICS = North American Industry Classification System; LCA = life cycle assessment; LCI = life cycle inventory; MRF = materials recovery facility; MSW = municipal solid waste; PRO = producer responsibility organization.

An emerging financing model, extended producer responsibility (EPR), alleviates financial burden on local governments by shifting residential recycling costs to producers who pay for their share of recycling collection and processing costs. These EPR financing rules differ from the original concept of EPR, which is most directly embodied in an "individual"—or "take-back"—policy, in which manufacturers are required to reclaim their own product packaging and eventually the product itself once it has reached the end of its useful life. Fullerton and Wu's (1998) economic model captures these incentives by demonstrating how market equilibrium—achieved when firms' production choices align with consumers' purchasing and disposal decisions—can drive optimal product design, output, and packaging choices, accounting for external disposal costs. Existing "collective" EPR laws provide financing but do not capture all these individual incentives. They also vary greatly by state, and the economic impact of EPR also varies depending on the scope of the law.

If properly designed, these systems may provide incentives for producers to reduce packaging volumes and increase recyclability of packaging and products.

Recommendation 4-1: The United States should increase reliance on extended producer responsibility (EPR), which should cover packaging and expand to other materials as appropriate. EPR policies should include eco-modulation to create economic incentives for manufacturers to design for recyclability, and funding streams for recycling systems and infrastructure. State governments should enact EPR policies to account for regional heterogeneity but should be supported and informed by a national framework with guidelines.

Summary

Key Policy Option 4-1: The U.S. Environmental Protection Agency (EPA), with appropriate funding and authority from Congress, could develop and facilitate a national extended producer responsibility (EPR) framework, as outlined in its 2024 report *National Strategy to Prevent Plastic Pollution*. If it pursues this framework, EPA should consult with state, local, and tribal governments; nongovernmental organizations; industry; and other relevant partners. This framework should provide guidelines on key elements of state-level EPR policies and recommend minimum state-level standards and best practices. A national framework should provide as much consistency across states as possible and support multistate efforts, while allowing for state-level variation in targets, fees, covered materials, and methods to reflect heterogeneity in costs and benefits across states.

Key Policy Option 4-2: State governments could enact extended producer responsibility (EPR) policies, informed by any minimum standards provided by the U.S. Environmental Protection Agency. State-level needs assessments should identify gaps in current services and programs and serve as a basis for setting EPR fees. Within an EPR framework, state governments could consider policies, such as recycled content standards, to enhance end markets for recyclable materials.

One of the primary objectives of MSW management programs is to make it easy and convenient for residents to recycle. Achieving this objective is a key reason why curbside collection services are provided on a regular basis to single-family residences. While this type of service can be provided more cost-efficiently to residents in urban and suburban communities, curbside collection may be cost-prohibitive for many tribal and rural communities because of their low population density and long distances between households. Alternative funding mechanisms for rural recycling areas include dedicated financing generated from statewide federal grants such as the Solid Waste Infrastructure for Recycling programs, state-level EPR policies that promote recycling programs in rural areas, and landfill tipping fee surcharges.

Conclusion 4-2: State-based landfill tipping fee surcharges can provide a dedicated revenue source to support recycling programs and can provide incentives for waste diversion from landfills (especially recyclable materials and organics). As such, landfill tipping fee surcharges can offset some of the costs of recycling, enhance social and environmental benefits associated with recycling, and provide stable financing for recycling systems.

Key Policy Option 4-3: State governments could implement mandatory surcharges on landfill tipping fees to provide incentives for recycling, support recycling and composting efforts, and divert waste from landfills. Moderate surcharges would minimize harmful responses (e.g., illegal dumping, increased contamination of recycling streams). State governments could collect and redistribute the funds to various recycling activities based on state and local priorities. Local uses of these revenues may vary with needs but could include grants for recycling infrastructure, shoring up enterprise funds for recycling operations, and funding local social modeling programs.

Conclusion 4-3: Relying on local government financing limits access to recycling programs, particularly for residents of rural areas, where recycling costs may be high. Alternative funding mechanisms, such as state or federal grants or EPR programs, would help distribute recycling costs across a broader population.

Key Policy Option 4-4: The U.S. Congress and state legislatures could authorize and appropriate funds for rural and tribal recycling. These funds could help communities overcome transportation distances and economies of scale through purchase of infrastructure such as trucks, drop-off and transfer facilities, and processing facilities. In parallel, the U.S. Environmental Protection Agency could continue to provide Solid Waste Infrastructure for Recycling grants for rural and tribal communities. State government funding could be derived from revenues generated from extended producer responsibility policies, landfilling tipping fee surcharges, or other state-based revenues.

Materials and Markets Considerations for an Effective Recycling System

In the United States, most communities focus on five material types that are collected curbside or at drop-off centers and processed at MRFs: plastics, paper, cardboard, glass, and metals. Less commonly collected in a separate stream are food and yard wastes. Unique or specific challenges and considerations arise for recycling each material type. For example, plastic is a ubiquitous component of today's manufactured items because of its strength, low cost, durability, and wide range in properties with dozens of types of plastic resins in use. In particular, recycling plastics is important because of their persistence in the environment, their generation from non-renewable sources, their contribution to litter problems, and more. However, the overall recycling rates for all plastics are low, partly because only certain resin types are accepted for recycling, as influenced by market demand and technological limitations.

Conclusion 5-1: A revenue-neutral policy that applies a fee for using virgin plastic resins and a reward for using recycled plastic resins would increase the cost-competitiveness of recycled materials relative to virgin inputs and would enhance end markets for recyclable materials.

Key Policy Option 5-1: The U.S. Congress could enact a new revenue-neutral fee-and-reward policy to increase the competitiveness of recycled materials relative to virgin inputs. It would encourage the use of recycled plastic resins in the manufacturing of plastic packaging and single-use products. This policy could comprise two levers:

- First, the Department of the Treasury, in partnership with the U.S. Environmental Protection Agency (EPA), could implement a new fee on the use of virgin plastic resins in product packaging and in the manufacturing of consumer products, and a corresponding reward for the use of postconsumer recycled plastic resins in those same manufacturing processes. If implemented, this new fee and reward should be paid and received by domestic manufacturers that use plastic resins in their manufacturing processes, should be weight-based, and should be of sufficient value to encourage the use of recycled plastic resins. Market parity can facilitate economic competition between recycled plastic resin and virgin resin.
- Second, the Department of the Treasury, in partnership with EPA and U.S. Customs and Border Protection could impose a new border adjustment fee on fully manufactured imported plastic packaging and single-use products, to be paid by the importer of those products.

If pursued, this policy should be revenue-neutral for the federal government, such that the total annual sum of fees collected equals the total annual sum of rewards distributed. Furthermore, the Department of the Treasury, in partnership with EPA and other relevant parties, would need to study and identify the appropriate levels of fees and rewards to fully encourage the use of recycled plastic resins while minimizing motivations for changing manufacturing locations.

End markets play a critical role in sustaining recycling systems, with higher-value materials such as aluminum containers, cardboard, and polyethylene terephthalate (PET) and noncolored high-density polyethylene (HDPE) plastics contributing the most reliable revenues. Sales of recyclable commodities in end markets provide revenues that reduce the expense of recycling for local governments and private parties with revenue-sharing agreements. However, the effect of end markets is not exclusively financial, because end uses also impact the environment as a benefit of recycling. The extent to which recycling improves environmental quality depends on the successful substitution of secondary materials for extraction or production of environmentally damaging primary materials and when these recycled materials can be incorporated into new products without requiring resource-intensive processing. To this end, improving product recyclability and developing new uses for recycled materials with consideration for reducing environmental costs and enhancing end markets is an area for further research. Thus, end markets and programs to support them need to be assessed for both their financial and environmental benefit.

Conclusion 5-2: Advancing research and development in technology areas relevant to recycling and adopting new technologies in the MSW recycling system can help achieve multiple policy objectives for recycling:
- *enhancing end markets for recyclable materials,*
- *increasing the cost-competitiveness of recycled materials relative to virgin inputs,*
- *improving the cost-effectiveness of recycling collection and processing,*
- *decreasing contamination of post-consumer recycling streams, and*
- *enhancing social and environmental benefits associated with recycling.*

Conclusion 5-3: Increased recycling collection may have little benefit without end uses for the collected materials that are environmentally sound and economically valuable. Thus, increased collection needs to be combined with support for end markets, with attention to the environmental implications of end uses.

Key Policy Option 5-2: Federal agencies that fund research related to recycling, including the U.S. Environmental Protection Agency, the Department of Energy, and the National Science Foundation, could enhance investments in research related to recycling systems and recyclable materials. When this option is pursued, the research should prioritize environmentally sound and economically valuable end uses for recycled commodities and other approaches to increase end use values nationally and internationally. Examples include recyclable design for consumer products, and technologies to reduce contamination of the recycling material stream. Funding from Congress to support this endeavor could include public–private partnerships in manufacturing innovation to increase opportunities for recyclable materials end uses.

Consumer and Social Impacts of Recycling Programs

Understanding how and why individuals engage or do not engage in recycling practices is crucial for designing policies that effectively increase participation rates and improve recycling outcomes. Household recycling behavior is shaped by various factors, including the availability and accessibility of recycling programs, convenience, public awareness, and economic incentives. While consumer surveys consistently find high support among respondents for recycling and its programs, they also highlight barriers—mainly a lack of convenience and confusion over what materials can be recycled. Inconsistent and misleading packaging labels, including the use of the chasing arrows symbol and resin identification codes, are significant causes of consumer uncertainty and misunderstanding.

Conclusion 6-1: Reforming product labeling regulations and practices to provide accurate information (i.e., to prevent mislabeling) on what products are or are not recyclable would achieve multiple policy objectives, including clarifying information for consumers, decreasing contamination, and increasing efficiency of recycling systems.

Recommendation 6-1: The Federal Trade Commission (FTC) should revise its *Guides for the Use of Environmental Marketing Claims* so that resin identification codes no longer use the chasing arrows symbol. Additionally, FTC should prohibit use of the chasing arrows symbol or any other indicator of recyclability on products and packaging unless the items are regularly and widely collected and processed for recycling across the United States. Furthermore, with or without a mandate to do so, producers should adopt and use updated resin identification symbols that do not include the chasing arrows symbol.

Key Policy Option 6-1: The U.S. Environmental Protection Agency (EPA), in partnership with producers could support and evaluate national recycling label standards—through education, outreach, and funding—such as the How2Recycle symbols created by the Sustainable Packaging Coalition. Additionally, the U.S. Congress, through EPA, could provide funding for small- to medium-sized companies that lack capability for transitioning to a new national recycling label standard.

Additionally, the presence of social norms and community engagement can further influence participation in recycling efforts. Social modeling programs or locally-organized programs that promote recycling norms and behaviors may facilitate this engagement. Social norms may be used to design communications that address the concerns and values of a target population. Behavioral interventions that promote recycling and decrease contamination are effective when they target a specific barrier to recycling for a given population of consumers. More regular collection and reporting of direct observations of household and commercial behavior related to recycling are needed to support recycling policy decision-making. As one example, new and more rigorously collected data on household time costs are needed to perform recycling cost-benefit analyses more accurately.

Conclusion 6-2: Social modeling programs are effective interventions for enhancing recycling behavior and establishing positive recycling norms in communities. Policies that promote social modeling programs can achieve various objectives for recycling. They can clarify information for consumers, decrease contamination, increase the cost-effectiveness of recycling collection and processing, and enhance the social and environmental benefits associated with recycling.

Recommendation 6-2: The U.S. Environmental Protection Agency should provide grants for state, municipal, local, and tribal governments for enhancing and expanding local social modeling programs, especially in disadvantaged communities and communities with high numbers of multifamily dwellings. Local governments, in turn, should implement or support social modeling programs, potentially through partnership with local nonprofits or other community-based groups, to engage directly with community members to promote positive social norms and recycling practices.

Key Policy Option 6-2: The U.S. Congress could reauthorize and further appropriate funds to the Consumer Recycling Education and Outreach Grant Program, authorized in the Infrastructure Investment and Jobs Act, to support social modeling programs.

Recycling rates vary significantly across the United States, with some cities achieving higher efficiency through mandates, specialized programs, and focused public policies; these cities demonstrate the potential for recycling efficiency to surpass national averages through targeted local actions. Different actors and pressures—including consumer behavior, economics, and available technology—govern the rate of recycling at different decision stages within the MSW system as well as the costs associated (Figure S-2). These represent key areas where policy choices can positively influence recycling rates.

Different contexts and recycling programs require tailored policy solutions, based on such factors as variations in materials, geographies, economies of scale, existing infrastructure and programs, and demographics and other social considerations. Guiding policies from higher levels of government can be appropriate, but it is necessary to consider and tailor policies for recycling based on local factors. Although they serve an important function, today's recycling programs sit at a crossroads. In recent years, challenges facing municipal solid waste recycling programs, especially economic-based challenges, have led some municipalities to stop funding their recycling programs altogether. This report explores the contemporary issues facing MSW recycling programs and lays out recommendations and policy options to chart a path forward.

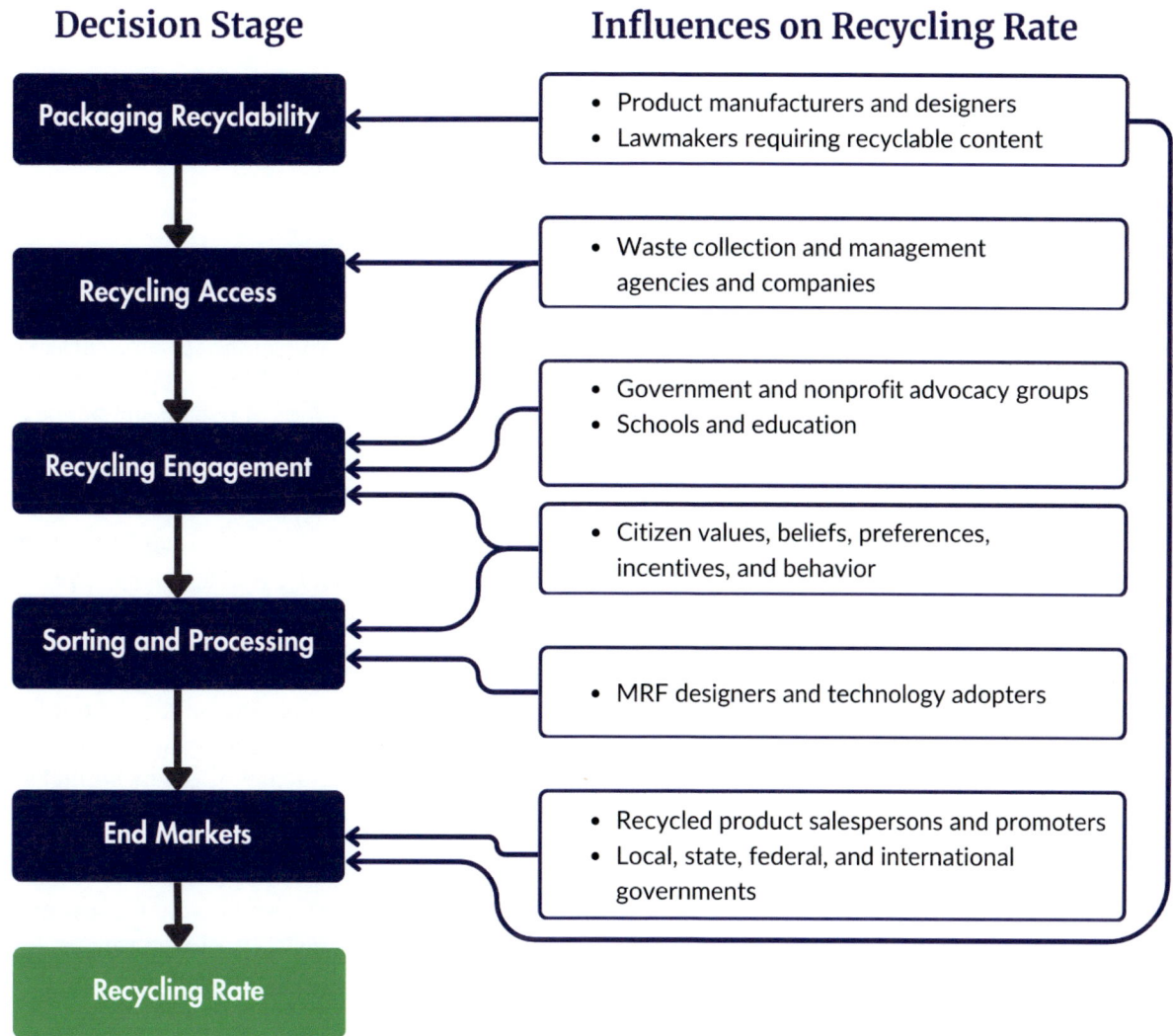

FIGURE S-2 Major technological or behavioral decision stages and influences affecting recyclability of solid waste.
NOTE: MRF = materials recovery facility.
SOURCE: Data from The Recycling Partnership, 2024.

1
Introduction

Municipal solid waste (MSW) management systems in the United States have undergone much evolution over time. The overall system has seen many changes, including changes in policy, the types of materials managed, infrastructure and technology, and the actors that play a role in the system. These evolutions have yielded the distributed waste management system found in the United States today: one that is made up of a complex and interconnected web of policies and distributed groups of people and public and private entities acting at local, regional, state, national, and international levels. Each of these actors holds a different motivation for recycling and responds to different incentives regarding the management of the near 300 million tons of MSW generated every year in the United States (EPA, 2020).

Likewise, MSW recycling programs in the United States have a long and evolving history, beginning in 1690 when the Rittenhouse Mill accepted linen and cotton rags (Robert C. Williams Paper Museum, n.d.). World War II saw a national recycling campaign for metal, rubber, paper, and other materials (Springate, n.d.). Environmental awareness grew in the 1960s and 1970s, leading to the rise of curbside recycling and the first Earth Day in 1970 (Eldred, 2020). Recycling was also seen as a method of decreasing the use of fossil fuels, one of which—namely oil—was suddenly in short supply due to the Arab oil embargo. Recycling provided a method of reducing oil usage due to the lower energy requirements associated with using some recycled materials rather than virgin materials for the manufacture of products and packaging. Also at that time was significant interest in recovering energy from waste through the construction of "Waste-to-Energy" plants, as waste was found to have about one-third of the energy value of coal (IEA Bioenergy, 2003). Finally, the 1970s saw predictions that the world would run out of material resources because of the growing world population. The combination of these developments laid the foundation for modern recycling programs and shifted national attitudes toward conservation and waste management.

While serving an important function for the manufacturing supply chain, public health, and pollution mitigation, today's recycling programs sit at a crossroads. In recent years, challenges facing MSW recycling programs, especially economic-based challenges, have led some municipalities to stop funding their recycling programs altogether (Waste Dive, 2023). This report explores the contemporary issues facing MSW recycling programs and lays out recommendations and policy options to chart a path forward.

1.1 THE COMPOSITION AND REGULATION OF SOLID WASTE

For the purposes of this report, municipal solid waste is defined as the non-hazardous solid waste generated by the residents, commercial businesses, and institutions of a community. The proper management of MSW is required on a regular and ongoing basis to protect the public health and the local environment of the community and to minimize litter and illegal dumping. MSW is sometimes defined to include construction and demolition (C&D) waste, although the U.S. Environmental Protection Agency (EPA) definition of MSW excludes C&D waste. Other solid wastes also typically excluded are junk automobiles and industrial process waste such as metal casting and combustion residuals. Although MSW is considered a non-hazardous waste, it is known to contain small quantities of hazardous waste that are discarded by residents and businesses.

Ensuring that MSW is properly managed in the United States is the responsibility of local governments, as assigned to them by their respective states. As a general practice, local governments are tasked to ensure that MSW—as well as recyclables sorted by the residents and businesses for separate collection—are properly contained and collected on a regular basis. Municipal governments are generally responsible for ensuring that the collected wastes and recyclables are properly processed for recovery of materials and/or energy or disposed in landfills; municipalities often contract for the collection, processing, and disposal of MSW with private companies. It is estimated that 85–90% of the permitted MSW landfill disposal capacity in the United States is owned by private companies (Karidis, 2018).

The costs associated with the proper management of MSW and its recyclable content are currently borne by the residents and businesses of the community, either directly through service charges or indirectly through taxes. Thus, MSW management services can be viewed as public utility services—similar to electricity, water, wastewater treatment, and stormwater management—that are vital to the ongoing well-being of the community. In the United States, managing MSW involves significant costs for the infrastructure, labor, and technology required for various disposal methods. As of 2018, the country generated approximately 292 million tons of MSW each year, with the majority going to landfills that are costly to build, operate, and maintain, especially as space becomes limited (EPA, 2020). Recycling and composting programs, which divert around 32 percent of MSW, also carry high expenses due to the collection, sorting, and processing needed to reclaim materials such as paper, metals, plastics, and glass (EPA, 2020). Separate organic waste processing adds additional costs for facilities that handle biodegradable waste and turn it into valuable products, a process that requires specialized handling and transportation. Waste-to-energy (WTE) facilities, which process about 12 percent of MSW, are among the most expensive due to high technology and operational costs, although they offset some expenses by generating electricity (EPA, 2020). These combined costs highlight the financial investment required to manage waste sustainably and reduce environmental impacts, with the expense often covered by a combination of taxpayer funds, fees, and public–private partnerships.

These costs have changed over time. In the early 1990s, EPA promulgated new regulations for the disposal of MSW in landfills (42 U.S. Code §§ 6941–6949a). Referred to as "Subtitle D" regulations, they were designed to minimize the environmental impacts of landfill disposal by requiring that composite liners be installed at the bottom of any new landfill to prevent leachate from leaking and contaminating the groundwater below. Subtitle D regulations also placed new requirements on the siting and operation of landfills. These requirements significantly added to the costs and technical expertise required to design, operate, and maintain landfills. These costs resulted in the decision of many cities and counties to get out of the landfill business and instead rely on privately owned regional landfills for the disposal of their wastes.

Alongside the implementation of Subtitle D landfill regulations, EPA and many state governments adopted MSW recycling goals for local governments to achieve in the management of the MSW generated by their communities. A key rationale for these goals was to conserve Subtitle D landfill capacity, which had become relatively expensive because of the liner requirements and other regulations. Many of these goals were stated in terms of an MSW recycling rate, calculated as the total weight of recycled MSW over the total weight of generated MSW (EPA, 2020). The definition of recycling that was commonly accepted during this time included the composting of yard waste and other MSW, as well as materials and products used in the manufacturing and packaging of consumer goods. However, because composting is a biological process that converts organic materials, it is considered by some to be a recovery process rather than a recycling process.

Although recycling advocates in the 1990s argued that sale of recyclable materials and cost savings from diverting materials from landfills could cover costs of the recycling of MSW, this argument has fallen short in practice. As a result, local governments are currently responsible for covering the added costs associated with the recycling of MSW as opposed to its disposal in landfills. The committee for this study was convened in part to determine the role that the federal government, states, and private companies should each play in covering the additional direct costs associated with the recycling of MSW versus its landfill disposal. Covering these costs acknowledges the higher indirect social and environmental costs associated with landfill disposal rather than recycling that portion of MSW.

1.2 THE IMPORTANCE OF RECYCLING

Solving contemporary challenges of MSW recycling programs is a critical mission for policymakers in the coming years. Policy choices across levels of governments will influence and impact the actors across the MSW management system. Evidence-based decisions regarding these choices can ensure an appropriate balance of the tradeoffs, helping to ensure that MSW recycling programs operate in a manner that is both economically and environmentally sound.

As this report lays out in further detail, well-designed and supported MSW recycling programs hold many benefits. In particular, recycling decreases reliance on non-renewable resources and virgin materials. Extending the useful lifespan of materials can avoid greenhouse gas emissions associated with the extraction of non-renewable virgin materials. While recycling and the remanufacture of materials have greenhouse gas emissions costs, well-designed recycling programs and the use of renewable energy sources in remanufacturing process can lead to an overall reduction in emissions. These concepts are explored further in Chapters 3 and 7 of this report.

In addition to their environmental benefits, recycling programs can lead to measurable economic gains. Recycling programs and the circular economies they enable create jobs, promote business development, and provide further positive social impacts for communities across the country. These concepts are further discussed in Chapters 3 and 6.

1.3 MSW RECYCLING COSTS AND THE ROLE OF PUBLIC POLICY

Before exploring specific challenges recycling programs encounter in various regions and materials, it is crucial to understand the role of costs and public policies. Markets alone do not provide the necessary incentives or information for households, businesses, or local governments to engage in effective recycling practices. Local, state, or federal government intervention through public policy can significantly improve the efficiency, affordability, and accessibility of recycling initiatives.

For instance, local governments are involved in waste collection not just for economic efficiency but also to maintain public health and cleanliness for the community's benefit. Without this service, some people might avoid disposal costs by dumping waste illegally, which creates broader environmental and health risks. While curbside waste collection is essential, it requires considerable public funding. Recycling helps reduce waste disposal costs by diverting materials away from landfills, yet recycling programs themselves are costly to run, involving collection, sorting, and processing of materials. Here, public policy can help balance costs by funding and supporting recycling efforts, especially where they benefit the community overall.

Government support can also make both landfills and recycling more cost-effective through "economies of scale." Processing plants, known as materials recovery facilities (MRFs), may need large volumes of recyclable material to be cost-efficient, which can require a significant initial investment in infrastructure. Government policy can make it possible to build and run these large facilities by centralizing services, reducing the cost per ton of recycled material, and allowing local programs to access the savings. Appropriate public policy can further reduce costs by creating streamlined markets for selling recycled goods, which helps make recycling programs financially sustainable.

Another area where public policy plays a critical role is in managing "external costs," where a sales contract between two parties in an exchange does not account for indirect costs on third parties, such as noise, traffic, and pollution. For instance, disposal of waste at a landfill may produce leachate that affects groundwater, produces odors, and generates greenhouse gases that impact entire communities, not just individual users or owners of the landfill. Policies, such as regulations on landfill emissions and incentives for recycling, can help reduce these external costs and improve economic efficiency. This report emphasizes that decisions and policies about recycling can be analyzed and optimized only in the context of the broader waste management system that includes alternatives to recycling, such as landfill disposal, composting, and use of WTE plants.

Recycling also helps to lessen environmental damage from mining and raw material extraction, which impose long-term costs on society. Government policy can help reduce these environmental costs of mining and materials extraction by providing subsidies or incentives to encourage recycling, which can reduce new materials extraction and its impact on natural resources. Thus, this report emphasizes that analysis of recycling policy needs to consider not only the broader waste management system (landfill disposal, composting, and WTE), but a circular economy that includes decisions and policies about mining, other extraction, product design for recyclability, repair, and reuse, along with recycling (Fullerton, 2025; Stahel, 2016).

Finally, public policies also can ensure that recycling efforts are fair and accessible to all communities. Well-designed policies can help ensure that the costs of recycling and waste management does not fall unevenly on different groups and that everyone benefits from cleaner, healthier, and more sustainable waste practices.

This report refers to these cost and policy issues as it examines the specific challenges and solutions for recycling across the United States, giving context to the discussions in later chapters.

1.4 STUDY SCOPE AND APPROACH

Recognizing the growing challenges in sustainable MSW recycling programs, Congress called on the National Academies of Sciences, Engineering, and Medicine to convene an ad hoc committee to study the programmatic and economic costs of these programs and to make recommendations on policy options for effective funding and incentives for recycling. The full statement of task can be found in Box 1-1. This report represents the final report of that committee.

BOX 1-1
Statement of Task

An ad-hoc committee of the National Academies of Sciences, Engineering, and Medicine will review available information on programmatic and economic costs of municipal solid waste (MSW) recycling programs in municipal, county, state, and tribal governments. The committee will provide advice in the form of options, including potential policy approaches, to facilitate the effective implementation of MSW recycling programs. (MSW materials and programmatic and economics costs are defined below.)

As part of its assessment, and to the extent sufficient data are available, the committee will address the following aspects with respect to MSW recycling programs. The committee will base its analyses on examination of several different case studies in the United States as exemplars. The case studies will represent a range of circumstances (e.g., some from each of the municipal, county, state, and tribal governments; environmental justice considerations such as different population sizes and demographics; different geographical locations; different economies; etc.):

- Describe differences in programmatic and economic costs across municipal, county, state, and tribal governments. Examples of possible considerations include:
  o Types and differences in programmatic elements and capabilities (e.g., urban vs. rural vs. tribal needs) across government types.
  o Types of recycling programs implemented (e.g., curbside collection programs, drop-off-only programs, commercial and residential programs, back-hauling programs).
  o Factors that impact a government agency's ability to fund and administer a recycling program.

- Examine the ways in which costs of MSW programs differ based on materials accepted for recycling. Examples of possible considerations include:
  o Infrastructure (including freight), technology, and end markets that exist for commonly recycled materials.
  o Costs of material-specific approaches such as single-stream vs. dual-stream residential recycling, curbside food and yard waste pickup services, and glass separation mandates.

continued

> **BOX 1-1** *continued*
>
> The committee will provide recommendations in the form of options to effectively incentivize and fund recycling activities in an economically and environmentally sound way. The development of options should include considerations of:
>
> - Supply-side policies (e.g., economic incentives for people to recycle) and demand-side policies (e.g., Extended Producer Responsibility, recycled content mandates, or tax credits for remanufacturing firms to prioritize use of recycled over virgin materials).
> - Estimates of the programmatic and economic cost implications and time frames for implementing the options.
> - Metrics for the evaluation of the effectiveness of different policies or other approaches.
> - Environmental impact and related climate change considerations that focus primarily on changes in greenhouse gas emissions, including emissions from transportation-related sources.
> - Uncertainties in the supply of and demand for recyclable materials that create complexity in cost-benefit analyses.
>
> The committee will review references pertaining to the costs of recycling programs, factors that impact a local government's ability to fund and administer a recycling program, and policies or other approaches that facilitate the implementation of recycling programs. The committee will characterize key limitations within the existing references on the costs of recycling programs and identify future research needs.
>
> MSW materials that are considered in-scope for this study include commonly recycled or composted materials, such as paper, metals (e.g., aluminum), glass, plastics (types #1 and #2), food scraps, and yard waste from the residential, commercial, and institutional sectors that are converted into raw materials and used in the production of new products. Materials that are specifically out of scope for recycling considerations include textiles, electronic waste, construction and demolition debris, household hazardous waste, auto bodies, municipal sludge, combustion ash, and industrial process wastes that might also be disposed of in municipal waste landfills or incinerators. Material management pathways that are considered in-scope for this report include mechanical recycling of MSW and composting of organic waste. Material management pathways specifically out of scope include any type of waste-to-energy process, incineration, or fuel substitute production.
>
> Programmatic costs are expenses needed to implement MSW recycling programs, such as purchase of collection trucks and operation and maintenance of materials recovery facilities. Economic costs may include opportunity costs of recycling vs. landfilling; fluctuations in the supply, demand, and price of recycled commodities; externalities (such as emission of greenhouse gases (GHGs), and a household's willingness to pay for recycling services based on marginal social costs and benefits.

Note that the committee was not tasked with exploring or making recommendations with respect to general waste management or diversion related to "reduce" and "reuse" efforts in the traditional "3R" framework. The committee was tasked specifically to focus on the policies and pathways that relate to materials for recycling—while bearing in mind that policy recommendations about recycling must depend on problems with alternatives to recycling listed above. In addition, the committee provides policy options as a way to effectively operationalize its recommendations.

The committee understood its task as primarily concerning the policies and systems in place that relate to the collection, sorting, processing, transport, and sale of recyclable materials, especially those that are traditionally processed in MRFs (although, the committee notes that expansion to other materials such as textiles and food waste would be possible and desirable). The committee considered recycling and composting as distinct but related components of MSW management. While C&D and hazardous waste are outside the study scope, the committee considered policies around disposal of specific materials, such as lithium-ion batteries, that often enter typical recycling processes, as they impact the direct (i.e., monetary) costs of recycling programs. Additionally, given the intertwined nature of the waste and recycling system

Introduction

and associated policies, the committee considered the policies and approaches that impact in-scope materials, even if they also relate to out-of-scope materials. For example, the committee detailed extended producer responsibility (EPR) policies in this report as relevant to in-scope materials, even though many EPR frameworks also relate to out-of-scope materials (e.g., household hazardous waste).

1.5 ORGANIZATION OF THE REPORT

This report describes the results of the committee's review and study of the information available regarding its statement of task. Chapter 2 presents the MSW management and recycling ecosystem from a systems perspective. Chapter 3 explores challenges with recycling programs and how policy can address those challenges. Chapter 4 provides an overview of the various costs and financing options available for recycling programs, and Chapter 5 reviews the types of materials commonly recycled and the markets involved in recycled materials. Chapter 6 details the social and behavioral considerations that are relevant for recycling programs. Finally, in Chapter 7, the committee identifies the environmental impacts of recycling.

REFERENCES

Eldred, S. 2020, April 14. *When did Americans start recycling?* History. Last updated May 28, 2025. https://www.history.com/articles/recycling-history-america.

EPA (U.S. Environmental Protection Agency). 2020. *Advancing sustainable materials management: 2018 fact sheet. Assessing trends in materials generation and management in the United States.* https://www.epa.gov/sites/default/files/2021-01/documents/2018_ff_fact_sheet_dec_2020_fnl_508.pdf.

Fullerton, D. 2025. The circular economy. In T. Lundgren, M. Bostian, and S. Managi (eds.), *Encyclopedia of energy, natural resource, and environmental economics*, 2nd ed., vol. 3, pp. 254–265. Elsevier. https://dx.doi.org/10.1016/B978-0-323-91013-2.00050-2.

IEA Bioenergy. 2003. *Municipal solid waste and its role in sustainability.* https://www.ieabioenergy.com/wp-content/uploads/2013/10/40_IEAPositionPaperMSW.pdf.

Karidis, A. 2018. *Why some landfills are becoming privatized, while others remain public.* Waste 360. https://www.waste360.com/landfill/why-some-landfills-are-becoming-privatized-while-others-remain-public.

Robert C. Williams Paper Museum. n.d. *Papermaking moves to the United States.* Georgia Institute of Technology. https://web.archive.org/web/20070814173933/http://ipst.gatech.edu/amp/collection/museum_pm_usa.htm.

Springate, M. n.d. *Material drives on the World War II home front.* National Park Service. https://www.nps.gov/articles/000/material-drives-on-the-world-war-ii-home-front.htm.

Stahel, W.R. 2016. The circular economy. *Nature* 531(7595):435–438.

Waste Dive. 2023. *Where curbside recycling programs have stopped and started in the US.* https://www.wastedive.com/news/curbside-recycling-cancellation-tracker/569250.

2
Municipal Solid Waste Management and Recycling Systems

> **Summary of Key Messages**
>
> - **Municipal solid waste (MSW) recycling programs are part of broader waste management systems:** These systems include those that manage materials destined for end-of-life treatment or disposal as well as commercial and industry recycling and remanufacturing systems.
> - **Actors and policies:** MSW recycling programs are impacted by actors and policies across the materials supply chain—from virgin material extractors to end-of-life waste managers—and by governments at all levels.
> - **Quantities of MSW generated and recycled annually:** The United States generates approximately 292 million tons of MSW annually. However, most MSW (68–79 percent) because of factors such as the material not being targeted for recycling, not being economically viable to recycle, limited access to recycling programs, and low participation rates (EPA, 2020; The Recycling Partnership, 2024).
> - **Changes in material streams over time:** The volume and composition of materials managed in the MSW system have changed and will continue to change over time with the evolution of consumer products (e.g., printed newspapers) and packaging materials (e.g., plastic versus glass containers).
> - **Technological advances may increase MSW system efficiency:** Increased use of automated sorting by households and at materials recovery facilities may decrease contamination issues and increase sorting efficiency. These technological advances likely will require significant financial outlay.
> - **Limitations of weight-based recycling rates:** Aggregated weight-based recycling rates (i.e., a ratio of recycling to total waste collected) and material-specific rates (e.g., percent of all discarded aluminum cans that are recycled) are incomplete metrics because they do not adequately account for all costs and benefits of using and reusing materials over their life cycle (e.g., economic, social, and environmental costs and benefits), changes in packaging material composition, and waste reduction efforts.
> - **Sustainable materials management:** A sustainable materials management approach provides a more complete picture of the costs and benefits of using and reusing materials across their life cycles than weight-based recycling metrics alone.

Today's municipal solid waste (MSW) management system is the product of countless iterative changes throughout the course of history, both in the United States and abroad. Changes in technology, economic activity (from local market development to international trade), demography, social norms and values, policy, packaging types, and other factors have all influenced contemporary MSW systems. Wilson (2023) provides a more detailed historical review of waste management, recycling, and composting.

2.1 TODAY'S RECYCLING SYSTEMS

As further detailed throughout this report, recycling systems are complex. Still, to understand the current state of recycling programs and to address this committee's charge, it is necessary to map a generalized system that considers the actors, processes, materials, and other components of typical municipal recycling programs (see Figure 2-1). This chapter describes the basics of MSW management and recycling; the actors, infrastructure, and processes involved; and methods for assessing the performance of a recycling system, as well as current concerns for and promising technological improvements to MSW processing.

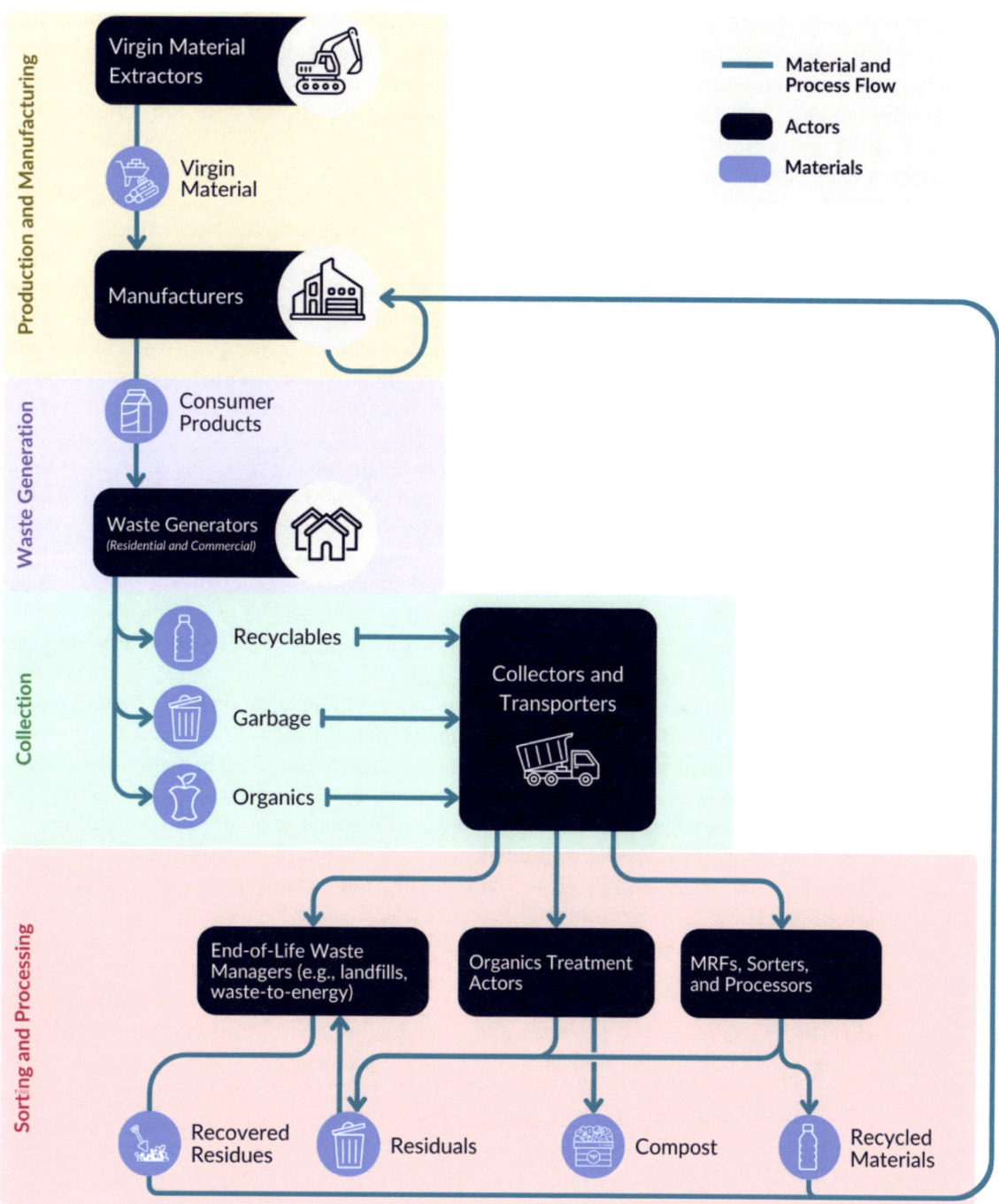

FIGURE 2-1 A simplified system diagram for municipal solid waste management systems primarily highlighting processes and materials discussed in this report.
NOTES: Other non-residual organic products (e.g., animal feed, energy) may result from treatment of organics but are not discussed in this study. MRF = materials recovery facilities.

Broadly speaking, the materials management system includes subsystems for (1) production and manufacturing, (2) waste generation, (3) waste collection, and (4) sorting and processing. *Production and manufacturing* include both virgin and recycled materials as inputs, which are valued differentially by markets that procure them for production. Also of note, manufacturer efforts to recycle their own materials play a critical role in supporting broader waste reduction goals, but these efforts typically occur separately from downstream MSW management systems. *Waste generation* has been studied in local, regional, national, and cross-national contexts, where differences in household size, income, education, values, waste ordinances, and information efforts are shown to affect household waste generation rates and composition (Firmansyah et al., 2024; He et al., 2022; Šomplák et al., 2023). Cecot and Viscusi (2022) conducted an extensive set of survey studies to identify the effect of these and other factors on household recycling decisions (see Chapter 6; see also Cecot and Viscusi, 2022; Huber et al., 2023; Viscusi et al., 2022). *Waste collection* methods influence the rate and composition of material flows. Single-stream, dual-stream, and source-separated or single-material methods include collecting and transporting recyclables separately, in contrast to traditional mixed-waste collection and transport. The degree of separation at collection also impacts quality of the outbound material. Contamination is lower and therefore quality is higher in material that is source separated or single stream, compared with mixed-waste or dual-stream collection, albeit at a higher cost. In the *sorting and processing* stage, materials recovery facilities (MRFs) receive sorted or treated waste; they then further sort and consolidate the waste to prepare recycled materials for use in remade products or other applications.

An important metric for the benefits and performance of a municipal recycling program is the cost-efficiency of collecting, sorting, and processing recycled materials (Bohm et al., 2010; Bolingbroke et al., 2021; Wilson et al., 2015). This efficiency is affected by the quality of both human (Jităreanu et al., 2023) and mechanical processes that create or reduce losses along the waste flow–recycle stream.[1] Much of these inefficiencies are attributed to contamination, which is the amount of nonrecyclables—15–20 percent on average—collected curbside with recyclable materials. Missed recyclables add to residue from MRFs and are often mislabeled as contaminants. These can create significant operational problems at recycling centers and increase processing costs (The Recycling Partnership, 2024).

Composting can be considered a form of recycling but involves very different operations. *Composting* is a biological process that converts recycled organic matter, primarily leaves and food waste, into a soil-like material that has value as a soil additive known as compost. Large-scale composting requires a controlled environment that includes, optimally, adequate oxygen levels, periodic compost turning, appropriate moisture levels, and temperature control. Composting can be done in homes or communities, or for large-scale operations, in engineered vessels, windrows, or aerated static piles. Composting plays a major role in meeting U.S. food waste reduction goals, recognizing that composting food waste has multiple advantages over landfilling. The U.S. Environmental Protection Agency (EPA) estimates that 98.5 million tons of food waste and yard trimmings were generated in 2018 (12.1 percent of total MSW generated was yard trimmings and 21.6 percent was food waste). Approximately 25 million tons of organic material were composted, representing 8.5 percent of MSW generated. Approximately 4.1 percent of food waste generated was composted (EPA, 2020). Interest is also growing in recycling food waste as animal feed, codigestion, donation, land application, sewering, anaerobic digestion, and thermal recovery for organic waste.

2.1.1 Actors and Decision-Makers in MSW Management

The MSW management system portrayed in Figure 2-1 includes the mass flow of products, materials, and wastes from production through use, waste generation, sorting, collection, and treatment, and on to reuse, recycling, or disposal. These operations are accomplished by actors in the MSW management system and their employees, as illustrated in Figure 2-2. In each stage, multiple actors influence each other through a combination of price and economics, education, persuasion, or directives supported by legal requirements. A fully effective recycling effort requires high-quality programs in each of these stages and strong planning and coordination among them.

---

[1] Some researchers have proposed variations of this metric (Bohm et al., 2010; Bolingbroke et al., 2021; Wilson et al., 2015).

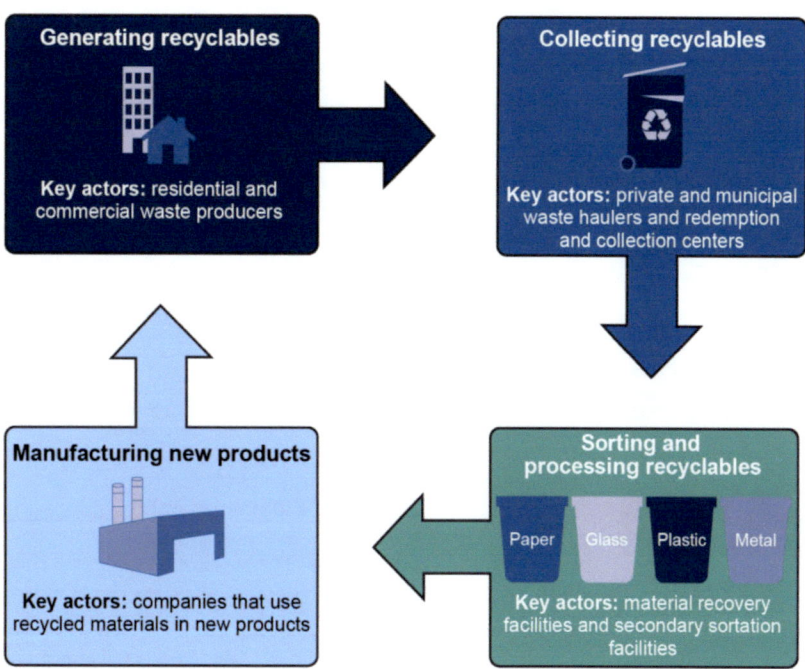

FIGURE 2-2 Key stages and actors in the U.S. recycling system.
SOURCE: GAO, 2020.

Each actor in this system is led by human decision-makers, who are influenced by their beliefs and values, technical and popular sources of information, social norms, economic costs and incentives, and constraints imposed by laws and regulations. These decision-makers include producers choosing technologies for product manufacturing and packaging; consumers choosing products and deciding between methods for waste separation, sorting, and collection; and waste management organizations choosing among technologies that support these behavioral decisions.

In a recent report, The Recycling Partnership (2024) presented an MSW systems approach that focuses on decision-makers and their choices. The model, adapted and simplified in Figure 2-3, depicts the five major stages in which critical decisions affecting the success of a recycling program are made. The overall recycling rate of solid waste is directly impacted by these actors and the choices they make at each stage. As such, these actors and decision stages represent key areas where policy choices can positively influence recycling rates.

The stages in Figure 2-3 can serve as a framework for assessing participant roles and decisions pertaining to recycling and the factors that affect these decisions. Waste collection and management agencies, whether private or public, play a central role in providing access to recycling; in coordinating policies and programs that encourage engagement; and in acquiring the best, most cost-efficient technologies for collection and MRF operation. They also must coordinate their operations with sales staff and go-between businesses seeking end markets for their recycled products. Throughout this process, waste management agencies must coordinate their activities with government policymakers and enforcement staff, educators and advocacy groups, customers, local communities that may be impacted by their facilities, and technology providers that wish to promote and sell their equipment and management services. While this section provides a brief overview of these stages, they will be discussed further throughout the rest of this chapter and report. In addition, many of the upstream influences in Figure 2-3 propagate downstream to affect concurrent or subsequent decisions. For example, end market performance is determined by the cumulative effects of system design decisions made by waste and recycling companies as well as regulatory decisions, incentives, and subsidies (e.g., to improve access to recycling) by government agencies at different local, state, national, and international levels.

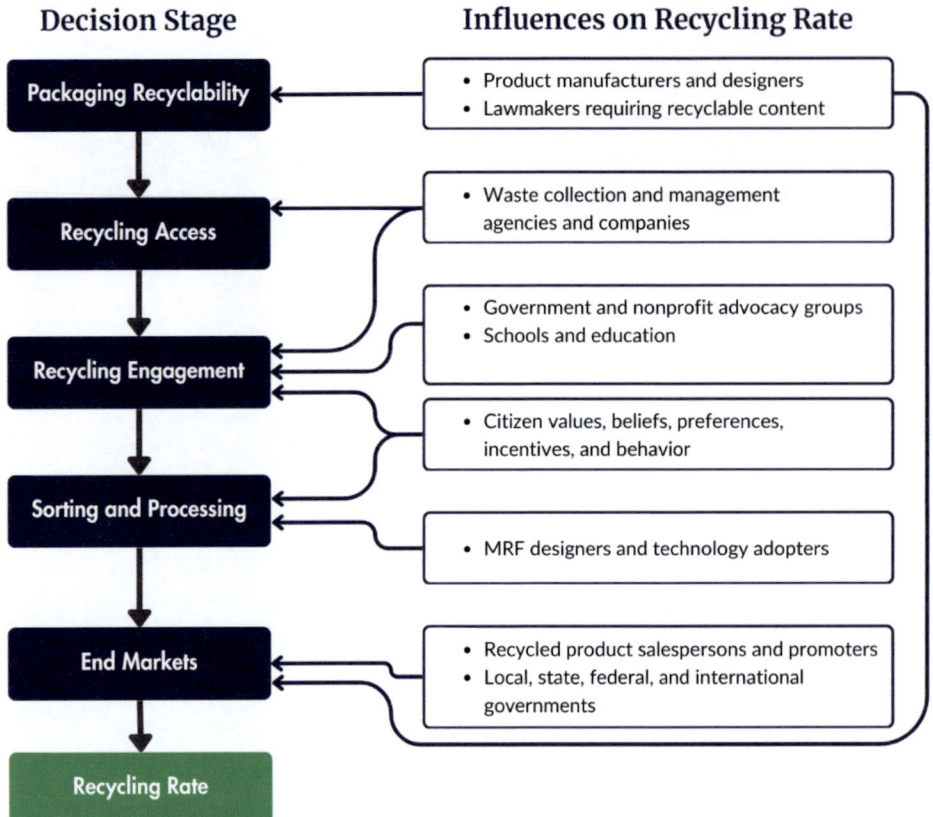

FIGURE 2-3 Major technological or behavioral decision stages and influences affecting recyclability of solid waste.
NOTE: MRF = materials recovery facility.
SOURCE: Data from The Recycling Partnership.

Packaging Recyclability

To improve recyclability, manufacturers need to design products and their packaging in such a manner that their delivery, use, sorting, reuse, collection, and processing are relatively easy for decision-makers to learn and implement, with limited added time and effort required for household waste management. Upstream, manufacturers need to design for and use recycled materials as inputs. These choices are increasingly motivated by state and local extended producer responsibility (EPR) laws and requirements for life cycle environmental impact assessments (see Chapter 4).

Recycling Access

Recycling access is generally provided to the public through curbside pickup programs or drop-off bins, which may be material specific. Access to recycling is generally high in the United States (73 percent); however, this rate is much higher for single-family homes (~85 percent) than for multifamily housing (~37 percent) where recycling dumpsters are more common (The Recycling Partnership, 2024). The comparatively lower rates of recycling access for residents who live in multifamily housing contributes to concerns that low-income neighborhoods face additional barriers to receiving the positive environmental impacts of recycling.

The collection of waste and recyclables typically represents about 75 percent of the costs of MSW management, while processing and disposal account for about 25 percent of those costs. Not all recyclables are created equal. Determining which materials to collect in a recycling program can be informed by a variety of parameters:

- Quantity of materials generated (or present in the disposed waste stream)
- Ease of collection and processing (e.g., adding the material to an existing recycling program)
- Public sentiment or resident demand for collection of specific materials
- Strong, stable market pricing with a positive demand outlook
- Proximity of markets to the MRF (reducing transportation costs).[2]

Recycling Engagement

A high level of consumer engagement is necessary to direct appropriately chosen, clean waste products from the home to the recycling facility. For those with access, recycling engagement is motivated by public outreach programs, educational materials, and presentations, and by influence from neighbors and friends as recycling becomes a social norm among the general population. States with higher access rates tend to have higher engagement rates, as multiple efforts to provide access signal the importance and support for recycling (Eunomia Research and Consulting, 2023). MSW planners and policymakers nationwide continue to target improvements in engagement (see Chapter 6).

Sorting and Processing

Sorting and processing are primarily provided at MRFs using a combination of optical scanning; air flotation; mechanical redirection; manual sortation; robotics, visual identification using artificial intelligence (AI), and size, magnetic, and weight sorting technologies. Ongoing improvements in these technologies have enabled a typical MRF to capture about 87 percent of its incoming accepted program recyclable materials (The Recycling Partnership, 2024). The highest capture rates are achieved for high-density polyethylene (HDPE) bottles and jars (93 percent) and steel cans (96 percent), while the lowest are estimated for non-bottle polyethylene terephthalate (PET) (60 percent) and film and flexible material, where those materials are accepted (40 percent). It is anticipated that new and better computerized or AI-driven air, optical, and mechanical systems will allow MRFs to approach 95 percent overall capture in the near future (The Recycling Partnership, 2024). Still, consumers will need to ensure that used products are free of contamination from food waste, non-recyclables, or potentially hazardous materials to allow MRF performance at this high level.

End Markets

Viable end markets are necessary to ensure that recycled products can be sold and reused at an economically sustainable price to create benefits for society. Over the long term, an end market that requires high government subsidies to stay in business will erode the confidence of investors, taxpayers, and recycling service consumers. Identifying and securing markets for collected materials are central to the cost-effectiveness, design, and operation of recycling programs.

One of the growing drivers of end-market demand comes from companies that have committed to using recycled materials in their products and packaging. These commitments may stem from increased consumer valuation for products made and packaged with recyclables, EPR requirements, or a general wish by manufacturers to be viewed as a green company (e.g., Iannuzzi, 2024).

As noted above, the location of end-market facilities and production operations affects transportation costs and the revenue paid for the materials. Having a variety of end markets for each material type is beneficial to help maintain demand for materials despite fluctuations in virgin material prices and other

---

[2] Transportation costs can be higher for MRFs in rural areas compared with those in more developed areas, which tend to be closer to highways.

changes in market conditions. When processing facilities have good access to major transportation routes (highway, rail, boat, or barge), they have more flexibility in selecting end markets. West Coast communities, for example, have much easier access to Asian markets for recyclables than do communities in the rest of the country.

2.1.2 Infrastructure and Access in the MSW Management System

The collection, processing, and marketing of recyclables are interdependent components, and each must be considered when designing and operating a recycling system as they affect one another. For example, changes in collection methods and materials collected will impact the design and operation of the MRF; how the MRF is designed and operated will determine whether materials will be produced that meet market specifications; and changes in market requirements may lead to changes in how materials are collected and processed. Furthermore, these systems must be designed and operated considering the material that is presented, which is dictated by manufacturing and purchasing decisions.

Collection Programs

A variety of approaches and configurations has been used to implement MSW recycling in the United States, principally curbside collection and drop-off programs. Collection services for recyclables are generally provided on a weekly or biweekly basis to single-family residences. Collection for businesses is often more frequent. Collection services are arranged in a variety of ways, including by contract, public entities, and franchise. In communities where residential waste collection services are organized on a franchise basis, one or a few entities may provide services to all residences in the franchise district. Municipal collection and, contracting is also common in portions of the United States.

In terms of facilitating the collection and processing of appropriate wastes, curbside collection programs are generally recognized as having the greatest potential for effective MSW recycling (Best and Kneip, 2019; Dahlen and Lagerkvist, 2010; Noehammer and Byer, 1997). Curbside collection of residential recyclables is provided in thousands of communities in North America and is the most convenient recycling option for residents. However, operations must be compatible with the processing capabilities of the MRF that will be receiving the collected material. For example, a program that collects materials in a single stream (several types of recyclables collected together) would be a poor choice if local MRFs are unable to process commingled materials. One of the most significant challenges facing managers of curbside programs is participation. Inconsistent participation can affect material volumes and equipment or labor needs, impacting operations (e.g., efficiency of truck routes) and costs. Other challenges include contamination, labor, equipment replacement, and fluctuating demand and pricing for recyclables and the ever-changing recycling mix (e.g., the significant drop in newspaper recycling).

Drop-off programs are the oldest form of public-sector-provided recyclables collection and are used by both residential and commercial participants. Drop-offs rely on individuals to bring recyclables to a designated drop-off location. For convenience, these locations are typically in public spaces, often at landfills or transfer stations where the public is already traveling to dispose of waste, as well as drop-off centers in retail locations. Drop-off collection is most often used to serve rural communities with low population density, where curbside recycling may be cost-prohibitive. Drop-off collection points may have higher contamination levels and engender illegal dumping if they are not staffed or monitored.

Single-day, or weekend collection events are often used to collect materials generated in smaller quantities that are not cost-effective to collect on a more frequent basis. They typically target materials that are banned from landfill disposal, such as used oil, tires, household hazardous wastes, electronics, and bulky items (e.g., appliances). Shredding events are becoming more prevalent as residents look for confidential recycling options for their personal documents. These events are often held in a large parking lot or public works yard to accommodate significant traffic flow. They also provide an educational opportunity for management of all types of recoverable and recyclable materials.

Commercial recycling collection may be performed in a manner similar to residential curbside recycling, using either wheeled carts or larger containers or designated collection vehicles and containers (e.g., trucks with large roll-off containers). For businesses with suitable storage space or large quantities of specific material (e.g., cardboard), collection may be provided by material brokers, with collected material being hauled to markets directly. In addition to storage space, having a baler on-site can facilitate direct-to-mill options for commercial materials. Space constraints are common, impacting the collection method and frequency selected by commercial operations.

Collection Methods and Equipment

Methods for collecting recyclables include segregated, dual stream, single stream, and mixed waste. The chosen method determines what type of equipment is needed for collecting and storing the recyclables.

- **Segregated (source separated):** Recyclable materials are manually sorted at the location where they are first discarded, such as homes or businesses, into designated containers or specific compartments within the collection vehicle. This multistream method is most often used in communities with limited processing capabilities and is found in both commercial and residential programs. Collection productivity is relatively inefficient, resulting in high collection costs relative to other methods. This method includes source-separated organic waste, which can be converted into fuel or compost.
- **Dual stream:** Fiber (newspaper, cardboard, office paper) and container (plastic, aluminum, bimetal, glass) recyclables are separated by the resident. Each stream (fiber and containers) is loaded into its own compartment in the collection vehicle, which may or may not be compacted. Dual-stream collection capitalizes on the initial labor provided by the participating residents. It produces cleaner recycled materials, especially paper, because it is not contaminated with broken glass and small plastics.
- **Single stream:** Recyclables are collected in a single, fully commingled form and subsequently separated and processed into marketable secondary materials at a MRF. Single-stream recycling is now the most commonly used method of collecting recyclables because of its convenience and popularity with consumers and its ability to accommodate a large and varying amount of recyclables. Its high efficiency and low costs are attributed to its ability to use automated collection trucks staffed by a single operator and the safety it provides the operator, who does not have to exit the cab to collect the recyclables. Programs that grow in size and volumes of material collected exert more pressure to consider additional commingling of recyclables because of the operational benefits provided and the ease of participation by generators. Many dual-stream programs have been or are being converted to single-stream programs where MRFs are equipped to process the single-stream material.
- **Mixed waste:** Relatively few facilities in the United States process mixed "residual" waste for the purpose of producing "spec fuels" for industries such as paper mills or cement kilns. Spec fuels generally consist of paper, plastics, and other organics that have energy contents that would contribute to the heating value of the fuel. These processing facilities generally include the sorting of recyclables remaining in the mixed waste for recovery and marketing to secondary materials markets.

Recycling containers provide material storage between collections and assist the vehicle operator during collection at the curb. In many recycling programs, bins or larger wheeled carts are provided at no direct charge to the residents, while others require residents to purchase containers or to purchase additional containers if one is insufficient. The choice of household container type must be consistent with the collection method, vehicle type, and material processing ability. For example, a 64-gallon wheeled cart for commingled recyclables will not work in a collection program with partitioned vehicles, where the local MRF does not have the ability to separate materials and requires them to be source separated. The size and number of recycling bins or carts needs to match the collection frequency and the projected volume of recyclables.

In commercial contexts and multifamily dwellings, recyclables are typically stored in five types of containers: (1) individual containers similar to those for single-family residences, (2) larger bulk containers collected by either front- or rear-load collection trucks, (3) roll-off compactor boxes, (4) open-top roll-off boxes, and (5) rear-load van-type trailers. Selection depends on space available for containers and the types and quantities of materials to be collected.

Materials Recovery Facilities

MRFs receive, sort, process, and market recyclable materials collected from municipal waste streams, and they are integral to most municipal recycling programs. In early recycling programs, when recyclable materials were sorted curbside to minimize contamination and maximize resale value, materials were merely densified before being sent to market. As recycling programs expanded and the types of materials collected increased, more programs initiated commingled collection of more recyclables, thereby increasing the need for sorting at a central location.

In addition to the space used for processing equipment, MRF design typically includes a tipping area for material storage and a space for processed materials, with load-out bays or docks for removal of consolidated, sorted materials. The tipping floor and load-out areas are typically of reinforced concrete construction to withstand use and travel by heavy equipment. Most MRFs are enclosed structures, providing protection of recyclable materials (particularly fiber materials) from weather effects and limiting the potential for litter.

The challenge of a MRF is to transform recyclable materials into marketable resources for future use. Therefore, the requirements of secondary materials markets need to guide those designing MRFs with respect to the types and quantities of recyclables they will accept. Issues to consider include:

- Level of sorting necessary before and after delivery to the MRF
- Screening for and removal of dangerous contaminants, most notably lithium-ion batteries, in separated materials (see Box 2-1)
- Size and sorting capacity of the equipment in relation to the volume of material to be processed and daily and downtime capacity storage requirements
- Balance of mechanical and manual sorting needed
- Degree of processing and consolidation required for end-market acceptance and economical transportation
- Amount of storage required for accumulating sufficient quantities of recyclables for transport and docks and loading capacity
- Revenue projections for the sale of recyclables.

Challenges in Rural and Urban Areas

Many factors contribute to the differences in recycling infrastructure, access, and behaviors in rural and urban areas. Population density and the built environment affect transportation costs, volumes of material collected, and options for collection equipment.

Rural Communities

Rural communities face unique challenges, including time-consuming and inefficient collection, longer hauling of materials, and low population density. Burning waste in barrels and piles occasionally is practiced in these areas, and therefore, may reduce the amount of waste to be collected.[3] However, burning has its challenges, including air pollution, the long-term disposal of ashes, and dealing with items that do not burn. Recycling participation increases during times of burn bans, as residents seek to dispose of waste. It is unknown whether these increases are sustained after the bans are lifted.

BOX 2-1
Lithium-Ion Battery Fires

Fires caused by lithium-ion batteries now pose considerable risk and raise costs for the recycling system, in collection vehicles and particularly in processing by materials recovery facilities (MRFs). Although these batteries ought to be managed separately, residents sometimes include them with materials collected in single-stream curbside recycling and with ordinary waste. Estimates suggest that MRFs currently average 18 of these fires per year each, with catastrophic fires (those causing more than $10 million in damage) affecting 1 percent of MRFs annually.[a] Insurance costs for MRFs have risen because of the contribution of these batteries to their fire risk. The insurance cost increase may be in the range of $8–$40 per ton recycled, representing a significant share of processing costs (which are about $100–$200 per ton; see Chapter 4).[a] Although landfills also suffer fires from these batteries, MRFs face higher costs because they have more workers and capital exposed to the risk (EPA, 2021a). The problem of lithium-ion battery fires has emerged largely within the last decade and is likely to grow worse with widespread adoption of lithium-ion batteries in short-lived consumer products, such as disposable e-cigarettes (EPA, 2021a).

Some public policies in the United States attempt to reduce the extent to which these batteries enter the municipal solid waste and curbside recycling streams, although the U.S. Environmental Protection Agency (2021a) does not have authority to address them. For example, New York City and New York State have banned disposal of rechargeable batteries, and the state requires that manufacturers fund a recycling program for these batteries and that retailers selling these products accept returns and provide information to consumers (Eunomia, 2021).

European countries have tried extended producer responsibility and deposit–return systems on lithium-ion batteries. Although these policies appear to increase separate collection of batteries, their effectiveness at reducing fires at MRFs has not been demonstrated. Longer-run improvements in battery technology may eventually reduce the flammability of these batteries (EPA, 2021a). However, until that occurs, the mismanagement of lithium-ion batteries is likely to continue to impose costs and hazards for recycling and to add to the challenges that communities face in funding curbside recycling programs.

[a] Anne Germain, presentation to the committee, April 2024.

Instead of hiring a municipal waste management company, rural communities often rely on drop-off locations for recycling, which results in lower recycling participation rates (Morawski and Wilcox, 2017). Additionally, many rural areas have "mom-and-pop" waste management businesses that collect and haul waste to transfer stations. Minimal or unorganized collection services can result in illegal dumping along roadsides or even burying of waste on one's property.

In areas with small populations and limited capacity for local collection and processing, as in some rural or isolated parts of the United States (e.g., parts of Alaska), backhauling is the main source of waste disposal. Backhauling occurs one to three times per year when a summer barge or truck makes deliveries to the community. After their goods are delivered, the vehicles are filled with bales of aluminum cans, containers of batteries, and other materials for proper disposal, with a focus on hazardous materials for environmental concerns. Backhauling is expensive, especially for residents in very small towns (e.g., <1,000 residents) (EPA Environmental Financial Advisory Board, 2019).

---

[3] Burn barrels are 55-gallon metal drums where trash is placed and burned. A burn pile is simply an area where trash is taken and burned.

Many small municipalities have tight budgets that restrict them from managing facilities independently and purchasing equipment to collect and manage recycling materials. County-wide collaborations are popular, allowing several small communities to work together. Hub-and-spoke recycling has been developed as a solution for helping rural communities work together on a regional level to consolidate larger volumes of recyclable materials. This model works by creating regional recycling processing centers within larger communities that serve as "hubs" and encourages smaller communities, or "spokes," to deliver their recyclables to these hubs. As shown in Figure 2-4, spokes could be a small municipality's recycling drop-off center, a town's curbside collection program, or other recycling services, all of which have material streams that feed into the centralized recycling infrastructure of the hub (Nebraska Recycling Council, 2024). Private provider hubs handle the transfer of small-volume recycling from the spokes. Waste transfer stations can sometimes use storage bins and top-loading tip or walking-floor trailers in their truck yards, which are contracted by public convenience centers or those who use these hubs to service subscription routes with recycling.

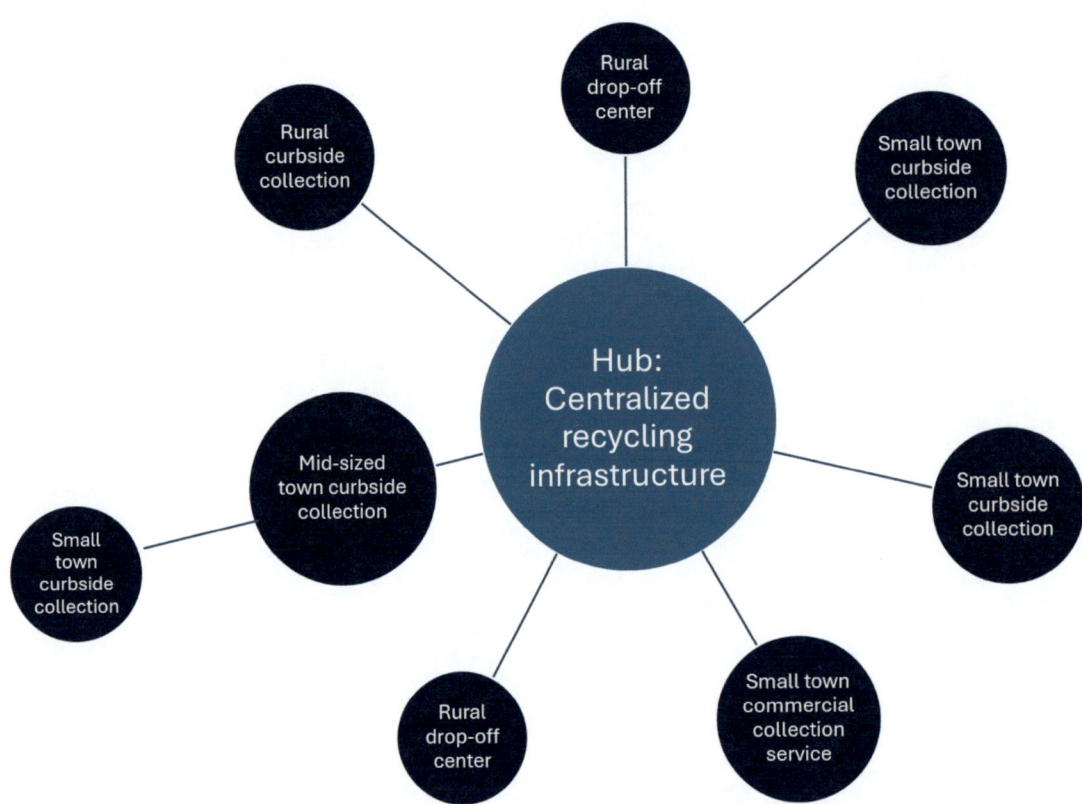

FIGURE 2-4 Hub-and-spoke recycling model.
SOURCE: Generated by the committee, modeled after Nebraska Recycling Council (2024).

Recycling hubs invest in or solicit grants for equipment and infrastructure needed to sort materials, then use this equipment to create and store bales of materials for end markets. Spoke communities invest in or solicit grants for recycling collection and transportation to the closest hub. Hub-and-spoke systems greatly reduce transportation requirements and increase overall efficiency of program operations. Spokes' transportation and operating costs decrease, while the hubs receive sufficient volume of materials to increase revenues, achieve economies of scale, and assist with operational costs. These types of programs also help small and remote communities implement recycling programs, reduce costs, and increase participation and recovery (Nebraska Recycling Council, 2019).

According to the Nebraska Recycling Council (2019), hub-and-spoke recycling can work in a variety of contexts, including:

- Recycling drop-off centers or trailers
- Public and private recycling operations
- Curbside recycling, either single or dual stream
- Towns in different counties or even across state lines
- Any type of recyclable material, including cardboard, plastic bottles and aluminum cans

Curbside Collection in Urban Communities

Compared with rural areas, urban areas tend to have more efficient and organized public and private programs. Despite this, urban areas face issues related to high volumes of waste and often limited space for bin placement.

Curbside collection of recyclables generally utilizes a large truck similar to those used for general waste, driving through the streets with a side-arm mechanism that picks up the cart and dumps its contents into the truck. Some trucks have cameras that allow the driver to review the contents as they are emptied into the truck and then provide feedback to the resident if the cart had materials in it that could be considered contamination. Thus, trucks with side-arm technology reduce labor costs, may result in cleaner communities, and may reduce contamination in the recyclables collected (Wise Guy Reports, 2024).

In older cities (e.g., New York City), narrow alleys and street parking preclude the use of side-arm trucks; in many of these areas, employees must pick up bags from the sidewalk or empty carts by hand. Using this method results in higher labor costs, employee injuries, litter in communities, and contamination in recyclables collected. In urban areas, safe curb-side waste collection has public health implications, as it reduces vermin populations and leakage of the materials (Budds, 2022).

2.2 MSW MANAGEMENT

MSW management is a complex and evolving industry. This section first reviews estimates of the quantity of MSW collected in the United States today, and the portion that is recyclable. It then describes objectives, frameworks, and performance metrics used to evaluate waste management systems.

2.2.1 Quantity and Composition of MSW

The tonnage of MSW collected in the United States annual has increased every year since it was first recorded (88.1 million tons in 1960). Comparing tonnage and population rates from 1960 and 2018, the per capita generation rate has increased by 82 percent in 60 years. While it is not known precisely what percentage of MSW is residential versus commercial, EPA estimates that residential waste makes up 55 percent of MSW, and commercial waste 45 percent (EPA, 2020).

The most recent national-level data on recycling as reported by EPA (for the year 2018) indicate that recycling plays a significant role in managing waste in the United States. For 2018, EPA estimated that 292.4 million tons of MSW were generated, of which 69 million tons (24 percent) were recycled and 25 million tons composted (EPA, 2024). More recent data by The Recycling Partnership (2024) estimates that 10 million tons of MSW—or 15 percent—were recycled through residential curbside and drop-off recycling programs in 2024, representing only 3.4 percent of the MSW generated (based on 2018's EPA estimate). These data suggest that commercial and multifamily recycling accounts for 85 percent of the recycling tonnage recovered from the MSW stream (see Table 2-1). However, since multifamily residential recycling rates are significantly lower than single family residential recycling rates, it is reasonable to assume that commercial recycling accounts for the major portion of the commercial/multifamily recycling tonnage.

The Recycling Partnership (2024) further reports recent data indicating that only 21 percent of residential recyclable material is being recycled, with 76 percent of this material lost at the household level. While a reported 73 percent of U.S. households had access to residential recycling in 2024, this number was significantly less for multifamily households (37 percent), and overall household participation in recycling was just 43 percent. Much of the material collected in recycling programs is directed to about 500 MRFs (National Waste and Recycling Association, 2018).

At the start of the COVID-19 pandemic, in March 2020, residential waste volumes increased abruptly, while commercial waste volumes decreased (Pinto et al., 2022). However, as stores started reopening and people resumed going out, commercial volumes substantially increased. E-commerce as a percentage of retails sales peaked in the fall of 2020, declined significantly, and only recently achieved the earlier percentage of retail sales, according to Department of Census data on e-commerce.[4] While these volumes may not have remained the same after the pandemic, they could reflect a shift in the types and amounts of materials used across different sectors. However, more important are the longer-term trends in packaging and product delivery, including glass losing market share to plastic and aluminum, adoption of flexible plastics, decline in printed paper, and increased home delivery. The evolution of packaging used in e-commerce continues to evolve.

TABLE 2-1 MSW Recycling in Residential and Commercial/Multifamily Sectors

Material	Total Recycled		Curbside Recycling from Single-Family Residences			Commercial and Multifamily Recycling		
	Tons[a]	% of Total Recycled MSW	Tons[b]	% Residential Single Family Recycling	% Total	Tons[c]	% Commercial/ Multifamily Residential Recycling	% Total
Paper and Paperboard	45,970,000	67%	5,808,858	54.1%	12.6%	40,161,142	68%	87%
Glass	3,060,000	4%	2,152,303	20.4%	70.3%	907,697	2%	30%
Metals	8,720,000	13%	–			8,720,000	15%	100%
Ferrous	6,360,000	9%	231,156	3.0%	3.6%	6,128,844	10%	96%
Aluminum	670,000	1%	393,488	3.4%	58.7%	276,512	0%	41%
Other Nonferrous	1,690,000	2%	–			1,690,000	3%	100%
Other Materials			–			–		
Plastics	3,020,000	4%	1,552,576	19.1%	51.4%	1,467,424	2%	49%
Rubber and Leather	1,670,000	2%	–			1,670,000	3%	100%
Textiles	2,510,000	4%	–			2,510,000	4%	100%
Wood	3,100,000	4%	–			3,100,000	5%	100%
Other	970,000	1%	–			970,000	2%	100%
Total MSW recycled	69,020,000	100%	10,138,381	100.0%	14.7%	58,881,619	100%	85%

[a] Data from EPA Advancing Sustainable Materials Management 2018 Tables and Figures, p. 2.
[b] Data from The Recycling Partnership – *The State of Residential Recycling 2024*, Figure 14.
[c] Tonnages in the Commercial Recycling column are the differences between the Total Recycled and the Curbside Recycling columns.
NOTE: MSW = municipal solid waste.

---

[4] See https://www.census.gov/retail/ecommerce/historic_releases.html.

The EPA study (2020) differentiates between "materials in products"—such as paper, glass, metals, and aluminum that can be recycled—and organic materials—such as food waste and yard trimmings that can be composted. In 2018, about 189.76 million tons of "materials in products" waste were generated, while 98.5 million tons of organic waste were generated (EPA, 2020).

The quantities and types of materials in the waste stream continually shift because of evolving technologies, consumer preferences, and product types. As a result, the composition of the waste stream has changed dramatically over the last 35 years. General trends have been the replacement of glass container discards with plastics, the reduction in newspaper discards, the increase in cardboard waste due to e-commerce, and the increase in plastic containers and other plastic products. These trends and the corresponding per capita increase in waste production have led to an increased effort to reduce waste generation through the design of products with fewer disposable materials, parts, and packaging, and to reduce, reuse, and recycle waste. These efforts, in conjunction with improvements in the design and performance of landfilling and thermal treatment (and to a lesser extent, mechanical or biological treatment) for that portion of the waste stream that is not reused or recycled, now constitute the principal elements of solid waste management programs in the United States and other countries (Awino and Apitz, 2024; Devi et al., 2024; Khan et al., 2022; Sharma and Jain, 2020, Sondh et al., 2024; Tsai et al., 2020).

2.2.2 Objectives for MSW Management

Objectives for MSW management typically include (1) providing the essential public service of waste management in an efficient and affordable manner; (2) reducing the health impacts, environmental damage, and contamination that results from mismanagement of solid waste; and (3) sustaining access to raw materials and other natural resources that are lost to the economy when they are disposed without reuse or recycling.

Several analytical frameworks have been developed for evaluating MSW management:

Flow tracking models:
- *Material flow analysis* evaluates the mining of raw materials through production, consumption, recycling, and disposal (e.g., Allesch and Brunner, 2015; Arena and Di Gregorio, 2014; Harder et al., 2014; He and Small, 2022; Makarichi et al., 2018).
- *Energy flow models* track the amount of heat, electricity, and other forms of energy generated, lost, or used for living requirements and economic activity in a region (Subramanyam et al., 2015).
- Measures of progress toward a circular economy (Chioatto et al., 2023; Dumlao-Tan and Halog, 2017; Meleddu et al., 2024; Salemdeeb et al., 2022; Tsai et al., 2020; Vines et al., 2023).
- Life cycle assessment of energy use and environmental and health impacts (Anshassi and Townsend, 2023, 2024; Olafasakin et al., 2023; Wang et al., 2021, 2022).
- Multicriteria analysis, including economic and social sustainability and resilience (Goulart Coelho et al., 2017; Gutierrez-Lopez et al., 2023; Jayasinghe et al., 2023; Makarichi et al., 2018; Taelman et al., 2020).
- Metrics of economic efficiency in achieving high diversion rates for reuse or recycling at low cost and with low environmental impacts (Hu et al., 2024; Laner and Schmidt, 2023; Mensah et al., 2023; Ng and Yang, 2023; Prenovitz et al., 2023; Wilson et al., 2015).
- Economic models of incentives and market failures within the waste management system (Acuff and Kaffine, 2013; Fullerton and Kinnaman, 1995; Palmer et al., 1995).

MSW management objectives are met through the activities and efforts of multiple participants, including product manufacturers, shippers, commercial enterprises, consumers, educators, local government solid waste managers, waste collectors, and workers and managers providing waste recycling and disposal services. Oversight and governance are provided by government agencies and regulators, ensuring effective formulation and implementation of MSW policies, rules, and regulations, including requirements for recycling (e.g., Baffoe-Bonnie and Ezeala-Harrison, 2023; Macauley and Walls, 2010; Ogieriakhi and Wang, 2024).

2.2.3 Performance Metrics for MSW Management Systems

Performance metrics play a valuable role in planning and assessing waste management program elements and outcomes. Across nations, states, and systems within a state or region, metrics help program managers and agencies benchmark and rank their programs and identify where changes and innovations would be most beneficial. Indicators can be designed to emphasize participation rates, the quantity of materials captured and recycled, program costs, and end-market performance.

Two case studies, presented in Boxes 2-2 and 2-3, describe system indicators for MSW management systems in New York and Florida. These cases studies are followed by a review of results from The Recycling Partnership's (2024) study of U.S. state programs, with a focus on metrics for statewide material capture and loss.

Anshassi and Townsend (2024) also assessed environmental return on investment—comparing the environmental benefits with the cost of maintaining a recycling program—for two contract structures (Contract A and Contract B) in each Florida region (see Figure 2-5). The environmental return on investment for recycling for a typical household was similar to or higher than that of switching from a gasoline vehicle to a hybrid or electric vehicle or using renewable energy. In most cases, except for three regions in 2021, discontinuing recycling led to net dollar savings, which can be viewed as the cost of maintaining recycling. However, the study found that recycling programs reduced environmental impacts, with 0.34–2.4 times lower greenhouse gas emissions. This study also found that focusing on high-value materials in recycling programs reduces both environmental impacts and system costs (Anshassi and Townsend, 2024).

BOX 2-2
Case Study: Evaluating Selected Waste Systems in New York State

Greene and Tonjes (2014) reviewed the structure and intent of several municipal solid waste (MSW) management performance indicators, noting a general historical preference for recycling rate as a metric, and raising arguments for additional measures that address household participation and environmental and economic impact. Green and Tonjes (2014) categorized the metrics into four tiers.

Tier 1 indicators track the mass flow of waste to recycling facilities and landfills, as well as that diverted from landfill and incineration over a unit of time (typically annually). Tier 2 indicators convert the Tier 1 mass flows to rates measured as percentages by dividing by the total mass of waste collected. Tier 3 indicators calculate these rates on a per capita basis. Tier 4 indicators address reductions in energy- and climate-related externalities achieved by recycling programs.

Greene and Tonjes's (2014) evaluated these indicators for ten municipal waste systems in New York State. They found that using different indicators yields significantly different rank ordering among systems; care and clarity is thus needed in indicator selection and presentation to avoid misunderstanding. Given such care, indicators can help support internal assessments of progress and program needs. Green and Tonjes (2014) found that several indicators were especially useful for general waste systems characterization:

- landfill disposal per capita
- diversion per capita
- diversion rate,
- landfill disposal rate

The authors noted however, that the choice of the best indicators depends on local and regional conditions and the waste management system design.

SOURCE: Greene and Tonjes, 2014.

BOX 2-3
Case Study: Assessing Recycling Programs in Florida

Anshassi and Townsend (2024) evaluated the costs, environmental impacts, and improvement strategies for municipal recycling programs in Florida's six regions, paying particular attention to indicators of greenhouse gas emissions based on life cycle assessment. The study predicts the counterfactual effects of eliminating all recycling in each of Florida's six regions.

The model by Anshassi and Townsend (2024) included costs for collecting and sorting and processing waste and recyclables. It accounts for composition of waste and recyclables, contamination rates, and end-market revenue generated from the sale of recyclables and it estimates the annual cost per household for curbside collection and management of recyclables and garbage for 2011, 2020, and 2021 under four illustrative recycling contracts. The four contracts account for variation in materials recovery facility (MRF) processing fees (in $/ton) and local government–MRF revenue-sharing agreements.

In addition to capital and operating costs, Anshassi and Townsend (2024) calculated the climate-related environmental impacts associated with alternative scenarios, including terminating recycling programs, increasing recycling rates, and targeting high-value commodity materials. They found that Florida's recycling programs reduce greenhouse gas emissions, energy use, water use, the potential harm to human health, the risk of damaging ecosystems, and the likelihood of increasing acid levels in the environment (Anshassi and Townsend, 2024).

As an example of a region-level counterfactual, in Central Florida, the model indicated that eliminating recycling would increase greenhouse gas emissions by 0.377 t CO_2 eq per household per year, which is equivalent to driving an extra 967 miles in a typical gasoline-fueled car (using 29 gallons of gas) (see Table 2-2 for estimated greenhouse gas emission increases). Note that the equivalent activity emissions of eliminating recycling are highest in Central Florida and lowest in Northwest Florida.

TABLE 2-2 Estimated Increases in Greenhouse Gas Emissions Based on Discontinuing Recycling Programs in Florida

Metric	Florida Region					
	Northwest	Northeast	Central	Southwest	South	Southeast
Miles driven by avg. gas passenger car	195	423	967	660	935	483
Gallons of gas consumed	9	19	43	29	41	21

NOTE: Factors for each metric were retrieved using the U.S. Environmental Protection Agency greenhouse gas equivalencies calculator.
SOURCE: Anshassi and Townsend, 2024.

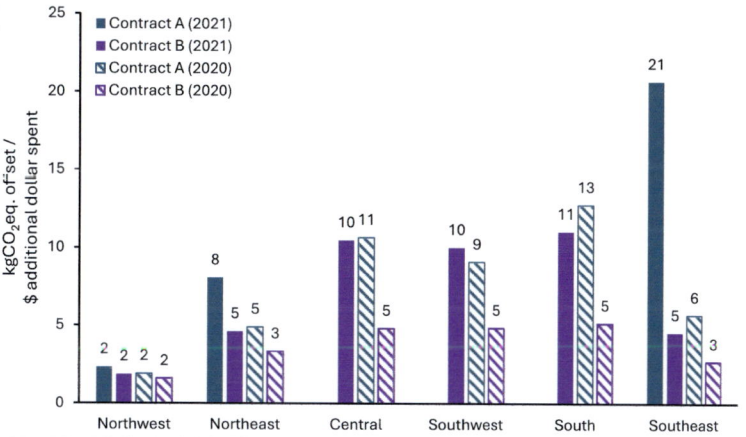

FIGURE 2-5 Environmental return on investment associated with the effects of keeping a recycling program, by region and two different contract structures (Contract A and Contract B).
NOTES: Calculated as the estimated potential greenhouse gas emissions offset by recycling divided by the cost of recycling. For Central, Southwest, and South regions under the Contract A scenario in 2021, dropping recycling would result in a net dollar cost increase, instead of a savings (which is assumed to be the added cost of maintaining recycling). Accordingly, no return on investment is presented for these scenarios.
SOURCE: Anshassi and Townsend, 2024.

Location-specific studies of municipal recycling programs' performance can allow a focus on specific features of a system and provide a basis for improved designs and operations. They may be based on detailed inputs and datasets, or they may have the ability to obtain such data as needed. In contrast, state evaluations often include approximations to account for incomplete datasets and uncertain aggregation. Despite these limitations, state evaluations provide a broadly based assessment of the factors that promote or limit the success of recycling programs and policies and allow citizens and their representatives to benchmark their state's performance relative to that of others.

As has been referenced throughout this chapter, The Recycling Partnership (2024) conducted an extensive national and state evaluation of the status of recycling in the United States. Among many metrics, it studied two of the most widely used indicators of recycling efficacy: residential recycling rates (see Figure 2-6) and residential recyclable material lost (in tons per year; see Figure 2-7).

Recycling rates ranged from 8 percent in the southeastern United States to 37 percent in California and Oregon (The Recycling Partnership, 2024; see Figure 2-6). Residential recycling rates remain the most widely used metrics of recycling performance and impact but provide a limited picture of recycling activity and impact as it does not distinguish between composition, environmental impact, and disposition of recycled waste. Other metrics that reflect important dimensions of composition and impact need to be used in conjunction with these.

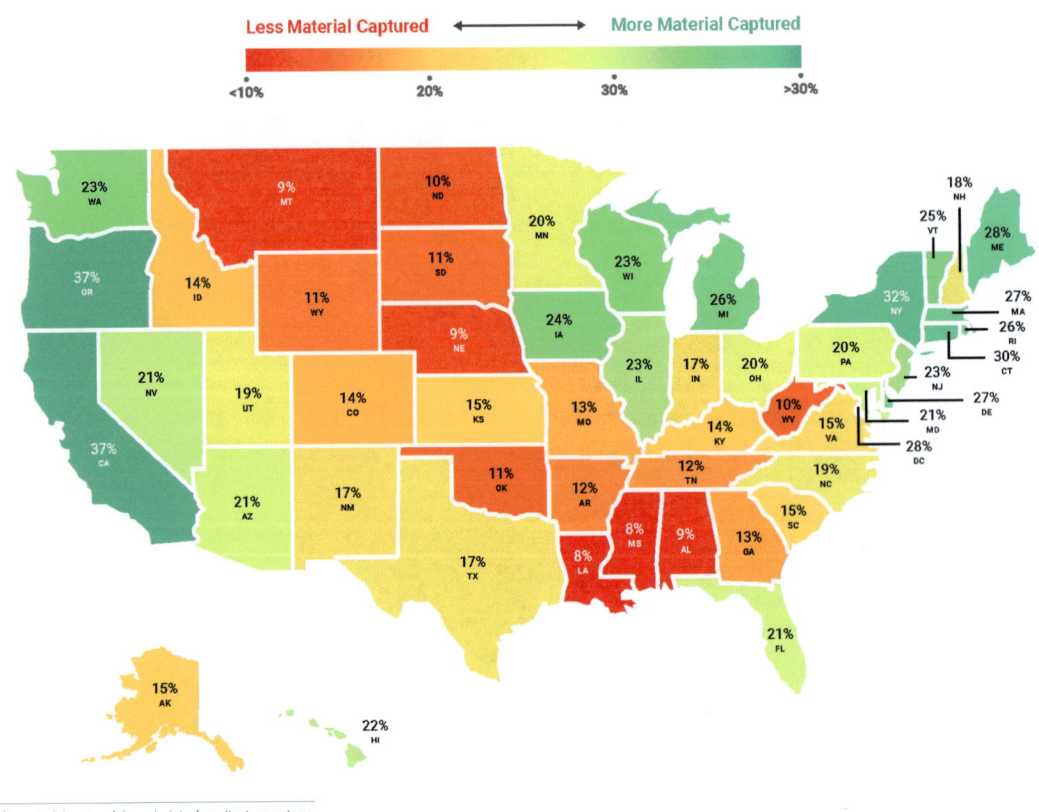

FIGURE 2-6 State-by-state residential recycling rates.
NOTE: Includes material captured through state deposit-return systems.
SOURCE: The Recycling Partnership, 2024.

An alternative metric (Figure 2-7) shows statewide estimates of residential recyclable material lost to disposal (e.g., in landfills). These values represent total flows of solid waste and are highly influenced by the state populations, with especially high estimates determined for California, Texas, Florida, New York, Pennsylvania, Ohio, and Illinois. While this metric, as assessed by recyclable material lost, provides a direct indication of a state's overall waste-related environmental impact, the improvements resulting from recycling are masked by population and the total overall tons of MSW generated. In contrast the residential recycling rate in Figure 2-6 expresses the recycling efficacy as the percentage of waste captured by the recycling program.

The residential recycling rate (Figure 2-6) and residential recyclable material lost metric provide complementary information but are not independent. One approach that allows a more direct comparison of these indicators is dividing the material lost tonnage by the state population, yielding a per capita estimate of the material lost from recycling (see Figure 2-8). As expected, the data indicate a negative correlation between a state's recycling rate and its tonnage of recyclable material lost—higher residential recycling rates tend to decrease the per capita residential recyclable material lost. This suggests that the trend shown in Figure 2-8 may be used for preliminary inference from state metrics such as estimating statewide recyclable materials lost from residential recycling rate, when the former data are unavailable. However, the spread in Figure 2-8 suggests that other factors may influence the rates as well and suggests a need for a more complete model.

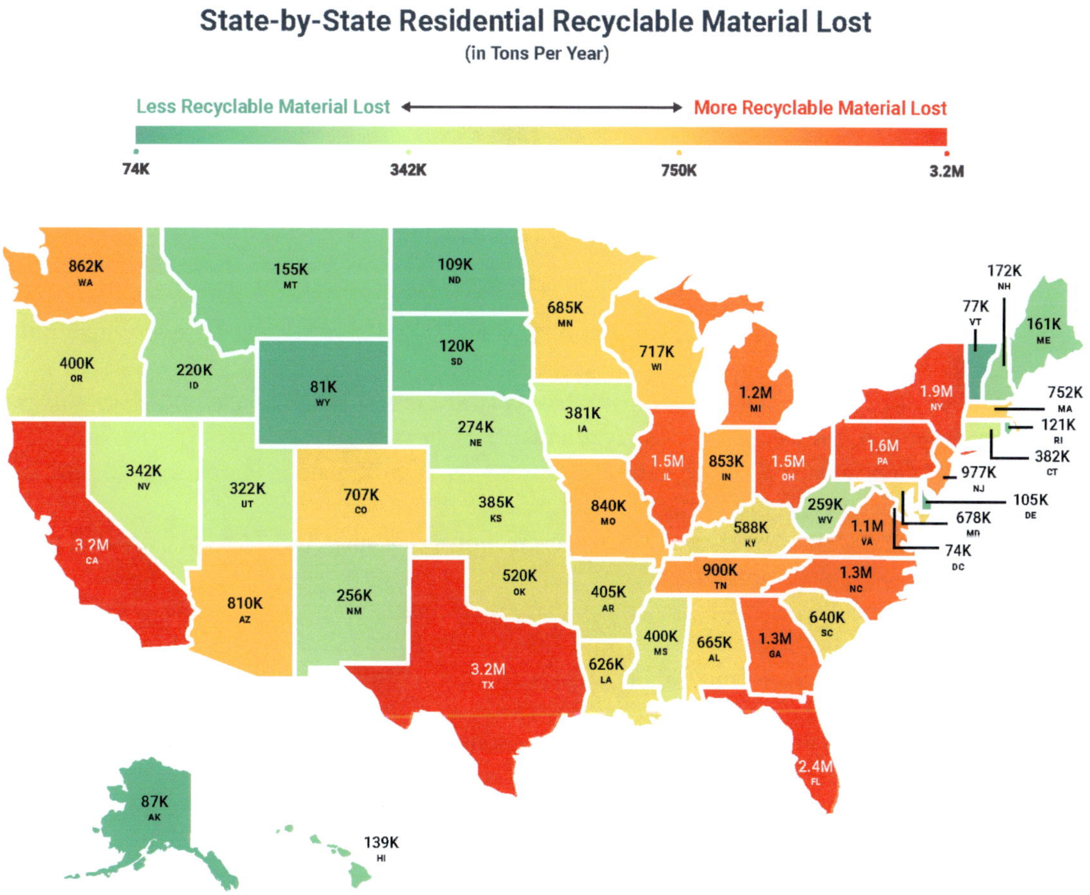

FIGURE 2-7 State-by-state residential recyclable material lost, in tons per year.
NOTE: *Recyclable material* here refers to all recyclables processed by materials recovery facilities and moved to end markets.
SOURCE: The Recycling Partnership, 2024.

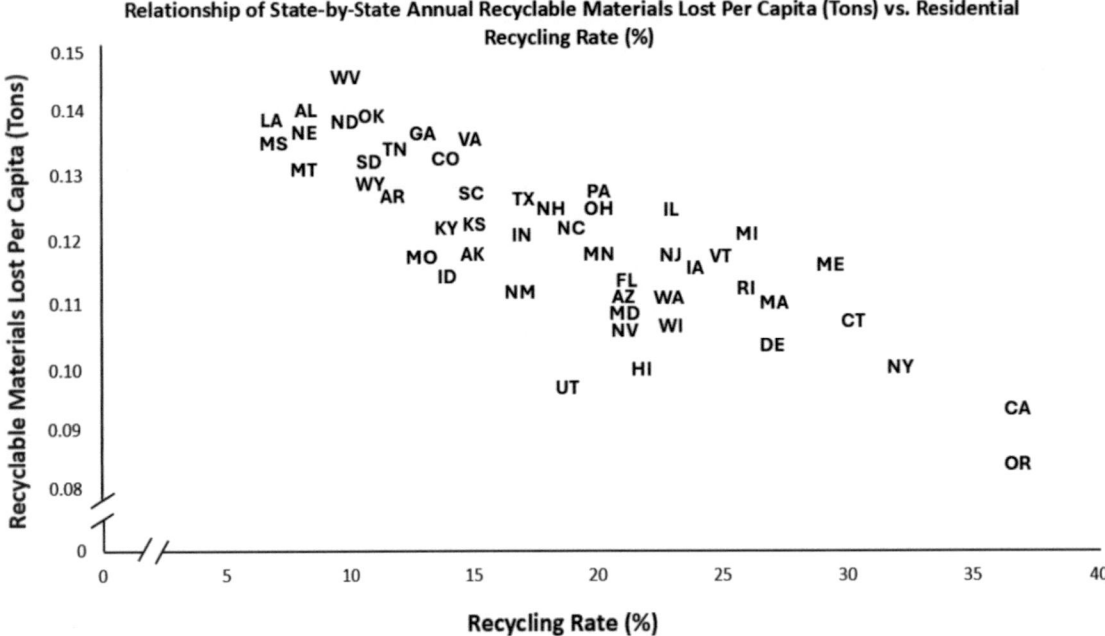

FIGURE 2-8 Relationship of state-by-state annual recyclable materials lost per capita versus residential recycling rate.
SOURCE: Data from The Recycling Partnership, 2024.

Nationwide Data on Status, Performance, and Engagement Metrics for MSW Recycling Programs

In addition to state-level data, The Recycling Partnership (2024) has identified the following key issues to be addressed using nationwide data:

- Twenty-one (21) percent of residential recyclables is being recycled—every material type is under-recycled.
- Seventy-six (76) percent of residential recyclables is lost at the household level, underscoring the importance of access and engagement.
- Only 43 percent of all households participate. Nonparticipation is due to both lack of access and insufficient communication and outreach.
- Of households that have access to curbside or drop-off recycling services, only 59 percent use their recycling service regularly, and of those that do, only 57 percent of recyclable material is put in recycling containers, meaning many households do not participate to the fullest extent possible. This participation rate is significantly lower than the 90 percent target benchmark that The Recycling Partnership (2024) sets for an effective recycling system.

Based on these data, it is often concluded that residents need more education, communication, and support to engage in recycling. However, a factor that is often not considered or discussed is that many residents—perhaps over 50 percent—do not feel that their participation in recycling is worth their time and effort. In other words, the results being achieved through residential curbside recycling programs (the diversion of 450 pounds per household per year) might be the maximum that can be expected from a program that involves the voluntary participation and effort of individuals on a regular basis. If system improvements needed to yield higher access, engagement, and overall recycling rates are not developed or implemented, then the fallback alternative may be to focus on more effective methods of recovering recyclables from the mixed solid waste stream. Meanwhile, improved mechanical capture and recovery make necessary the creative efforts to educate and influence citizens to build a culture of participation.

2.3 TECHNOLOGICAL ADVANCES IN MSW RECYCLING

Technological improvements are being developed for multiple decision stages to increase the rate of municipal recycling, from product redesign for recyclability to machines to facilitate collection and processing (see, e.g., Box 2-4). While the latter tend to add upfront cost to MRFs, they promise to increase efficiency and the quality of produced recyclables. In the last 4 years, the fleet of U.S. single stream MRFs has spent close to $2 billion in retrofits and have announced that another $1 billion is to be invested in the next 3–5 years to take advantage of advancements in automation.

2.3.1 Intelligent Waste Bins

Another recent area of investigation is the use of intelligent waste bins. These bins sort recyclables into appropriate containers using sensors, Internet of Things networks, and data analytics (Kaverina, 2018). They may also be used to monitor quantities of material in large containers. Sensors could notify waste collectors that a bin is full, minimizing trips to pick up a partially full load and reducing spillover and littering. Sensors can monitor gas, temperature, odor, sound, and humidity, which may be useful in planning collection and disposal activities.

Interest is also growing in the use of technology that is AI-based and provides real-time feedback to individuals about recyclability of individual items (e.g., via mobile apps) prior to bin placement. However, these systems are not widely available because they require complex setup, controlled environments, and significant computational resources. Zhang and colleagues (2021) proposed a mobile-friendly waste classification model based on recyclable waste images and deep learning, with highly accurate results.

BOX 2-4
ReMADE Institute

The ReMADE (Reducing EMbodied-energy And Decreasing Emissions) Institute focuses on supporting efforts to use less energy in making and processing materials and products, which helps lower carbon emissions in manufacturing. This institute, funded by the Department of Energy's Office of Energy Efficiency and Renewable Energy (EERE), is one of a national network of 17 manufacturing institutes under Manufacturing USA. ReMADE has developed a Technology Roadmap that covers precompetitive research, industry-directed projects, large-scale initiatives, and strategic interest groups, all of which support the scaling-up of promising technology.

ReMADE-funded projects address such areas as circular design, digital remanufacturing, full electric vehicle reuse and remanufacturing, next-generation materials recovery facilities, plastics recycling, recovery and recycling of e-scraps, and textile recycling. The overarching goals include reducing the need for primary materials and increasing use of secondary feedstocks, in part by developing technologies that make the secondary feedstocks economically competitive with or even preferable to virgin materials. These goals are bolstered by institute support for the adoption of new technologies and education materials for the attendant workforce.

The institute was designed as a public–private partnership and had over $65 million obligated from EERE through December of fiscal year 2026, with expected private contributions of $70 million through the first grant period. As of November 2024, ReMADE membership comprised 88 industry members, 38 academic partners, 34 affiliate organizations, and 8 national laboratories, and had 39 completed, 41 active, and 13 pending projects.

SOURCE: More information available at https://www.usaspending.gov/award/ASST_NON_DEEE0007897_8900 and https://remadeinstitute.org/.

2.3.2 Advanced MRF Technologies

2.3.2.1 MRF Automation

Although data on MRF age are not collected, many MRFs were built several decades ago when MSW had a different composition and was generated at lower rates. The introduction of automation in the 2010s has been crucial for producing high-quality secondary materials for recycling. However, automation significantly increases capital expenditures, meaning larger MRFs with higher processing capacity and broader service areas are needed to justify the investment. At the same time, automation helps lower operational expenditures, reducing ongoing processing costs. With automation and robotics, MRF operations can expand to 24 hours per day with minimal downtime. Automation reduces risk by reducing injuries associated with manual labor. The addition of automation may also allow recycling to reach higher levels at MRFs primarily using manual labor. Automation can eliminate screening, which is subject to clogging and requires frequent maintenance, and can reduce contamination of sorted materials (The Recycling Partnership, 2024; see Figure 2-9). That said, while isolated communities with long spoke distances to a hub may have lower processing costs, their collection system is effective. Small-volume MRFs that use less automation can allow for processing lightweight packaging and increase recycling rates competitively through freight savings. Many of these MRFs are independent or municipally owned and have persisted for decades.

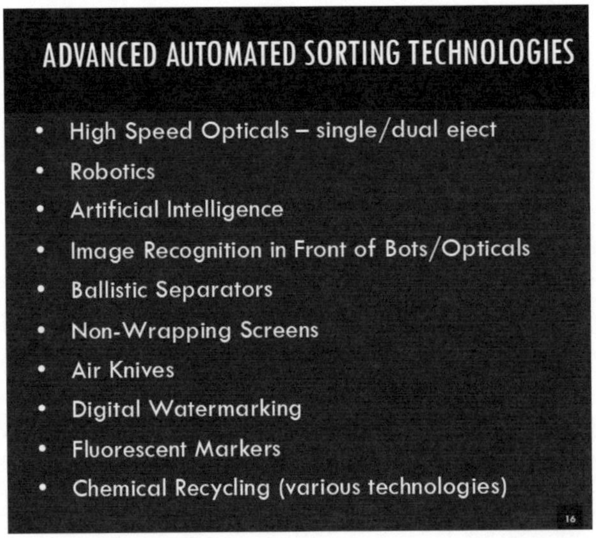

ADVANCED AUTOMATED SORTING TECHNOLOGIES

- High Speed Opticals – single/dual eject
- Robotics
- Artificial Intelligence
- Image Recognition in Front of Bots/Opticals
- Ballistic Separators
- Non-Wrapping Screens
- Air Knives
- Digital Watermarking
- Fluorescent Markers
- Chemical Recycling (various technologies)

FIGURE 2-9 Examples of automation sorting technologies.
SOURCE: Presentations to the committee by Nathiel Egosi, June 11, 2024, and Jim Frey, June 11, 2024.

Optical sorting uses a spectrometer or AI-assisted camera image recognition to identify recyclable material and uses high pressure air systems to separate it from the waste stream. Optical sorters can sort two materials away from the stream at the same time instead of one. They are highly accurate but rely on optical properties rather than shape. For example, they can detect material composition, such as the difference between PET and HDPE plastic, based on how they reflect infrared light. However, they cannot distinguish dark-colored materials. Some packaging manufacturers are adding radio-frequency ID tags to recyclables that can be recognized by sorting sensors.

Financing automation in a MRF is a challenge and its impacts on collection programs is substantial. Automating MRFs requires private and public financing through issuance of bonds, savings from reduced landfill or incineration costs (avoided cost of disposal), funds produced from selling recyclables, and grants from state and federal agencies. These funding dynamics have fundamentally changed the business model of recycling—widespread MRF processing charges are now the norm, the main source of capitalization, and a central feature in the discussion of system financing. This creation of a consistent revenue stream has fueled massive investments in new and revamped MRFs, but it has also added a cost burden to collection programs in delivering materials to MRFs ($100/ton is typical). Some state programs respond to these significant shifts. For example, one of the objectives of Oregon's EPR[5] program and the proposed Program Plan by the Circular Action Alliance in Colorado[6] is to have these "gate rates" covered by EPR fees, reducing the direct cost burden to municipalities and haulers in transacting with MRFs.

While automation may be more practical in urban areas than in smaller communities, the reduced cost of operation may justify transporting recyclables longer distances. Although automation reportedly can increase MRF efficiency, a remaining concern is that, instead, increased automation leads to escalating electricity demand and associated costs. As targets for recycling efficiency increase and more advanced technology is developed and implemented for improved MRF performance, greater expectations and costs may be shifted to upstream sorting and collection to meet MRF specifications. This shift will likely lead to changes in financing agreements, with coverage of shortfalls by collectors possibly addressed by other mechanisms, such as by EPR fees (Bradshaw et al., 2025; Liu et al., 2024). See Box 2-5 for a case study of one company using automation in a MRF.

BOX 2-5
Case Study: Rumpke Recycling & Resource Center

Rumpke Waste & Recycling Services provides waste management for 50 Ohio counties; it serves a population of around 3 million, and 96 percent have access to a curbside recycling program. Rumpke recently opened a state-of-the-art facility, including a highly advanced materials recovery facility (MRF) and education center. The MRF can process 250,000 tons per year at speeds from 30 to 60 tons per hour, and it includes sufficient redundancy to have 97–98 percent uptime. The MRF is equipped with ballistic separators, trommel screens, 19 optical sorters, magnetic separators, eddy current processors, balers, and artificial intelligence–assisted features for material tracking and characterization. Optical sorters are used for two- and three-dimensional separation of plastics and paper fiber, minimizing manual sorting. The process includes recirculation of waste material to ensure that capture of the maximum amount of recyclables. In fact, Rumpke expects to recover 98 percent of recyclables moving through the MRF. The company has invested $100 million in the facility.

SOURCES: Weiker, 2024; see also https://www.rumpke.com.

2.3.2.2 Artificial Intelligence

AI is expected to become an industry standard to address the complexity and expense of recycling collection and processing. It can play multiple roles in improving efficiency, increasing throughput, and reducing contamination. For example, AI can monitor waste materials through the MRF to power an analytical dashboard. It can optimize truck routes and logistics with a combination of GPS and GIS. And it may produce data streams that could be useful to future modelers.

The addition of AI to optical sorting is particularly useful, as it can differentiate colors and shapes through computer visioning, image recognition, and convolutional neural networks (Fang et al., 2023). Combined with machine learning, the systems can continuously adapt to changes in waste streams and recognize dirty, deformed, or damaged materials that are still recyclable.

---

[5] See https://www.oregon.gov/deq/recycling/Pages/Modernizing-Oregons-Recycling-System.aspx.
[6] See https://circularactionalliance.org/circular-action-alliance-colorado.

2.4 KEY POLICY OPTIONS AND RECOMMENDATIONS

Effective recycling policies require a combination of clear goals; strategic investments; and coordinated efforts across government, industry, and communities. This section outlines key policy options and recommendations for enhancing recycling systems, improving material recovery, and supporting sustainability objectives.

2.4.1 Setting Recycling Goals

Policymakers at all levels of government and other actors across the waste management and recycling system set recycling goals to identify desirable outcomes of policies and programs. Recycling goals may be used to identify benchmarks, measure progress, evaluate success, and simplify the communication of a policy's or program's purpose to important stakeholders (e.g., constituents and citizens of a community, businesses, company shareholders). Some entities assess progress toward sustainability using a broad set of metrics, including recycling rates, diversion rates, greenhouse gas reduction targets, cost savings, and job creation. This range of metrics reflects economic, environmental, and social goals related to sustainability.

EPA's MSW Recycling Rate Goals

The most widely used metric for evaluating recycling progress remains EPA's (2020) MSW recycling rate calculated as the total weight of recycled MSW divided by the total weight of generated MSW. The popularity of this metric stems from its simplicity and applicability across states and regions, making it accessible to a range of stakeholders. In general, however, methodologies for calculating recycling rates vary dramatically, according to which materials, processes, and sources of materials are included in the calculation.

Many states and provinces established recycling goals, such as a 25 percent MSW recycling rate, 20 or more years ago. In recent years, as some state-level goals have been attained and surpassed, some states are reevaluating and increasing recycling goals. Furthermore, some communities have adopted a "zero waste" or "circular economy" framework to organize and communicate their waste and recycling goals. At the federal level, EPA recently established the National Recycling Goal to increase the MSW recycling rate to 50 percent by 2030.[7]

However, for several reasons, the overall aggregation of weight-based recycling is insufficient to set a recycling goal as a policy. Limitations include lack of differentiation between material types, social goals for recycling, changing rates of total MSW, and the need to distribute responsibility across the system of actors.

First, the sum of the weights of recycled materials in the numerator does not adequately address important factors related to the heterogeneity of materials, including the relative environmental benefits of recycling each type of material. For example, weight does not account for how recycling a given material decreases natural resource extraction and landfill disposal. Recycling 1 ton of aluminum offers significant environmental advantages, such as reductions in energy use, greenhouse gas emissions, and human ecotoxicity. In contrast, recycling 1 ton of glass has much smaller environmental benefits. The weight-based approach overlooks these distinctions.

Second, social goals for recycling have often been overlooked or regarded as indirect factors. Although EPA's (2021b) national recycling strategy advocates for a more equitable and inclusive approach to waste management, it lacks indicators to track access and inclusion.

Third, the relevant environmental and social goals relate to recycling itself, in the numerator of the recycling rate, not total MSW as the denominator. While use of the ratio helps put recycling in context, it

---

[7] See https://www.epa.gov/circulareconomy/us-national-recycling-goal.

would be more appropriate if the denominator (total MSW generated) were fixed. However, many factors will alter total MSW before EPA's target year (2030), potentially distorting the ratio. For instance, shrinking overall MSW generation rates would artificially increase recycling rates. Conversely, changes that increased total MSW generation may be perceived as decreasing the recycling rate when in fact it is associated with economic growth and improved lives. For example, a reduction in virgin materials prices would help consumers, but it could lead to more MSW and reduce the recycling rate, even if consumers increase their quantity of recycling in the numerator.

Fourth, the onus to reach recycling goals sometimes falls largely on households, ignoring upstream and downstream decisions that also affect the recycling rate. For example, simple MSW recycling rate goals often ignore whether particular wastes can be recycled at all (e.g., how much plastic waste fits the definition recyclability of the Association of Plastic Recyclers[8]). They may also ignore whether adequate end markets exist for materials collected. Considering these factors would yield more realistic recycling goals and would better support holistic policy decisions across the MSW management system.

Expanding from the MSW Recycling Rate

For these reasons, the committee asserts that traditional recycling rate goals need to be augmented with new goals established around sustainable materials management (SMM). This framework focuses on enhancing the environmental, social, and economic benefits through all life stages of a material, concepts that are discussed throughout this report. Environmental, social, and economic benefits can be realized across the value chain of a product's extraction, manufacture, use, and end of life. This concept is driven by robust data and can provide transparent and quantifiable progress toward societal betterment.

A common framework for effective goal setting is the "SMART" mnemonic that lists five critical components of effective recycling goals, or any other goals (Hammond et al., 2015):

- **Specific:** Is the goal clearly defined? Does it address "what," "why," and "how"?
- **Measurable:** Is it possible to track progress and measure the outcome?
- **Achievable:** Is the goal realistic and attainable?
- **Relevant:** Does the goal align with long-term objectives?
- **Time-bound:** Does the goal have a deadline or a time frame to work within?

Using a SMART goal-setting method will support more standardization among the range of stakeholders while retaining the simplicity and applicability of the current metric.

Recommendation 2-1: Goals for recycling policy should expand from weight-based recycling rates to include informative metrics for sustainable materials management. To support these efforts, the U.S. Environmental Protection Agency should study how to combine multifaced sustainability goals into an overall policy framework, provide guidance for state and local governments to set and measure progress toward those goals, and use this information to evaluate progress. National recycling goals should be material specific but flexible to account for heterogeneity across regions and municipalities. These goals should include environmental, social, and economic targets, including cost-effectiveness. Goal-setting should be leveraged to design a policy framework and set new national recycling goals using best practices such as life cycle assessment and SMART (specific, measurable, accessible, relevant, and time-bound) metrics.

---

[8] See https://plasticsrecycling.org/apr-design-hub/apr-design-guide/pet-rigid.

2.4.2 Identifying and Filling Data Gaps and Needs

Data collection, reporting, and analysis are essential for guiding recycling policies and improving overall recycling program effectiveness. Reliable data sources are also necessary to properly assess and evaluate the impacts of specific policy choices, including those recommended by this committee. Unfortunately, recycling data are outdated, incomplete, or inaccessible. Throughout this report, the committee identifies data problems that hinder informed decision-making in waste and recycling systems at the national, state, local, and tribal levels. To underscore this problem, a primary public source of waste and recycling characteristics in the United States—EPA's facts and figures webpage[9]—is 6 years out of date at the time of writing. Furthermore, some of these data are inconsistent with estimates reported by private entities, perhaps because of the lack of standardized definitions of various components of the recycling system.

EPA can play a pivotal role in filling these gaps by continuing and expanding its efforts to develop standardized definitions and measurement methodologies for collecting crucial data on recycling in partnership with key stakeholders. For one important example, a centralized EPA platform or dashboard could provide streamlined access to current information.

Consistent federal funding is necessary for adequate and reliable data collection, analysis, and reporting relevant to environmental, social, and economic dimensions. A structured framework would support the prioritized collection of such data, allowing stakeholders to address technical aspects and support data-driven policy development systematically.

Data collection requires cooperation of actors across all levels of product life cycles, from extraction of virgin materials to end-of-life disposal. It requires partnerships among governments at all levels, nongovernmental organizations, and industry.

Recommendation 2-2: The U.S. Environmental Protection Agency (EPA) should enhance data collection and reporting efforts related to municipal solid waste (MSW) and MSW recycling programs to fill significant data gaps, to ensure sufficient and contemporary data are available to inform policy decisions, and to aid in developing and evaluating recycling goals based on sustainable materials management. These efforts should include appropriate input from stakeholders including other federal partners; state, local, and tribal governments; and industry partners. Additionally, EPA's efforts should include:

- **Updating its publicly available website on at least a biennial basis with national-level facts and figures about materials, waste, and recycling. Where possible, this information should expand from input-output modeling figures to include direct observation data. Where necessary, EPA should continue to provide sufficient funding for collecting and reporting these data.**
- **Developing standard definitions of recycling and methodologies on data collection and reporting for recycling and MSW. These definitions and collection methodologies should distinguish between pre- and post-consumer recycling and differentiate between open- and closed-loop recycling. This public information should include, at a minimum, material-specific data on MSW generation, recycling, composting and other food and yard waste management, combustion with energy recovery, and landfilling. To the extent possible, these data should also be reported at regional, state, and local levels.**

A summarized list of data needs and their uses is provided in Table 2-3. This list is not exhaustive but is representative of the need to improve data availability for decision making.

---

[9] See https://www.epa.gov/facts-and-figures-about-materials-waste-and-recycling.

TABLE 2-3 Summarized List of Data Needs and Their Uses

Domain	Data/Units (where applicable)	Purpose and Use of Data	Primary Actors to Collect and Report Data
Product characteristics	• Product recyclability, composition, recycled content • Aggregate producer sales records by NAICS code and region	Ensure that related policies (e.g., EPR/PRO, interventions, product bans) are working, support recyclable labeling, project future material flows	EPA, manufacturers, U.S. Census Bureau, Federal Reserve Board, Bureau of Labor Statistics
Waste generation and composition	• Timely solid waste generation estimates, bin survey results for major composition categories (tons/year)	Help advance the understanding of how the recycling system is performing, estimate level of contamination in recycling streams, complete LCA and LCI, evaluate recycling goals	EPA, states, local governments
MSW- recycling systems	• Number and capacity of recycling programs • Capital costs, operating costs, revenue	Evaluate the availability of programs, recycling capacity, level of consumer access and participation, and fiscal stability	States, municipalities, local MRF owners
System costs	• Distribution of system capital costs and operating costs • Performance versus cost histories for reporting MRFs • Consolidated fossil fuel and other virgin material taxes and subsidies	Estimate consumer cost of recycling more completely and accurately	Local government, compiled by each state
State policies and rules	• MSW facility operating rules and reporting requirements • Economic Incentives (taxes/subsidies) • Listing and brief description of state recycling targets, tipping fee surcharges, recyclable content goals, recycling rates, public participation rates • EPR data	Identify the objectives, economic impact, and constraints of government policy	EPA, state, nongovernmental, and industry experts on policy and regulation
Technological innovation	Descriptions and inventories of new MRF technologies, sorting technologies for consumers, and new patents	Understand how recycling performance could improve in the future	MSW research and development experts from industry
Environmental impacts and improvements	Input data for LCA and LCI; greenhouse gas emissions; air, water, land pollutants; from waste and recycled material transport and processing; exposures and health impact estimates (environmental, economic, and social metric units)	Ensure LCA and LCI are updated and accurate	EPA
Inventory of recyclable and recycled materials	Performance metrics, including fraction recycled (tons/day) and other impact-based indices	Improve markets and enable potential buyers and sellers of materials to be matched more easily	Local governments, MRF owners and operators, states
Consumer knowledge and behavior	• Household and establishment survey on waste and recycling • Summaries of survey studies in literature (links to key studies and papers)	Provide regular direct observations of household and commercial behavior; to improve the ability to evaluate the empirical impacts of public policies; measure social impacts of recycling, including health, distribution of programs; evaluate true cost and benefits of recycling	EPA - Surveys and bin audits measuring behavior and contamination; local governments

continued

TABLE 2-3 *continued*

Domain	Data/Units (where applicable)	Purpose and Use of Data	Primary Actors to Collect and Report Data
Macroeconomic impacts	• Recycling process data describing inputs to recycling supply chains • Jobs associated with recycling and composting • Commodity values over time (i.e., price of scrap and recycled materials per ton)	Survey recyclers to enable estimates of material flowrates, enabling estimations of material availability; evaluate recycling impacts on economy	States, manufacturers

NOTES: Data needs not described in Chapter 2 are further described throughout this report. EPA = Environmental Protection Agency; EPR = extended producer responsibility; NAICS = North American Industry Classification System; LCA = life cycle assessment; LCI = life cycle inventory; MRF = materials recovery facility; MSW = municipal solid waste; PRO = producer responsibility organization.

Key Policy Option 2-1: The U.S. Environmental Protection Agency (EPA) could support studies to update or otherwise fill missing data gaps to ensure sufficient data are available to inform policy decisions on recycling. These studies could include:

- Tracking household time spent by single and multi-family households on recycling to support more complete and accurate estimates of the economic and social costs of recycling and ensure that life-cycle assessment models are as updated and accurate as possible.
- Regularly collecting and reporting direct observations of household and commercial behavior related to recycling. In addition to filling knowledge gaps, these data would complement top-down modeling in the recycling system and enable empirical study of the impact of public policy. As part of these efforts, EPA could consider a periodic household and commercial survey for waste and recycling akin to the Energy Information Administration's Residential Energy Consumption Survey.

REFERENCES

Acuff, K., and D.T. Kaffine. 2013. Greenhouse gas emissions, waste and recycling policy. *Journal of Environmental Economics and Management* 65(1):74–86.

Allesch, A., and P.H. Brunner. 2015. Material flow analysis as a decision support tool for waste management: A literature review. *Journal of Industrial Ecology* 19(5):753–764.

Anshassi, M., and T.G. Townsend. 2023. The hidden economic and environmental costs of eliminating kerbside recycling. *Nature Sustainability* 6(8):919–928.

Anshassi, M., and T.G. Townsend. 2024. Residential recycling in Florida: A case study on costs, environmental impacts, and improvement strategies. *Resources, Conservation and Recycling* 206:107627

Arena, U., and F. Di Gregorio. 2014. A waste management planning based on substance flow analysis. *Resources, Conservation and Recycling* 85:54–66.

Awino, F.B., and S.E. Apitz. 2024. Solid waste management in the context of the waste hierarchy and circular economy frameworks: An international critical review. *Integrated Environmental Assessment and Management* 20(1):9–35. https://doi.org/10.1002/ieam.4774.

Baffoe-Bonnie, J., and F. Ezeala-Harrison. 2023. Analysis of optimal solid waste recycling policy: Evidence from US using panel data. *Advances in Management and Applied Economics* 13(3):1–2

Best, H., and T. Kneip. 2019. Assessing the casual effect of curbside collection on recycling behavior in a non-randomized experiment with self-reported outcome. *Environmental and Resource Economics* 72:1203–1223.

Bohm, R.A., D.H. Folz, T.C. Kinnaman, and M.J. Podolsky. 2010. The costs of municipal waste and recycling programs. *Resources, Conservation and Recycling* 54(11):864–871.

Bolingbroke, D., K.T.W. Ng, H.L. Vu, and A. Richter. 2021. Quantification of solid waste management system efficiency using input–output indices. *Journal of Material Cycles and Waste Management* 23:1015–1025.

Bradshaw, S.L., H.A. Aguirre-Villegas, S.E. Boxman, and C.H. Benson. 2025. Material recovery facilities (MRFs) in the United States: Operations, revenue, and the impact of scale. *Waste Management* 193:317–327.

Budds, D. 2022. It's time to end our "rat nutrition" program. The case for containerized trash. *Curbed*. https://www.curbed.com/2022/09/new-york-city-waste-containerization.html.

Cecot, C., and W.K. Viscusi 2022. The hierarchy and performance of state recycling and deposit 6643 laws. *Vermont Journal of Environmental Law* 23(4):319–348.

Chioatto, E., M.A. Khan, and P. Sospiro. 2023. Sustainable solid waste management in the European Union: Four countries regional analysis. *Sustainable Chemistry and Pharmacy* 33:101037.

Dahlén, L., and A. Lagerkvist. 2009. Evaluation of recycling programmes in household waste collection systems. *Waste Management & Research* 28(7):577–586. https://doi.org/10.1177/0734242X09341193.

Devi, R., A.K. Singh, A. Kumar, R. Kumar, S. Rani, and R. Chandra. 2024. Development of technologies for municipal solid waste management: Current status, challenges, and future perspectives. In *Integrated waste management: A sustainable approach from waste to wealth* (pp. 37–62). Singapore: Springer Nature.

Dumlao-Tan, M.I., and A. Halog. 2017. Moving towards a circular economy in solid waste management: Concepts and practices. In S. Goel (ed.), *Advances in solid and hazardous waste management*. Springer. https://doi.org/10.1007/978-3-319-57076-1_2.

EPA (U.S. Environmental Protection Agency). 2020. *Advancing sustainable materials management: 2018 fact sheet. Assessing trends in materials generation and management in the United States*. https://www.epa.gov/sites/default/files/2021-01/documents/2018_ff_fact_sheet_dec_2020_fnl_508.pdf.

EPA. 2021a. *An analysis of lithium-ion battery fires in waste management and recycling*. https://www.epa.gov/system/files/documents/2021-08/lithium-ion-battery-report-update-7.01_508.pdf.

EPA. 2021b. *National recycling strategy: Part one of a series on building a circular economy for all*. https://www.epa.gov/system/files/documents/2021-11/final-national-recycling-strategy.pdf.

EPA. 2024. *National overview: Facts and figures on materials, wastes and recycling*. https://www.epa.gov/facts-and-figures-about-materials-waste-and-recycling/national-overview-facts-and-figures-materials.

EPA Environmental Financial Advisory Board. 2019. *Revenue options for a waste backhaul service program in rural Alaska*. https://www.epa.gov/sites/default/files/2019-12/documents/revenue_options_for_a_waste_backhaul_service_program_in_rural_alaska.pdf.

Eunomia Research and Consulting. 2021. *Cutting lithium-ion battery fires in the waste industry*. https://eunomia.eco/reports/cutting-lithium-ion-battery-fires-in-the-waste-industry.

Eunomia Research and Consulting. 2023. *The 50 states of recycling: A state-by-state assessment of US packaging recycling rates*. https://eunomia.eco/the-50-states-of-recycling-a-state-by-state-assessment-of-us-packaging-recycling-rates.

Fang, B., J. Yu, Z. Chen, et al. 2023. Artificial intelligence for waste management in smart cities: a review. *Environmental Chemistry Letters* 21:1959–1989. https://doi.org/10.1007/s10311-023-01604-3.

Firmansyah, F., I. Park, M. Corona, O. Aphale, A. Ahuja, M. Johnston, K.L. Thyberg, E. Hewitt, and D.J Tonjes. 2024. Variation in municipal solid waste generation and management across time and space. *Resources, Conservation and Recycling* 204:107472.

Fullerton, D., and T.C. Kinnaman. 1995. Garbage, recycling, and illicit burning or dumping. *Journal of Environmental Economics and Management* 29(1):78–91. https://doi.org/10.1006/jeem.1995.1032.

GAO (Government Accountability Office). 2020. *Building on existing federal efforts could help address cross-cutting challenges*. https://www.gao.gov/assets/gao-21-87.pdf.

Goulart Coelho, L.M., L.C. Lange, and H.M. Coelho. 2017. Multi-criteria decision making to support waste management: A critical review of current practices and methods. *Waste Management & Research: The Journal for a Sustainable Circular Economy* 35(1):3–28.

Greene, K., and D.J. Tonjes. 2014. Quantitative assessments of municipal waste management systems: Using different indicators to compare and rank programs in New York State. *Waste Management* 34(4):825–836. https://doi.org/10.1016/j.wasman.2013.12.020.

Gutierrez-Lopez, J., R.G. McGarvey, C. Costello, and D.M. Hall. 2023. Decision support frameworks in solid waste management: A systematic review of multi-criteria decision-making with sustainability and social indicators. *Sustainability* 15(18):13316.

Hammond, J.S., R.L. Keeney, and H. Raiffa. 2015. *Smart choices: A practical guide to making better decisions.* Harvard Business Review Press.

Harder, R., Y. Kalmykova, G.M. Morrison, F. Feng, M. Mangold, and L. Dahlén. 2014. Quantification of goods purchases and waste generation at the level of individual households. *Journal of Industrial Ecology* 18(2):227–241.

He, R., and M.J. Small. 2022. Forecast of the US copper demand: A framework based on scenario analysis and stock dynamics. *Environmental Science & Technology* 56(4):2709–2717.

He, R., Sandoval-Reyes, M., Scott, I., Semeano, R., Ferrao, P., Matthews, S., and Small, M.J. 2022. Global knowledge base for municipal solid waste management: Framework development and application in waste generation prediction. *Journal of Cleaner Production* 377:134501.

Hu, J.L., Honma, S., and Chang, T.M. 2024. Recycling efficiencies in Japan's administrative regions: Findings from network data envelopment analysis. *Circular Economy and Sustainability* 4(1):1–21.

Huber, J., W.K. Viscusi, and J. Bell. 2023. Using objective characteristics to target household recycling policies. *Environmental Law Reporter* 53:10804.

Iannuzzi, A. 2024. *Greener products: The making and marketing of sustainable brands*. CRC Press.

Jayasinghe, P.A., S. Derrible, and L. Kattan. 2023. Interdependencies between urban transport, water, and solid waste infrastructure systems. *Infrastructures* 8(4):76.

Jităreanu, A.F., G. Ignat, M. Mihăilă, and C.L. Costuleanu. 2023. Real-time feedback and education: The key to reducing contamination in home recycling waste. *Environmental Engineering & Management Journal* 22(6).

Kaverina, S. 2018. Sort It Out: TrashBot, a Smart Bin developed by CleanRobotics, uses AI to help improve waste management. *Medium*. https://medium.com/makersbootcamp/sort-it-out-a-smart-trash-bin-from-cleanrobotics-66a46bedff11.

Khan, A.H., E.A. López-Maldonado, N.A. Khan, L.J. Villarreal-Gómez, F.M. Munshi, A.H. Alsabhan, and K. Perveen. 2022. Current solid waste management strategies and energy recovery in developing countries-State of art review. *Chemosphere* 291:133088.

Liu, M., S. Grimes, K. Salonitis, and L. Litos. 2024. Modular model and simulation for process optimisation in advanced material recovery facilities (MRFs). *Procedia CIRP* 129:250–255.

Macauley, M.K., and M.A. Walls. 2010. Solid waste policy. In *Public policies for environmental protection* (pp. 261–286). Routledge.

Makarichi, L., K.A. Techato, and W. Jutidamrongphan. 2018. Material flow analysis as a support tool for multi-criteria analysis in solid waste management decision-making. *Resources, Conservation and Recycling* 139:351–365.

Meleddu, M., M. Vecco, and M. Mazzanti. 2024. The role of voluntary environmental policies towards achieving circularity. *Ecological Economics* 219:108134.

Mensah, D., N. Karimi, K.T.W. Ng, T.S. Mahmud, Y. Tang, and S. Igoniko. 2023. Ranking Canadian waste management system efficiencies using three waste performance indicators. *Environmental Science and Pollution Research* 30(17):51030–51041.

Morawski, C., and J. Wilcox. 2017. Is access everything? *Resource Recycling*. https://resource-recycling.com/recycling/2017/03/10/is-access-everything.

National Waste & Recycling Association. 2018. *Material recovery facilities issue brief*. https://wasterecycling.org/wp-content/uploads/2020/10/Issue_Brief_MRFS.pdf.

Nebraska Recycling Council. 2019. *Hub and spoke recycling: Regional cooperation reduces costs, increases efficiency*. https://nrcne.org/wp-content/uploads/2019/05/Nebraska_Hub-and-Spoke.pdf.

Nebraska Recycling Council. 2024. *State of recycling in Nebraska*. https://storymaps.arcgis.com/stories/2b3b3e1dd15d40daadabe023cb7d57f4.

Ng, K.S., and A. Yang. 2023. Development of a system model to predict flows and performance of regional waste management planning: A case study of England. *Journal of Environmental Management* 325:116585.

Noehammer, H.C., and P.H. Byer. 1997. Effect of design variables on participation in residential curbside recycling programs. *Waste Management & Research* 15(4):407–427.

Ogieriakhi, M.O., and X. Wang. 2024. Do mandatory environmental policies really work? A case study of California's mandatory commercial recycling law. *Economic Analysis and Policy* 81:915–930.

Olafasakin, O., J. Ma, S.L. Bradshaw, H.A. Aguirre-Villegas, C. Benson, G.W. Huber, V.M. Zavala, and M. Mba-Wright. 2023. Techno-economic and life cycle assessment of standalone Single-Stream material recovery facilities in the United States. *Waste Management* 166:368–376.

Palmer, K., H. Sigman, and M. Walls. 1995. The cost of reducing municipal solid waste. *Journal of Environmental Economics and Management* 33(2):128–150.

Pinto, A.D., H. Jalloul, N. Nickdoost, F. Sanusi, J. Choi, and T. Abichou. 2022. Challenges and adaptive measures for U.S. municipal solid waste management systems during the COVID-19 pandemic. *Sustainability* 14(8):4834.

Prenovitz, E.C., P.K. Hazlett, and C.S. Reilly. 2023. Can markets improve recycling performance? A cross-country regression analysis and case studies. *Sustainability* 15(6):4785.

Psyrri, G., M.Z. Hauschild, T.F. Astrup, and A.T.M. Lima. 2024. Recycling for a sustainable future: Advancing resource efficiency through life cycle assessment resource indicators. *Resources, Conservation and Recycling* 209:107759. https://doi.org/10.1016/j.resconrec.2024.107759.

Salemdeeb, R., R. Saint, F. Pomponi, K. Pratt, and M. Lenaghan. 2022. Beyond recycling: An LCA-based decision-support tool to accelerate Scotland's transition to a circular economy. *Resources, Conservation & Recycling Advances* 13:200069.

Schmidt, S., and D. Laner. 2023. Environmental waste utilization score to monitor the performance of waste management systems: A novel indicator applied to case studies in Germany. *Resources, Conservation & Recycling Advances* 18:200160.

Sharma, K.D., and S. Jain. 2020. Municipal solid waste generation, composition, and management: The global scenario. *Social Responsibility Journal* 16(6):917–948.

Šomplák, R., V. Smejkalová, M. Rosecký, L. Szásziová, V. Nevrlý, D. Hrabec, and M. Pavlas. 2023. Comprehensive review on waste generation modeling. *Sustainability* 15(4):3278.

Sondh, S., D.S. Upadhyay, S. Patel, and R.N. Patel. 2024. Strategic approach towards sustainability by promoting circular economy-based municipal solid waste management system-A review. *Sustainable Chemistry and Pharmacy* 37:101337.

Subramanyam, V., D. Paramshivan, A. Kumar, and Md. A.H. Mondal. 2015. Using Sankey diagrams to map energy flow from primary fuel to end use. *Energy Conversion and Management* 91:342–352. https://doi.org/10.1016/j.enconman.2014.12.024.

Taelman, S., D. Sanjuan-Delmás, D. Tonini, and J. Dewulf. 2020. An operational framework for sustainability assessment including local to global impacts: Focus on waste management systems. *Resources, Conservation, and Recycling* 162.

The Recycling Partnership. 2024. *State of recycling: The present and future of residential recycling in the US*. https://recyclingpartnership.org/wp-content/uploads/2024/01/Recycling-Partnership-State-of-Recycling-Report-1.9.23.pdf.

Tsai, F.M., T.D. Bui, M.L. Tseng, M.K. Lim, and J. Hu. 2020. Municipal solid waste management in a circular economy: A data-driven bibliometric analysis. *Journal of Cleaner Production* 275:124132.

Vines, V., M. Pasquali, S. Ganguli, and D.E. Meyer. 2023. Understanding the trade-offs of national municipal solid waste estimation methods for circular economy policy. *Journal of Cleaner Production* 412:137349.

Viscusi, W.K., J. Huber, and J. Bell. 2022. Quasi-experimental evidence on the impact of state recycling and deposit laws: Household recycling following interstate moves. *American Law and Economics Review* 24(2):614–658. https://doi.org/10.1093/aler/ahac006.

Wang, D., Y.T. Tang, Y. Sun, and J. He. 2022. Assessing the transition of municipal solid waste management by combining material flow analysis and life cycle assessment. *Resources, Conservation and Recycling* 177:105966.

Wang, Y., J.W. Levis, and M.A. Barlaz. 2021. Development of streamlined life-cycle assessment for the solid waste management system. *Environmental Science & Technology* 55(8):5475–5484.

Weiker, J. 2024. August 6. Rumpke opens what it calls the largest recycling center in North America on Joyce Avenue. *The Columbus Dispatch*. https://www.dispatch.com/story/news/environment/2024/08/06/new-rumpke-recycling-center-is-largest-in-north-america-company-says/74685159007/.

Wilson, D.C. 2023. Learning from the past to plan for the future: An historical review of the evolution of waste and resource management 1970–2020 and reflections on priorities 2020–2030—The perspective of an involved witness. *Waste Management & Research* 41(12):1754–1813.

Wilson, D.C., L. Rodic, M.J. Cowing, C.A. Velis, A.D. Whiteman, A. Scheinberg, R. Vilches, D. Masterson, J. Stretz, and B. Oelz. 2015. "Wasteaware" benchmark indicators for integrated sustainable waste management in cities. *Waste Management* 35:329–342.

Wise Guy Reports. 2024. *Global arm type garbage truck market research report*. https://www.wiseguyreports.com/reports/arm-type-garbage-truck-market.

Zhang, Q., X. Zhang, Z. Mu, R. Tian, X. Wang, and X. Liu. 2021. Recyclable waste recognition based on deep learning. *Resources, Conservation and Recycling* 171:105636.

3
Benefits and Challenges of Recycling Programs and the Role of Policy

Summary of Key Messages

- **Economic constraints on recycling programs:** Recycling programs face significant economic challenges, particularly related to high operational collection and infrastructure costs, limited local budgets, and unstable resale values for recyclables. Smaller municipalities, especially in rural areas, struggle to achieve economies of scale, making recycling programs financially unsustainable without external support.
- **Balancing environmental benefits and external costs:** Recycling offers substantial environmental benefits by reducing landfill use and multiple polluting emissions associated with virgin material production and use. While recycling processes also introduce external costs, including emissions from transportation and energy use during manufacturing, the environmental benefits of recycling can outweigh these costs. Effective policies balance these trade-offs to optimize environmental and social welfare.
- **Heterogeneity is a significant factor in recycling programs:** The implementation of recycling policies is complicated by heterogeneity—that is, a broad range of differences across costs, benefits, existing capabilities, different material volumes, transportation distances, access to end markets, and cultural norms across regions.
- **Recycling programs require policies that are region specific, flexible, targeted, enforceable, and easy to understand:** A one-size-fits-all approach to recycling policy may be ineffective, given differences across U.S. regions and municipalities in recycling capabilities, population densities (rural versus urban), economies, and environmental concerns. Tailored national policies that address regional and local constraints and provide targeted support can enhance the effectiveness and sustainability of recycling programs across the United States.
- **Market failures from incorrect incentives:** Even with access to recycling, households and businesses that decide which items go to the landfill or to recycling do not face all of the social and environmental costs of their choices (e.g. external costs from virgin material extraction and from landfill disposal). Policy interventions need to account for these inherent market failures by sufficiently encouraging recycling at the system, commercial, and household levels.

Recycling involves choices by households, businesses, and many levels of government. This chapter takes a step back to first consider the role of government in the recycling system. In order to identify major goals for waste and recycling policy, the chapter considers the fundamental problems that recycling may address and how governments might be able to reduce those problems. It then describes the logic of how each alternative public policy can help, including the advantages and disadvantages of each.

3.1 WHAT PROBLEMS SHOULD RECYCLING POLICIES FIX?

Recycling is a concern not only for households and businesses but also for policymakers at local, county, state, and federal levels. Later chapters detail the funding provided by municipal governments and regulations from higher levels. But what goals should governments set when choosing waste and recycling policies and financing? If consumers and firms took into account all the relevant costs and benefits of their waste and recycling decisions, government activity may be unnecessary. But in reality, policymakers need to intervene to address a well-identified *market failure*—a misalignment between private incentives and social costs and benefits (the term applies even where no formal market exists). Policymakers may also intervene when the distribution of costs and benefits across households does not align with social goals.

Two categories of market failure are particularly relevant for waste and recycling. The first is "public goods," where government provision can increase social and economic welfare. This category can encompass basic waste collection, economies of scale, information, and transaction costs. The second category of market failure is ignoring external costs such as from waste collection, recycling, and virgin materials extraction. Markets may also ignore costs for future generations if resources are depleted, especially nonrenewable resources.

3.1.1 Public Goods

When the public values a benefit such as a park, bridge, or wildlife preserve, the government may provide this public good because it would not be cost-effective for a private firm to do so. A firm would need to charge a price from every user, which may be impractical to enforce (e.g., charging to cross a pedestrian bridge or to enter a huge wildlife preserve). If people value the public good by more than its cost, then a government can use tax revenue to buy the land, build the infrastructure, provide beneficial use for free, and raise total *social welfare*, which is defined as including economic value, environmental value, and social justice. Similarly, a waste collection and disposal firm, which charges a price high enough to cover its costs, cannot prevent people from disposing of waste at no cost by dumping it in a remote area.

Municipal waste and recycling collection is inherently a public good because individuals cannot be charged for their benefits from environmental and public health protection (i.e., the reduction of the external environmental costs from waste disposal). Health problems associated with improper waste management have confounded cities back to ancient Athens and Rome. Regarding colonial America, Melosi (2004) writes, "In eastern cities, where crowding became a chronic problem as early as the 1770s, the streets reeked with waste, wells were polluted, and deaths from epidemic disease mounted rapidly." Early policymakers recognized waste collection as an essential public service that government must provide (or must ensure provision by regulated private firms). They later recognized that waste can be controlled more sustainably by recycling materials that can be sold to offset some costs.

Even when governments rely on private entities—including households and waste management firms—they can support effective waste management in their jurisdiction by taking on some fixed costs, namely providing information, and by addressing economies of scale.

Governments may support waste management in their jurisdiction by taking on the fixed costs of gathering, verifying, and disseminating information. Households need information about what material can be recycled, how to sort these items, what day they are collected, and how to store them until collection. And each materials recovery facility (MRF) needs information on preparing recycled materials for sale, finding buyers for each material, negotiating prices and contracts, and generating forecasts about future prices. Additionally, new market entrants can benefit by making use of the government-provided information, brokerage services, and standardized contracts. Thus, free provision of information can improve recycling behavior and recycling markets, thereby reducing environmental damage.

Governments can also address challenges associated with economies of scale through tax dollars. Landfills, for instance, certainly exhibit economies of scale. They require large pieces of equipment or facilities to operate (e.g., scale house, compactors, scrapers) regardless of the landfill size or the quantity of waste to be disposed. Smaller facilities will therefore have a higher capital cost per unit of waste disposed, while larger facilities can spread these costs over a higher quantity of waste, to achieve lower per unit costs. Thus, in areas of lower population density, governments may need to use tax dollars to support landfills.

For similar reasons, MRFs also exhibit economies of scale that cause market failures in some cases, especially in small towns and rural areas. Industry experts report uniformly that operating costs per ton are high for rural, less densely populated areas using labor-intensive technology, and that costs per ton are much lower for urban, more densely populated areas using capital-intensive technologies (Pressley et al., 2015). Bradshaw and colleagues (2025) concluded, "Small MRFs report minimal equipment use, relying primarily on picking lines for sorting. Advanced sorting technology such as optical sorters, robots, and infrared spectroscopy are used exclusively in large MRFs, reflecting the importance of quantity of inbound material to justify capital investment in equipment." Another challenge for rural areas is that they tend to be located at long distances from a MRF, and the MRFs they use tend to have long travel times to end markets for recycled materials. In some locations, a hub-and-spoke system, as discussed in Chapter 2, could help reduce costs for multiple, small-population, rural areas that together incur the up-front cost to build one MRF of sufficient size to achieve cost-effectiveness.

Economies of scale are not a market failure in densely populated areas with competition among several large MRFs. In remote areas where only one MRF can achieve cost-efficient size, however, a private firm could take advantage of market power and raise prices. This market failure can justify local government ownership of the large cost-efficient MRF, or possibly local government regulation of the price charged by a private, cost-efficient MRF.

In addition, hybrid facilities can partially process materials in locations with lower population densities; these limited sorting facilities are already a key part of an efficient system in many parts of the country. Other areas, with even greater transportation distances and lower throughput, will benefit from public policy that uses tax revenue to help build a MRF that is government owned or regulated. Instead of allowing a private firm to charge a price high enough to cover all costs, this policy can cover the cost of recycling while encouraging more quantity by charging a low price per additional ton.

In summary, governments can support waste management activities by providing information to the public and to waste management firms, and by leading and subsidizing efforts to achieve cost-effectiveness. Alternative sources of funding are discussed in Chapter 4.

3.1.2 External Costs

Recycling and other aspects of waste management are part of a broad process that extends from mining minerals in the most remote parts of the world to manufacturing food and household goods, to consuming those goods, to disposing of the resulting waste. Decision-makers throughout these stages may perceive some costs and benefits but not perceive or experience other costs; these are known as *external costs*. Ignoring external costs can lead to private decisions that are not socially optimal; it can lead to a view that virgin materials are economically favorable, despite the overall advantages of using secondary materials. This market failure can be addressed by public policy or regulations that balance private and external costs and benefits and thus raise social welfare.

A household that sorts wastes for recycling and disposal receives no private compensation for these actions. The household also receives no benefits from buying items that are cheaper for MRFs to process or that yield higher-value recyclables.[1] Although many households still choose to recycle and buy recyclable products for social reasons, not all households have this value structure, nor do all have the luxury to spend time on activities that yield no financial reward.

Public policies may address this misalignment of incentives for households in two ways: First, policies may create a financial incentive for recycling. Examples of such policies include a deposit-return system and programs that charge a price per bag of garbage. All such policies are further discussed in Chapter 4. Second, in light of the lack of compensation for households, public policies may attempt to increase the convenience of recycling through single-stream curbside recycling and weekly rather than biweekly services.

Municipalities face key tradeoffs. Incentives for households to recycle may include lower costs for recycling than for disposal, but those incentives may also affect litter, dumping, and contamination in the recycling stream. Also, municipalities' recycling may create external costs such as noise and air pollution from recycling trucks. Most municipalities will not account for the external costs of their recycling activities that are imposed on other jurisdictions.

---

[1] A lack of incentives for households to care about recyclability translates into a lack of incentives for product producers and retailers as well.

Existing regulations may help reduce external costs. For example, landfill regulations require protective linings to capture leachate, and can require methane capture; some states provide incentives for using that methane for heat or electricity generation. These policies reduce external costs to the environment but place burdens on municipalities. Several states, including California, Washington, and New York, have carbon fees or cap-and-invest programs that include landfill and vehicle emissions, transforming the leachate and air pollution costs from external to internal. These fees are typically paid by municipalities, waste management operators, and ultimately consumers through waste collection fees, landfill tipping fees, and carbon credit purchases. Some costs are passed to businesses and producers, in locations where extended producer responsibility (EPR) programs apply. Still, remaining indirect and external costs may justify policies that encourage recycling.

Recycling policies and regulations aim to balance economic and environmental trade-offs, shifting costs and benefits across different stakeholders. While municipalities may benefit from lower landfill costs and reduced litter, they also face external costs such as increased truck emissions and noise pollution. In some cases, existing regulations help internalize these costs—for example, landfill fees and carbon pricing programs place financial burdens on municipalities, waste operators, and consumers to account for environmental impacts. However, not all external costs are fully addressed. Gaps in enforcement, inadequate infrastructure, and insufficient incentives can lead to unintended consequences, such as illegal dumping, unsustainable resource extraction, and environmental harms that extend across jurisdictions and generations.

Littering and Dumping

Regulations on haulers and landfills may lead to higher fees for legal waste disposal, which may inadvertently provide incentives for litter or improper dumping. Most jurisdictions have rules against these practices that are not enforced and are easily ignored. The social or environmental costs of littering and dumping can be substantially higher than those of landfills or recycling facilities, but those costs are external to those who commit these acts. Litter is a visual pollutant, and its removal to a landfill is much more costly than curbside collection of waste. Illegal dumping has even higher external costs, including substantial damages to ecosystems. Many policymakers have addressed this market failure by enacting deposit-return systems—for example, for beverage containers (to avoid visual pollution from litter) and lead-acid batteries (to avoid the serious health costs of lead exposure).

Virgin Materials Extraction

Research shows that the external costs per ton of virgin materials extraction are substantially higher than those per ton of waste or recycling. Advocates of a circular economy point out that recycling provides secondary materials for use as inputs for production and thus can reduce the need for mining new metals and harvesting new timber (Fullerton and Kinnaman, 2025; Stahel, 2016). The use of secondary materials for product manufacturing and infrastructure is advantageous from environmental, sustainability, and materials security perspectives. But mining and other primary sources continue to dominate material supply chains for several reasons, including the greater ability of mining sources to meet high-volume demand in a reliable manner, the history of existing infrastructure and contracts for primary sources, and the variability (and low predictability) of recycled material prices (Moore et al., 2024; Schmidt, 2021).

For 100 years, the Eagle Mine in Colorado extracted virgin materials including gold, silver, and zinc. The U.S. Environmental Protection Agency named it a Superfund site in 1986, pointing to soil contamination from large quantities of arsenic, zinc, cadmium, and lead. The mine also killed fish in the Eagle River and threatened downstream drinking water. Resulting environmental damages were estimated using empirical models of residential house prices, finding reductions in property value. Within 6 miles of the mine, these devaluations were about $25,000 per house, in 1985 U.S. dollars (USD) (Damigos, 2006). Current mining activities face stronger regulations than in those past years, so their damages are lower but not necessarily small. What is more, external costs can be substantial in developing countries with less regulation.[2]

Many studies of external costs from virgin materials extraction use life cycle assessment models to estimate overall environmental costs associated with final products sold to consumers. Kinnaman (2014) summarizes external costs per ton of virgin materials extraction and also external costs per ton of waste at the end of a product's life. His estimate of the external cost of waste disposal is only about $10–$15 per ton, and the external cost of recycling might be similar; but the external costs of some virgin materials extraction can be over $200 per ton. If mining policy does not directly address external costs, and if recycled material is a good substitute for virgin material to make a new product, policy interventions by U.S. and state governments may be justified. Such policies include those that encourage recycling; make recycled inputs cheaper; and reduce use of virgin materials, thus reducing environmental damages from extraction.

Interstate and International Pollution and Harms

Because municipal solid waste (MSW) recycling budgets are often limited, municipal policymakers may not worry about environmental damages from extraction of virgin materials elsewhere in the United States or in other countries. The same is true for environmental and public health harms associated with pollution due to exporting recyclables and other waste (e.g., electronics) to other countries. External costs that cross many jurisdictional boundaries could be addressed by financial support and other policies at the federal level that can maximize U.S. economic well-being by taking into account all direct, indirect, and external costs on all U.S. citizens. And external costs can become more internal and direct when U.S. citizens care about damages to ecosystems and wildlife around the world; this concern may increase support for policies for reducing interstate and international costs of recycling and waste disposal.

Future Generations

Using natural resources has costs for future generations that may not be felt by present generations. For instance, funds can be raised from private investors for major investments to build new sanitary landfills. Those investors earn a return by charging fees sufficient to cover the reduction in the value of their investment as the landfill becomes full. Indeed, higher fees that reflect the full costs of landfills can provide incentives for recycling. And landfill owners can use the extra fees to reinvest in future assets. But these dynamics do not address the reduction in the remaining value of the land from resource depletion. For another example, public lands are leased to private companies for mining, forestry, and other resource extraction. Economic studies show that these public lands and even private lands are often leased at rates that are too low to cover environmental damages or the reduction in land value from using up the resources on the land (Prest, 2022). These activities impose a cost on future generations, as well as people alive today who will live long enough to feel the reduction in national wealth.

Sustainability

To some extent, finite resources can be depleted and still satisfy the definition of sustainability, if reductions in natural resources are offset by new investments that ensure that future citizens are at least as well off as current citizens (Solow, 1991). Those new investments can be in the form of physical infrastructure, technology, or intangible assets. However, if investments are insufficient to maintain future well-being, then using up natural resources imposes an external cost on future generations across the

---

[2] One study of mineral mining in Indonesia estimated damages from mining noise, dust, decreased quality and quantity of groundwater, and various air pollutants. From interviews with 50 households, Furoida and Susilowati (2021) collected a combination of objective data and other qualitative information about costs of illnesses and of replacing losses to water supply. They estimate that the average family's loss is about 143 USD/year, a substantial figure in an area where the average annual income is 205 USD.

country, including long-term pollutants such as greenhouse gas emissions, and depletion of land, minerals, and other natural resources.

The local governments and private parties that fund most recycling in the United States (see Chapter 4) cannot be expected to account for the external costs such as littering and dumping or virgin material extraction that are felt outside their jurisdictions (across states and internationally) and in future generations. State and federal governments must help if these external costs are to be taken into account. This help may come in the form of support for local policies with statewide and national benefits that justify funding from higher levels of government. So far, state and national policymakers have enacted targets and even mandates for MSW recycling programs (see Chapter 7). Most of these mandates are not funded, however, so they impose costs on municipalities for the sake of statewide or nationwide benefits. Finally, moreover, accounting for external costs of waste disposal and recycling would also likely affect product design and prices.

3.2 THE HETEROGENEITY OF RECYCLING

Different materials have different optimum recycling levels, which change over time; thus no one public policy is ideal for all materials. Both waste management in general and recycling in particular have costs and benefits that are extremely heterogenous. This heterogeneity affects both private household costs and benefits and the public policies that will be most effective.

First, material type is key to determining whether items should be recycled, landfilled, or incinerated. Each household or commercial recycling bin could include bottles and jugs made of polyethylene terephthalate (PET) and high-density polyethylene (HDPE), cans made of aluminum and steel, paper, cardboard, and glass bottles and jars. These materials are typically processed cost-effectively by a MRF; other materials that could technically be recycled cost more to collect and process than they can earn in end markets. In addition, many materials are not accepted at MRFs because they do not have critical mass in generation to link to regular truckload shipments of those materials to markets. These factors are affected by available technology and market fluctuations (Table 3-1). In addition to immediate financial cost considerations, some inert materials could remain in landfills with low external costs, while other materials (e.g., lithium-ion batteries [see Box 2-1]) are dangerous for MRFs to handle.

Second, as discussed in earlier sections, waste management in general and recycling in particular differ greatly across locations. Large cities usually have good access to recycled commodity markets, but, in some cases, remote locations can have extremely high costs of recycling. Collection trucks must travel further to the MRF, which raises costs even before that material can be cleaned, crushed, and baled. Then the bales may require longer and more expensive transport to end markets. Geography, community size, and access to markets are complex in the United States and result in different experiences with recycling. Locations also differ by demography and preferences, which can increase the costs of promoting participation in recycling programs. Table 3-1 shows how private and external costs and benefits together can make recycling socially worthwhile, even when private costs exceed benefits, for some material types. Accounting for these external costs can justify public intervention to encourage or require recycling. Other material types have lower total costs and higher net social benefits associated with disposing of them in a landfill.

Third, waste disposal choices might need to differ across time because of changes in technology, market prices, preferences, and other conditions. An older MRF may rely on labor-intensive technology to collect and sort materials and to clean or bale each material. Newer facilities have capital-intensive technologies where the truck can dump mixed materials onto a conveyor belt that uses video cameras trained by machine learning to identify each material. Then a puff of air or a robotic arm can cost-effectively move each item to the appropriate pile. Other examples appear in the third column of Table 3-1. End-market prices for each recycled material change over time and are often quite volatile.

TABLE 3-1 Examples of the Wide Variation Across Components of Cost for Waste and Recycling

Costs	Dimensions of Variation		
	Different Materials	Different Locations	Changes Over Time
Private cost of recycling	• Secondary aluminum, PET and HDPE have high market values, offsetting cost at MRF • Glass can cross-contaminate other recyclables and raise facility maintenance costs • Glass has low value in end markets • Higher value for clear glass or plastic than for colored glass and plastic • MSW that contains lithium-ion batteries creates high risk of fires, raising internal cost at a MRF	• MRFs in some locations face high wage costs • Rural areas have higher transportation costs • Manual, low-volume MRFs can be more expensive than automated MRFs • Household storage of recyclables is more costly in areas with high housing costs per square foot	• Improved sorting technology reduces MRF labor costs and improves quality and value of commodities • Easier international trade raises the value of sorted recyclables but can be subject to volatile political tensions for some commodities such as mixed paper, metals, and mixed plastics • End market prices are volatile, so MRF planning is risky and difficult if based on commodity value
External cost of recycling	• Because of mixed components and cross-contamination, processing paper can yield high levels of residues • MSW that contains lithium-ion batteries creates high risk of fires, raising external costs from a MRF	• Rural areas with longer distances to a MRF or end markets use more fuel, raising emissions • MRFs in a populated area impose larger damages from traffic, noise, odors	• Sorting improvements reduce MRF residuals • Paper manufacturers incorporate more secondary fiber, reducing timber harvest activity • Technology improvements allow better control of MRF odors
Private cost of landfilling	• Bulky items cost more to transport and to landfill • Plastic bags are costly to collect, but small and light to transport	• Landfill tipping fees vary across the country • Local opposition can raise private costs of a new landfill	• Technology changes can facilitate compacting of waste in landfills, and restoration upon closing
External cost of landfilling	• Plastic waste may result in the release of microplastics • Food waste creates more leachate and gas • Other materials create methane (a greenhouse gas) or leachate (which may contain other pollutants)	• Soil attributes in some areas make leachate or emissions more damaging • Areas relying on waste combustion create air pollution, with higher costs in populated areas	• Improved methane capture from landfills reduces climate harms from waste • Technologies include sanitary lining, leachate collection, and venting
External cost of litter or dumping	• Some dumped items can pose risks to wildlife through entanglement, ingestion, and habitat disruption • Toxic liquids from improperly disposed wastes get into groundwater	• Dumping in areas with sensitive ecosystems has higher external costs • Litter in areas with larger population has higher external costs	• Technology can create new materials that have higher external costs of dumping • New biodegradable products can reduce external cost of dumping

NOTE: HDPE = high-density polyethylene; MRF = materials recovery facility; MSW = municipal solid waste; PET = polyethylene terephthalate.

To further explore the role of volatility in end-market prices, Table 3-2 shows the change in price for each of seven materials from 2020 to 2021. The lowest price increase was 69 percent, but most prices in this table tripled between those years. Any large jump in price may be followed the next year by a larger price crash. This price volatility increases risk to planners trying to ensure that MRFs are profitable. Thus, strategies are needed for dealing with volatile prices, even if the average price over time is adequate. This risk can be shifted away from MRFs by long-term contracts. For example, MRFs are now ensuring cost recovery and margins through processing fees.

TABLE 3-2 National Average Prices per Ton

	Dollars per Ton		
Commodity	September 2020	September 2021	Ratio 2021/2020
Corrugated cardboard	$60	$171	2.85
Mixed paper	$18	$96	5.33
HDPE	$1,100	$2,169	1.97
PET	$130	$511	3.93
Polypropylene	$105	$663	6.31
Aluminum cans	$915	$1,550	1.69
Steel cans	$78	$250	3.21

NOTES: Prices based on materials sold after they are cleaned, crushed, and baled. Glass is not listed because it is near zero or negative, depending on location. HDPE = high-density polyethylene; PET = polyethylene terephthalate.
SOURCE: SWANA, 2021.

Moreover, the question of whether an item ought to be recycled does not depend on the private and external costs of recycling alone. It also depends on external costs for other forms of disposal—sending to a landfill, littering, or dumping. While a particular material might not be recycled through a MRF because private costs are high or its sale price is low, that same material may have much higher external costs in a landfill, and even higher if it is littered or dumped. In that case, environmental damages will differ across these various materials, their toxicity, their persistence in the ecosystem, and their threat to wildlife. Damages for each material type will also differ by location, the social value of the ecosystem, and the fragility of the ecosystem. Box 3-1 displays a mathematical formula for assessing the net social costs of recycling versus other forms of disposal. For an application of those formulas, Box 3-2 considers a case study for determining the net value of recycling a specific material in a specific place—namely, glass jars or bottles in Fargo, North Dakota.

3.3 POLICY APPROACHES THAT RESPOND TO HETEROGENEITY

While some municipalities may choose to recycle many materials, a few might be wise to recycle fewer or even no items (e.g., a town in a very remote location with low access to commodity markets and low landfill costs). A state or national policy that requires a uniform list of recycled materials may result in costs of recycling that exceed its social and environmental benefits.

Trade-offs arise between the simplicity of the recycling system and the balancing of costs and benefits. Consistent rules about what can or cannot be recycled are easier for consumers to understand and can reduce the information burden of the system; however, as mentioned above, consistent rules may create too much uniformity.

Several state EPR laws for packaging aim to create more uniform curbside recycling programs across the state. However, the requirements vary not only between states but sometimes within a single state. For example, Oregon's EPR law includes a Uniform Statewide Collection List, but it makes an exception for glass collection in the Portland Metro region. California's program also includes a standard list of materials that must be accepted in curbside recycling, but local jurisdictions are allowed to add more items if they choose. These examples show that while uniformity is often a goal, states still recognize the need for flexibility in how recycling systems are implemented.

> **BOX 3-1**
> **Is Recycling Worth It?: A Formula for Assessing Net Social Costs**
>
> A municipal solid waste authority must decide whether a particular item is to be collected for recycling, "R" or sent to a landfill, "G." Suppose the private cost of recycling exceeds the private benefit from selling it, so the *net* "private cost of recycling" is positive. Call it PCR_{ilt} (for item i in location l, at time t). The external cost of recycling it is ECR_{ilt}. Those private costs plus external costs are the social cost of recycling. In addition, the external cost of mining a new source of the material is ECM_{ilt}. Then the *net social cost of recycling* this item, on the left side of the comparison below, is the net private cost and external cost, minus the avoided external cost of the amount of mining that would be needed if this item were not recycled:
>
> $$PCR_{ilt} + ECR_{ilt} - ECM_{ilt} < PCG_{ilt} + ECG_{ilt}$$
>
> To enhance environmental and economic welfare (i.e., overall social welfare), the net social cost of recycling on the left must be less than the social cost of placing that item into a landfill or incinerator instead (PCG_{ilt} plus ECG_{ilt}, on the right). The heterogeneity in those variables means that welfare-maximizing decisions depend on measuring all five kinds of costs for each material in each location at each point in time. In other words, optimal recycling policy is complicated. No single policy is best everywhere for all recycling, and these local policies may need to change as technology improves with time.
>
> SOURCE: Based on Fullerton and Kinnaman, 2025.

> **BOX 3-2**
> **Case Study: Recycling Glass Jars or Bottles in Fargo**
>
> To assess the value of recycling glass jars in Fargo, North Dakota, several key pieces of information need to be researched and measured:
>
> 1. Private costs, which might be higher than for other items because glass is dense and requires extra care to avoid hazardous breakage during collection.
> 2. Capital and labor costs, which are specific to the area of Fargo.
> 3. External emissions costs, which might be lower than other items, as clean glass gives no odor or toxic emissions in the materials recovery facility.
> 4. External extraction costs, which are not trivial for acquiring new virgin material if this jar is not recycled. The principal source of silicon dioxide for making glass is sand, usually taken from beach, river, or lake deposits (Corning, n.d.).
> 5. Private and external costs of landfilling the recycled jar instead. Direct private costs of landfilling depend on cost of transport or distance to the landfill versus the materials recovery facility and the price of land—which is lower in North Dakota than in many other states. And glass is relatively inert in the landfill, causing no leachate or methane emissions.
>
> This initial overview is not intended to be conclusive but to demonstrate the difficulty of research and measurement even to determine whether the total net social cost of placing this item in the landfill is higher or lower than the total net social costs of recycling the glass jar. In other words, recycling in this example may or may not reduce total costs or enhance social welfare.

With differentiated rules over time and space, recycling information is bound to be imperfect and sometimes outdated. Consumers may receive conflicting information when they talk to people in other jurisdictions that have a different list of recyclable items. MRF operators often complain about contamination by nonrecyclables in the recycling bin, but some amount of improper sorting may be a cost worth paying to allow flexibility in the rules. Future technology that improves sorting of single-stream recycling (e.g., using artificial intelligence) may thus have the benefit not only of improving the quality of secondary materials, but also of making complexity less costly.

3.3.1 "Ideal" Waste Management Policy

It is helpful to sketch out an "ideal" or "optimal" policy (although its implementation would be infeasible) as a benchmark for comparison with real policies. One version of such a policy relies on the economic idea of using corrective taxes (known as "Pigovian" taxes, from Pigou, 1920) to make every external cost into a cost borne by private parties. The Pigovian tax system would place a tax on every activity that generates an external cost, which could include a tax not only on garbage heading to the landfill, but also on dumping and virgin materials extraction (e.g., mining). Household waste going to landfills imposes external costs on society that households do not consider when sorting their waste. Market-based initiatives can internalize the cost of the externality by increasing the relative price of sending waste to landfills. The effectiveness of these instruments depends on the demand elasticity with respect to the tax or price. Low price elasticity makes these instruments either environmentally ineffective or costly.

If every polluting activity was discouraged appropriately by the right tax rate, then recycling policy would not have to take on goals related to those other activities. In that case, the ideal system would only tax recycling at a rate that reflects its own external cost (see Box 3-3).[3] However, the possibilities of illegal littering, dumping, or burning of waste present major obstacles to the Pigovian tax approach. Because a tax on garbage collection would raise the cost of legal waste disposal, it might encourage substitution with illegal waste management. Illegal littering and dumping have very high social costs but are nearly impossible to tax and very difficult to observe, regulate, or punish. However, the job of discouraging litter might be shifted to other policies that apply to activities that *are* market transactions, such as the deposit collected on beverage containers upon purchase in some states and the refund paid for their observable and documented return after use. The right-hand column of the table in Box 3-3 shows the optimal deposit-return system (DRS) that is equivalent to the ideal tax system. Real-world policies do not need to work perfectly, of course, but this column clarifies usefully that the perfect DRS would need to be tailored to the toxicity of the item and to the ecosystem fragility of its possible dumping location, as well as to changes from year to year as technology and costs evolve.

This discussion can be summarized in three key points. First, an ideal set of tax incentives on every form of disposal might be infeasible, but the same outcomes can be re-created by a feasible combination of appropriately chosen tax and subsidy rates (i.e., deposits and returns). Second, this optimal DRS can include many items in addition to beverage containers and lead-acid batteries. It can include any item for which disposal choices can be redirected away from improper disposal. Third, heterogeneity is relevant not only to the ideal tax system but also to the ideal DRS; the optimal deposit rate and refund can depend on the waste material and the location where dumping is most costly.

In a different model, Palmer and Walls (1997) calculated optimal rates for a DRS, finding that the deposit must equal the refund, and that both must be set equal to the marginal external damage per unit of disposal. Because of transactions costs, Numata (2011) argued that the refund should be equal to the sum of the following three components: (1) the suppliers' marginal net revenue from collecting and treating used, returned goods; (2) the marginal external cost; and (3) the deposit multiplied by the share of the unredeemed deposits that the government and the recycler collect from the supplier. Porter (1983) also disputed the perhaps intuitive-seeming notion that higher deposit and refund amounts necessarily lead to higher return rates. He noted that, although Michigan had relatively high deposit-refund amounts of 5–10 cents per aluminum can (other U.S. states at the time had deposit-refund amounts of 2–5 cents per can) recycling rates in Michigan were not higher than other states. Chapter 4 discusses DRS in detail and examines available data for assessing their effectiveness.

---

[3] These "ideal" taxes are useful for a conceptual discussion about how to fix problems with external costs, but do not fix other market failures, such as monopoly power, or public goods, such as information or economies of scale. Also, this conceptual discussion pertains only to the optimal pricing of each activity like landfill disposal, recycling, or illegal dumping. It does not deal with the practical issues of implementation, enforcement, administration, financing, political feasibility, or the distribution of burdens from such taxes. These topics are discussed elsewhere in this report.

> **BOX 3-3**
> **Optimal Incentives via Taxes or Deposit-Return System**
>
> A list of heterogeneous waste material costs is indexed by item i in location l, at time t. Suppose the external cost of placing it in the garbage is ECG_{ilt}, the external cost of recycling it is ECR_{ilt}, and the larger external cost of dumping it is ECD_{ilt}. A simple model like the one in Fullerton and Kinnaman (1995), where firms and households are only interested in their own private costs and benefits, can be used to show how two policies can achieve the same first-best optimal recycling increases and dumping reductions.
>
> The first column of Table 3-3 shows the optimal tax system of Pigou (1920), with no tax upon purchase of the item, but with a set of positive tax rates: the tax is ECG_{ilt} if it is placed in the garbage and ECD_{ilt} if it is dumped. However, if the tax on dumping is infeasible, then the exact same optimal outcomes can be achieved by a deposit-return system (DRS).
>
> **TABLE 3-3** Comparing Pigovian Tax Rates with Deposit-Return Systems Costs
>
> | | Pigovian Tax Rates | Item-Specific DRS |
> | --- | --- | --- |
> | **Tax on Purchase** | 0 | $ECD_{ilt} > 0$ |
> | Tax on Garbage | $ECG_{ilt} > 0$ | $ECG_{ilt} - ECD_{ilt} < 0$ |
> | Tax on Recycling | $ECR_{ilt} > 0$ | $ECR_{ilt} - ECD_{ilt} < 0$ |
> | Tax on Dumping | $ECD_{ilt} > 0$ | 0 |
> | Tax on Mining | $ECM_{ilt} > 0$ | $ECM_{ilt} > 0$ |
>
> The optimal DRS charges a tax (i.e., deposit) upon purchase of any item, at a rate equal to the external cost of dumping that item in that location that year (ECD_{ilt}). Then the "tax" per unit of garbage is equal to its external cost (ECG_{ilt}), minus the refund of the original deposit (ECD_{ilt}) because it was not dumped illegally. That net tax is negative (a subsidy to garbage collection), because the refunded amount (damage from dumping) is larger than the damage from putting in in the garbage. Garbage collection is indeed subsidized using local tax revenue.
>
> The optimal tax per unit of recycling is equal to its external cost (ECR_{ilt}), minus the refund of the original deposit (ECD_{ilt}). This net tax is also negative (a subsidy), because the refunded amount (damage from dumping) is larger than the damage from recycling. If the external cost from garbage is higher than from recycling, then the optimal rate of subsidy to recycling is larger than the subsidy for garbage collection.

3.3.2 Recycling Subsidies

Alternative economic choice models can help gain general insights into behavior and the factors likely to influence it. More detailed models can allow extensive evaluations of system constraints, feasibilities, and actual performance, but they may require more extensive data collection, calibration, and verification. Including such information in MSW data collection efforts is thus an important part of improved nationwide recycling assessment and design.

Fullerton and Stechuk (2024) solved for an alternative tax system for when the ideal is infeasible. They suggested a tax on garbage and negative tax on recycling (a subsidy for recycling) that can help divert waste from other untaxable forms of disposal with high external costs. This subsidy might be difficult to provide per household recycling cart, but similar incentives can be provided with easier administration if the subsidy for each recycled material is paid to a MRF per ton of recycling that is cleaned, crushed, and baled. It could even be set for each material at a different rate that reflects its damage to the environment if not recycled. To obtain more of this subsidy, the MRF has incentives to encourage household participation. In general, the ideal subsidy per cart or per ton has the logical advantage of getting people to recycle additional carts or tons, but similar effects might be achieved in other ways. Cities can get more people to recycle and achieve additional quantities simply by devoting additional municipal expenditures to aid recycling generally, including collection, transportation, and processing at MRF facilities. This logic provides the strongest rationale for most existing recycling finance and policy (Fullerton and Stechuk,

2024). Moreover, since many external costs from waste disposal spill over to other counties and states, this logic also provides a rationale for funding from higher levels of government for proper MSW management.

3.3.3 Quantity Regulations

This discussion of externalities can also explain the conceptual equivalence between tax incentives and quantity regulations. For example, a recycled content standard (RCS) requires that a certain fraction of each produced item must be composed of recycled material. And an EPR rule can require that the producer be responsible to pay for disposal of their packaging or even the eventual disposal of the item they produce. These regulations are discussed in more detail both in this chapter and in Chapter 4.

3.3.4 Trade-Offs

Each tax or regulatory approach has trade-offs. For example:

- The tax or fee approach may have lower information and administrative requirements. To set each tax rate (or deposit-return rate), a policymaker needs to know only the external cost of disposal. To set the optimal behavior or quantities, however, they also need to know how households and MRFs would react to those tax rates in adjusting their behavior and the chosen disposal quantities. Imperfect information means imperfect regulations.
- Uncertainty can be a challenge in either approach. If policymakers require a specific quantity reduction, the costs of compliance could end up being too high. But if they set a tax rate instead, they provide no guarantee it will lead to the intended reduction in waste, since households and MRFs might not adjust their behavior as expected.
- Voters tend to oppose new taxes, so adding a tax might be politically infeasible; on the other hand, voters may support a new tax if it provides revenue that can be used to cut other tax rates, or for some other useful purpose.
- The distribution of tax burdens will likely differ, depending on the chosen policy, but not always in the ways that might be expected. A tax on waste disposal is a very visible policy with high salience to households, and its burden might be a higher fraction of income for those with low income. Perhaps the revenue from the tax can be used to help these households. In contrast, regulations can have similar effects, but their burdens on households are less transparent. Levinson (2019) showed how and why regulatory burdens on low-income households can be higher than tax burdens on low-income households. For example, "a policy that targets energy-efficient appliances, either by subsidizing efficient ones or taxing inefficient ones, will favor richer households because they already spend more on energy efficiency" (Levinson, 2019).
- Either policy type can be difficult to change as circumstances change. New information about external costs might suggest a change to either disposal taxes or regulations. But each would also require an update to account for changes in economic growth, behavior, or technology.

In some ways, the above distinctions between taxes and regulations are overstated. While a waste tax or DRS is a price-incentive policy, and regulations are not, a requirement to undertake particular activities or to reach certain targets will affect relevant prices in the economy, and these price changes themselves are incentives. For example, a regulation may require producers to pay for packaging waste (i.e., EPR). Knowing they must pay for this waste, producers will charge higher prices to cover future disposal costs, with greater price increases for those products with high-quantity packaging and/or packaging with expensive disposal. Thus, the cost to consumers for such regulations may be equal to the amount of a tax. Similarly, requiring that producers use more recycled materials in production (i.e., RCS) increases the demand for recycled materials, which may drive up the price for recycled materials and raise the supply of recycling—just as would a price incentive such as a recycling subsidy.

3.3.5 Belt and Suspenders Approaches

In light of these pros and cons, policymakers could consider a combination of policies that support each other (i.e., a "belt and suspenders" approach). If a deposit-return system generates additional supplies of recycled material, without additional demand, then much of that additional recycled material may find no market and get relegated to a landfill. However, the deposit-return system can be combined with a recycled content standard that requires producers to use more recycled materials in production. Then the extra demand may better match the extra supply (see Basuhi et al., 2024; Lifset et al., 2023).

3.3.6 Addressing Economies of Scale

This chapter discussed economies of scale for landfills and MRFs. Governments can adopt policies to address this market failure and reduce total costs of recycling materials. This government provision differs from a tax or regulation. For instance, a small municipality can subsidize or otherwise encourage recycling in a way that generates enough total volume to build a MRF with reduced processing costs per ton. If the environmental benefits of that additional recycling spill over to other towns or states, then a higher level of government could improve statewide or nationwide economic and environmental welfare by helping small towns build MRFs of sufficient scale to take advantage of the lower cost per ton of processed material. County or state governments can help pay for a hub-and-spoke system in which multiple rural towns build one MRF of sufficient size to achieve cost-effectiveness (see Chapter 2).

3.4 CONCLUSIONS FOR RECYCLING PROGRAMS AND POLICY CHOICES

Tremendous heterogeneity and variability across time, materials, and geography impact the cost and benefits of recycling. The committee draws the following conclusion based on its assessment of these factors.

Conclusion 3-1: Effective MSW recycling programs help meet waste management needs, save resources, improve the environment, and benefit society. These economic, social, and environmental benefits can outweigh the costs of well-designed recycling programs.

Different contexts and recycling programs require tailored policy solutions, based on variations in materials, geographies, economies of scale, existing infrastructure and programs, demographics. and other social considerations. While guiding policies from higher levels of government can be appropriate, it is important to consider and tailor policies for recycling based on local factors.

Conclusion 3-2: Because of the significant heterogeneity in local conditions, no one-size-fits-all nationwide recycling policy can effectively fund and encourage recycling for all municipalities, waste generators, dwelling types, or materials. Understanding community differences is important when tailoring location-specific programs for consumers and recycling operations.

Regardless of the specific context, however, it is helpful to identify and articulate the objective(s) that effective recycling policy is designed to achieve.

Conclusion 3-3: Effective recycling policy would target some or many of the following objectives:
1. *Enhance end markets for recyclable materials*
2. *Provide stable financing of recycling systems*
3. *Clarify information for consumers, including what is recyclable, how to recycle, and which products best support recycling goals*
4. *Track and evaluate recycling activities through improved data collection and distribution*

5. *Increase the cost-competitiveness of recycled materials (relative to virgin material inputs) and of recycling (relative to landfilling)*
6. *Improve access to recycling collection and processing*
7. *Increase the cost effectiveness of recycling collection and processing*
8. *Decrease contamination of postconsumer recycling streams*
9. *Enhance social and environmental benefits associated with recycling*
10. *Maintain affordability, without undue burdens on low-income households*

REFERENCES

Basuhi, R., K. Bhuwalka, R. Roth, and E. A. Olivetti. 2024. Evaluating strategies to increase PET bottle recycling in the United States. *Journal of Industrial Ecology* 1–12. https://doi.org/10.1111/jiec.13496.

Bradshaw, S.L., A.H. Aguirre-Villegas, S.E. Boxman, and C.H. Benson. 2025. Material recovery facilities (MRFs) in the United States: Operations, revenue, and the impact of scale. *Waste Management* 193:317–327.

Damigos, D. 2006. An overview of environmental valuation methods for the mining industry. *Journal of Cleaner Production* 14:234–247.

Fullerton, D., and T. Kinnaman. 2025. The economics of recycling heterogeneity. In *Oxford Research Encyclopedia of Economics and Finance*. Oxford University Press.

Fullerton, D., and S.L. Stechuk. 2024. *The good, the bad, and the ugly: Second best taxes on recycling versus disposal, with no tax on dumping.* Presentation at the Association of Environmental and Resource Economists (AERE) meeting in Washington DC, June 2024.

Kinnaman, T.C. 2014. Understanding the economics of waste: Drivers, policies, and external costs. *International Review of Environmental and Resource Economics* 8(3–4):281–320. https://doi.org/10.1561/101.00000071.

Levinson, A. 2019. Energy efficiency standards are more regressive than energy taxes: Theory and evidence. *Journal of the Association of Environmental and Resource Economists* 6(S1):S7–S36.

Lifset, R., H. Kalimo, A. Jukka, P. Kautto, and M. Miettinen. 2023. Restoring the incentives for eco-design in extended producer responsibility: The challenges for eco-modulation. *Waste Management* 168:189–201. https://doi.org/10.1016/j.wasman.2023.05.033.

Melosi, M.V. 2004. *Garbage in the cities: Refuse reform and the environment.* Pittsburgh: University of Pittsburgh Press. https://upittpress.org/books/9780822958574.

Moore, K.R., E. Marquis, K. Shanks, and F. Wall. 2024. Mining of primary raw materials as the critical foundation of "sustainable" metals: A wicked problem for technology innovation clusters. *Philosophical Transactions A* 382(2284):20230241.

Numata, D. 2011. Optimal design of deposit–refund systems considering allocation of unredeemed deposits. *Environmental Economics and Policy Studies* 13(4):303–321.

Palmer, K., and M. Walls. 1997. Optimal policies for solid waste disposal taxes, subsidies, and standards. *Journal of Public Economics* 65(2):193–205.

Pigou, A.C. 1920. *The Economics of Welfare.* London: Macmillan and Company, Limited.

Porter, R.C. 1983. Michigan's experience with mandatory deposits on beverage containers. *Land Economics* 59(2):177–194. https://doi.org/10.2307/3146047.

Pressley, P.N., J.W. Levis, A. Damgaard, M.A. Barlaz, and J.F. DeCarolis. 2015. Analysis of material recovery facilities for use in life-cycle assessment. *Waste Management* 35:307–317. https://doi.org/10.1016/j.wasman.2014.09.012.

Prest, B.C. 2022. Supply-side reforms to oil and gas production on federal lands: Modeling the implications for CO_2 emissions, federal revenues, and leakage. *Journal of the Association of Environmental and Resource Economists* 9(4):681–720.

Schmidt, M. 2021. The resource-energy nexus as a key factor for circular economy. *Chemie Ingenieur Technik* 93(11):1707–1716.

Solow, R.M. 1991. *Sustainability: An economist's perspective.* Lecture to the Marine Policy Center, Woods Hole Oceanographic Institution, Woods Hole, MA, June 14, 1991.

Stahel, W.R. 2016. The circular economy. *Nature* 531(7595):435–438.

SWANA (Solid Waste Association of North America). 2021. Recycling markets have strongly recovered since national sword. https://www.stpaul.gov/sites/default/files/2021-11/recycling-markets-have-strongly-recovered-final.pdf.

4
Direct Costs and Financing of Recycling Programs

Summary of Key Messages

- **Costs of recycling:** Although local governments and private parties pay for most costs of recycling, they experience only a small fraction of its global and regional benefits; thus, incentives are weak for these parties to maintain or increase recycling.
- **Volatility and market disruptions:** Volatility in end markets for recyclables can undermine the economic sustainability of recycling programs.
- **Traditional financing approaches:** Financing for residential recycling in the United States relies largely on local governments, with little contribution from state or federal sources. When businesses recycle, they typically pay for recycling services that are provided by private waste management companies.
- **Extended producer responsibility (EPR):** As an emerging financing model, EPR alleviates financial burden on local governments by shifting residential recycling costs to producers and, indirectly, to households that purchase their products. If properly designed through "eco-modulation," these systems may provide incentives to producers to reduce packaging volumes and increase recyclability of packaging and products. Existing EPR laws vary greatly by state.
- **Deposit-return systems (DRSs):** DRSs encourage direct recycling by consumers and can lead to reduced litter and high recycling rates of the products covered by the policy (e.g., beverage containers). However, their effectiveness is limited to covered products. The adoption of new DRS policies may decrease revenues of existing curbside programs and materials recovery facilities (MRFs), which could undermine the economic viability of those programs and therefore the recycling of materials *not* covered by the DRS. This tradeoff could be managed with alternative financing approaches for existing curbside recycling programs (e.g., via EPR).
- **Investments in MRFs:** Investing in MRF technology is important for cost-effective recycling, as advanced sorting and processing equipment reduces per-ton processing costs and increases the quality of recyclable materials.

Local governments and private entities bear the primary financial burden of recycling in the United States, yet they receive only a fraction of the broader environmental and economic benefits. Spending on curbside and other recycling systems involves costs of collection, such as trucks, labor, and fuel, and costs of processing, sometimes at capital-intensive materials recovery facilities (MRFs). This spending often exceeds the revenue earned from selling recyclable commodities, straining municipal budgets and discouraging program expansion. Emerging financing approaches, such as extended producer responsibility (EPR), show promise in redistributing costs and stabilizing material quality. This chapter examines the costs and financing sources for recycling programs nationwide, with particular emphasis on curbside recycling as the dominant mode of collection (see Figure 2-1 in Chapter 2).

4.1 CURBSIDE RECYCLING PROGRAMS

Many different waste and recycling systems operate throughout the United States, which entail multiple collection and processing stages. The most common recycling system for households is curbside recycling. An estimated 53 percent of the U.S. population—about 176 million persons living in approximately 69.8 million households—has curbside recycling services provided to their homes automatically as a part of the residential solid waste management systems in their communities (The Recycling Partnership, 2020a). Furthermore, most of these residents receive single-stream recyclables collection service, meaning that they do not separate recyclable materials from each other but place them in a commingled fashion in a single recycling collection container, which is wheeled or carried to the curb by the resident on the collection day. In contrast, recycling systems for commercial establishments may be single stream or have different collection for specific types of materials. In many rural communities, households do not have curbside collection but may have access to recycling drop-off locations. The discussion of local government costs in this chapter focuses on curbside recycling because it is the dominant collection method and because limited cost data are available for other forms of collection.

Curbside recycling involves the provision of two services: (1) collection by municipal governments or private companies and (2) processing of the collected recyclables by municipal governments, solid waste authorities, and/or private companies. This chapter first distinguishes between these collection and processing costs and then considers the revenues from recycled commodities that may help offset the processing costs. Finally, these categories are combined to compare curbside recycling costs with costs of disposing of these materials in other ways.

4.1.1 Curbside Recycling Collection Costs

Curbside collection costs include expenses for bins provided to households and the costs of owning, operating, and maintaining trucks for collection. These costs (e.g., labor, fuel prices, equipment purchases) vary by state and within states across the country. Rates are calculated based on the number of households served per crew per day and/or tons collected per day by each crew. Communities with high participation rates may achieve lower average costs per ton because of efficient truck and labor use. Labor costs are lower in the South than in many urban areas of the Northeast and the West Coast, which are often unionized. Capital costs associated with trucks and equipment yards make up a significant portion of the cost of collecting waste and recyclables from households and businesses in the United States. While most regions of the country have some cities with extensive recycling programs, recycling services in rural areas may not be integrated into broader solid waste programs.

A full picture of these costs across the country is not available, but a few studies provide details on these costs at specific locations. In North Carolina, a long-term survey provides detailed data on recycling and refuse collection expenditures for a set of municipalities (UNC School of Government, 2024). Most of the communities studied provide single-stream recycling collection services to single-family residents on a biweekly basis. In 2022–2023, the total cost associated with these services—including the capital costs of the collection vehicles and containers—averaged $45 per household per year. The North Carolina data also provide the opportunity to calculate costs per ton of recyclables collected (and not just per household). These data show an average collection cost of $275 per ton in 2022–2023 (UNC School of Government, 2024). Reflecting economies of scale, communities with higher recycling set-out rates had lower average costs per ton. For Florida, Anshassi and Townsend (2024) provide similar figures. Across the six counties they studied in 2021, collection costs for curbside recycled averaged $49 per household, similar to the North Carolina values. The Florida costs ranged from $23 per household to $81 per household (Anshassi and Townsend, 2024).

Additionally, Anshassi and Townsend (2023) used Florida data to estimate the average costs for a typical U.S. residential household cost for separate recyclables collection. They found an average cost of $45 per household per year. Among the factors they considered are the number of households at one stop, participation rate, collection frequency, number of working days per week, and working hours per day per vehicle. The collection costs change as a function of mass and composition (for both garbage and recyclables stream) as well as the frequency of the collection service (which is typically weekly or biweekly). The primary factors that vary by region include the waste compaction density, recycling participation rate, total masses collected and recycled, and number of households participating (Anshassi and Townsend, 2023).

4.1.2 Curbside Recycling Processing Costs

Recyclables collected at the curb from single-family residences are typically processed at MRFs. Figure 4-1 shows recent trends of processing costs per ton on waste from MRFs in the Northeast region (10 states).

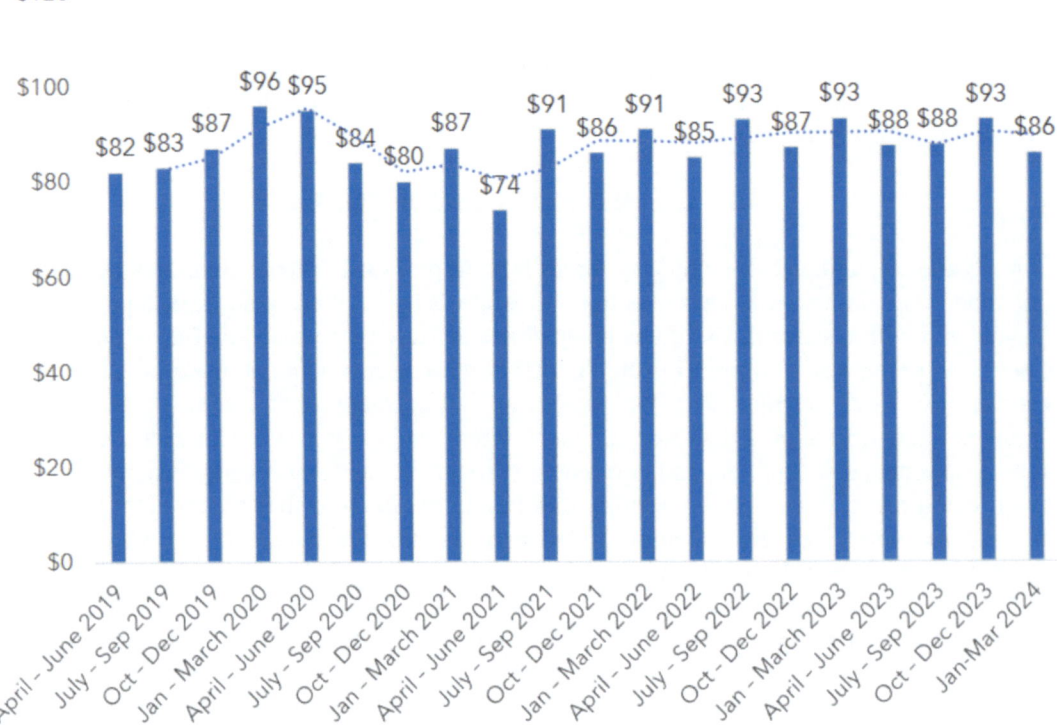

FIGURE 4-1 Processing costs for materials recovery facilities in the Northeast, 2019–2024.
NOTE: Data are as reported by facilities and underlying methodologies may vary.
SOURCE: NERC, 2024.

A review of commingled recycling costs completed for the Oregon Department of Environmental Quality audited a small number of MRFs in Oregon using their actual 2022 fixed and operating costs (Crowe LLP, 2024). These MRFs ranged from small, labor-intensive facilities to larger MRFs with investments in some newer technologies, although none are as advanced as the newest MRFs being built in the United States (see Box 2-5 in Chapter 2). The average cost of processing commingled recyclables through the single-stream MRFs in Oregon in 2022 was $129/ton (Crowe LLP, 2024).

Processing prices can also vary in response to international policy changes. China and other countries have increasingly imposed restrictions on the levels of contamination allowable in imported recyclables (Resource Recycling, 2022). The private company WM estimated that these restrictions have increased processing costs by about 15 percent—or about $13 per ton—across its 43 single-stream MRFs. These increased costs cover labor needs and investments in sorting equipment to meet the new contamination standards.

Investments in new technology have large initial costs that enable MRFs to process each additional ton of recycling less expensively and more efficiently (see Chapter 3 for more on economies of scale). Over time, advancements in recycling technologies may increase sales of materials by raising the quality of the sorted recyclables.

4.1.3 Curbside Recycling Revenues

The net costs of the curbside recycling programs depend not only on the direct costs of collection and processing but also on the revenues received from the recyclable commodities at the end of processing. These values fluctuate with the macroeconomic conditions, product demand, material quality, transportation costs, and price of competing resources. Demand for recycled commodities reflects their value to manufacturers who use them, so prices indicate some of the current value of the recycled material. Other values, such as environmental improvements, are outside of the market and may not be captured by prices. Chapter 5 presents more information on individual recycled commodity prices.

Figure 4-2 shows estimated revenues that MRFs in the Northeast received from selling processed materials.[1] Comparing the revenues in Figure 4-2 with the processing costs in Figure 4-1 demonstrates that revenue sometimes covers processing costs (e.g., in 2021 and 2022), but other times it does not. Periods when end-market revenues did not cover costs include 2019, shortly after China's National Sword Policy was instituted (see Chapter 5), and much of 2023, which did not feature these trade disruptions. MRFs may still make profits in years when the materials revenue is lower than their processing costs. because of the processing fees paid to them by local governments and third-party collectors (Anshassi and Townsend, 2024).

FIGURE 4-2 Revenues per ton recyclables for materials recovery facilities in the Northeast, 2019–2024.
NOTE: Data are as reported by facilities and underlying methodologies may vary.
SOURCE: NERC, 2024.

---

[1] Although these values are specific to this region and time period, they appear to be representative. Using the average prices of recyclables reported by the U.S. Environmental Protection Agency for 2009–2018 (see Figure 5-6 in Chapter 5) and typical material shares reported by The Recycling Partnership (2004) yields a revenue estimate of $167 per ton (not inflation-adjusted), within the range shown for the Northeast in Figure 4-2.

4.1.4 Overall System Costs for Curbside Recycling

Ignoring external costs for now, the direct costs above include collection plus processing costs minus revenue received from recycling. Putting these direct costs together provides an overall picture of the net financial costs of curbside recycling. A review of rates in several states highlights the variability of these overall costs across the country.

National Estimates

Anshassi and Townsend (2024) estimated government expenditures (net of commodity revenues) on waste and recycling systems in regions across the United States (see Figure 4-3). They found that these expenditures vary based on MRF processing fees, revenue-sharing structures with MRFs (described in the figure by different contract letter types), and market conditions for recyclables. The lowest annual household waste management expenditures were in 2011, during peak recycling markets, while the highest expenditures occurred in 2020, during a market low. Expenditures ranged from $124 to $241 per household per year in 2011, $166 to $267 in 2020, and $154 to $243 in 2021 (Anshassi and Townsend, 2024). Across all years and regions, the lowest expenditures were associated with Contract A, which featured a high revenue share of nearly 97 percent and a processing fee of $85 per ton (see Figure 4-3). The Northwest and Northeast regions had the lowest expenditures, while the Southeast region had the highest. In these estimates, garbage and recyclables collection costs accounted for 53–78 percent of total household waste management costs (excluding recyclables revenue), while landfill and waste-to-energy disposal made up 19–34 percent, and MRF processing (once commodity revenue in subtracted) ranged from less than 1 percent to 23 percent (Anshassi and Townsend, 2024).

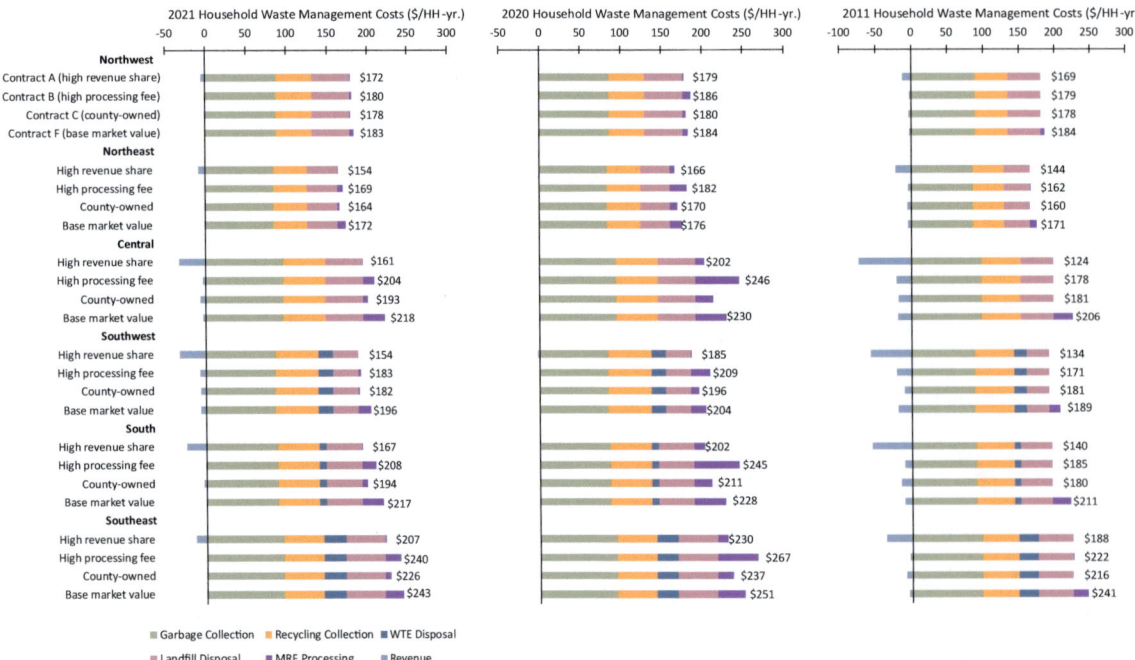

FIGURE 4-3 Estimated annual waste management costs for 2021, 2020, and 2011 by region.
NOTES: Costs for each region based on four types of contracts with MRFs: Contract A (referred to as high revenue share), Contract B (high processing fee), Contract C (county-owned), and Contract F (base market value). HH = household; MRF = materials recovery facility; WTE = waste-to-energy.
SOURCE: Anshassi and Townsend, 2024.

California. The state of California recycles 20 percent of the volume of recyclables collected in the entire United States. Regulations require separate diversion of waste, recycling, and organics, which means that multiple trucks service each house each week. Recycling requirements and fees on waste disposal increase the overall cost to the customers. With few exceptions, all costs are bundled into a monthly garbage charge that includes recycling at no additional visible cost to customers.

The cost of waste and recycling in California is higher than in other areas of the country because of this regulatory environment. These costs appear in fees for consumers. For example, in San Jose, households are charged $53 per month, or $641 per year, for residential solid waste management and recycling.[2] In San Mateo, this service costs $42 per month.[3]

Washington. Washington is one of only two states that regulate waste and recycling rates (West Virginia is the other). While incorporated cities can contract with private companies for services or provide municipal services, customer rates in unincorporated areas of the state are set by Washington's Utilities and Transportation Commission. The annual cost for all waste and recycling services in Washington State ranges from $406 to $500/year across various urban and suburban areas of the state. The recycling portion of these rates is highly scrutinized, and regulated rates range from $103 to $128 per year per household.

Although, existing literature does not provide a systematic comparison of the costs of recycling and disposal. Table 4-1 shows the per ton costs for disposal and for recycling in North Carolina (not per household monthly costs as above). As reported in the table, collection accounted for most of the cost of both recycling (73 percent) and disposal (80 percent) (EREF, 2024; NERC, 2024). In North Carolina in 2023, average costs per ton recycled exceed the cost of disposing of a ton, even when revenues from end markets were high for the year. These cost differences do not indicate the savings from shifting materials from recycling to disposal (or the costs of shifting them from disposal to recycling), because they do not capture possible increasing or decreasing returns to scale for either activity. As with the per-household costs, they only capture average costs of current levels of recycling and disposal.

These cost estimates are context and time dependent and do not apply to all curbside recycling programs. In some areas of the United States, such as the Northeast, where landfill costs are high, curbside recycling may have lower average costs than landfilling. In addition, as Figure 4-2 shows, recycled commodity revenue is sometimes much higher than in 2023, which would make the net per-ton cost of recycling more favorable. However, even when average costs (net of commodity revenues) are lower per ton for recycling than for disposal, curbside recycling may not save towns money because of the duplication involved in sending both a garbage and a recycling truck to households every week or every other week.

TABLE 4-1 Costs of Curbside Recycling and Residential Refuse Collection in North Carolina, 2023

Curbside Recycling	Average Cost ($/ton)
Collection Costs	$270
Processing Costs	$100
Revenue from Sale	–$70 to –$110
Net Recycling Costs	**$260–$300**
Refuse	
Collection Costs	$160
Disposal Costs	$40
All Refuse Costs Total	**$200**

SOURCES: Data from North Carolina Benchmark Project 2.0 for collection and processing costs; NERC (2024) for 2023 revenues; EREF (2024) for disposal costs in North Carolina.

---

[2] See https://www.sanjoseca.gov/your-government/departments-offices/environmental-services/recycling-garbage/residents/garbage-recycling-rates-billing.
[3] See https://www.cityofsanmateo.org/DocumentCenter/View/89316/2022-Solid-Waste-Notice-of-Public-Hearing.

The cost discussion thus far considers only the "on-budget" costs of recycling. Even single-stream recycling requires input of household time and effort not included here, although it is a real resource cost to consider in assessing these programs. Table 6-1 in Chapter 6 presents estimates that suggest household time costs may be either on par with or several times higher than the local government's collection costs for recycling. Although households may also have time costs for setting out waste, those costs are likely much lower. In addition, none of these costs are full social costs of recycling or waste management because they recognize market costs from energy and materials use and do not account for other external environmental impacts (see Chapter 7). Box 4-1 describes an effort to calculate this broader definition of costs by the State of Oregon.

BOX 4-1
Case Study: Incorporating Social and Environmental Costs in Oregon

As the result of a long stakeholder process, Oregon adopted a 2050 Vision for Materials Management in 2012 (ODEQ, 2012), taking a unique approach to prioritizing material management in the state. As part of the State's extended producer responsibility program (EPR) implementation, the Oregon Department of Environmental Quality (ODEQ) evaluated a range of materials, considering their commodity value and end markets, as well as the cost of handling each material. Recyclables included on Oregon's new Uniform Recycling List (required for curbside recycling) were all subject to a review of direct (e.g., infrastructure, labor, fuel) and indirect costs. These indirect costs include impacts from climate change; loss of ecosystem services; and illness, disability, and death (and associated health care costs). Reducing the need for extraction and production, recycling—if done well—creates "savings" of negative indirect costs, in the same way the commodity revenues create savings with negative direct costs (ODEQ, 2023a).

Direct costs of disposal other than recycling ranged from $983 to $1030 per ton as indicated by the modeling, while indirect social and environmental costs were $495–$595 per ton (ODEQ, 2023b). Recycling can perform better than disposal criteria for the material to be added to the Uniform Recycling List and included in Oregon's new producer-funded recycling programs.

Once implemented, the state's EPR program will pass along many of the financial costs of recycling to the producers. In Oregon, the cost of processing and marketing, rural drop-off collection, contamination management, education and outreach, and other one-time infrastructure costs are part of the EPR program covered by producers.

The discussion of recycling costs above concerns curbside programs, which are not universally available to households. When households have access only to drop-off programs, the on-budget costs of the programs are probably much lower, although estimates are not available. However, households' time and effort costs are much higher for drop-off programs than for curbside recycling, so the overall costs of collection under these programs may be higher and participation lower.

Recycling costs fluctuate over time and vary over space for many reasons, including the cost of infrastructure, labor, and fuel, and variations in housing density and transportation distances. In addition, different municipalities offer different services and impose varying participation requirements, that also generate heterogeneity in costs. The regulatory environment may also be an important source of variation in costs, because some cities keep recycling costs low while burdening disposal with fees. Expanding household participation in recycling can help reduce the costs per ton of material recycled and thus help make recycling more competitive with landfilling.

While the data presented in this chapter thus far display recent, contemporary, and potential near-term costs of recycling programs, it is important to note potential future changes (such as developments in renewable energy seen in Box 4-2) that could have implications for direct costs of recycling and overall waste management systems.

> **BOX 4-2**
> **Waste and Recycling Streams from Renewable Energy Technologies**
>
> Some installments of renewable energy technology have already reached the end of their useful life, which comes with attendant waste that largely ends up in landfills. The projected scale of waste from solar and wind energy entering landfills will be on the order of 10 million tons in the United States by 2050 (Cooperman et al., 2021; Duran et al., 2021). This waste may have ancillary economic impacts on future municipal solid waste management as it may affect the capacity and operating costs of landfills.
>
> Solar panels contain materials (e.g., silver) that can be valuable but are present in relatively low concentrations. Some of these components can be toxic and problematic in landfills but may also be problematic for recycling. Life cycle assessments that can evaluate the environmental benefits of recycling versus landfilling and accounting for several different factors and scenarios (Smith et al., 2024).
>
> While some materials in wind turbines can be recovered relatively easily, the blades are made of composite materials that are not currently easily recycled. They can be used as filler in concrete but are otherwise burned or landfilled. In landfills, the turbine blades can cause other complications, as their size and rigidity make them difficult to pack efficiently.
>
> To increase the recovery of materials from renewable energy technology waste streams, more infrastructure is needed for recycling; incentives or bans are needed to prevent landfilling; and, more broadly, proactive planning is needed for end of life when developing a renewable energy installation (Duran et al., 2021).

4.2 TRADITIONAL FINANCING APPROACHES

Financing of the recycling system in the United States comes from both private and public sources. Typically, local governments, households, and commercial establishments pay for recycling collection and processing. MRFs and other recycling processors receive revenue from end markets for recycled materials and from gate tipping fees (fee per ton paid for processing by governments and collectors). This section discusses traditional financing approaches to cover the costs of recycling collection and processing that exceed the money earned from end markets, which covers only a small share of the total costs of recycling and processing.

4.2.1 Public Financing

Almost all government financing of residential recycling in the United States comes from local governments (cities, towns, and counties), with at most small contributions from higher levels of government. Some local governments use a "general fund" approach, where recycling does not have a dedicated revenue source and is funded along with other categories of expenditure. Other municipalities rely on an "enterprise fund" approach, whereby the municipality collects fees for recycling and garbage collection, perhaps as an item on property tax bills or utility bills or as an explicit charge for businesses. For residential recycling, the general fund approach is more common than the enterprise fund approach (Sheahan, 2024). Governments may use general funds for residential collection but collect fees from commercial establishments.

Local governments pay for recycling at all stages of the process. For recycling collection, they may use their own employees or contract with a recycling collector. They pay gate tip fees for processing the recyclables they collect and implicitly or explicitly pay these fees as part of their contract with a recycling collector. In addition, when recycling drop-off programs are available, either in place of or as supplement to curbside collection, they are likely to be funded by local governments. Recent data are lacking, but in 1995, governments funded the programs in 77 percent of communities with drop-off services (Walls et al., 2003). Finally, about 20 percent of MRFs were publicly owned in 2020, often by county governments (Anshassi and Townsend, 2024). These public MRFs often collect processing fees from private collectors and other governments that use their services. However, the governments that own them may use general funds to cover some costs.

In the United States, state and the federal government provide little financial support for recycling. For all solid waste management, average state spending in 2021, was $4 per capita compared with $86 per capita for local government spending (U.S. Census Bureau, 2024). Figures for recycling specifically are not available but are likely to have a similar breakdown. Recent years have seen some of the first significant federal expenditures on recycling (see Box 4-3).

4.2.2 Private Financing

Commercial establishments and households may also pay privately to have their waste collected and processed. This private funding for recycling collection falls into two categories: (1) With a "franchise" provision, the municipality requires households or establishments to subscribe directly with private firms that provide recycling services at rates and service characteristics that follow a contract with the government. Household participation in these services may be mandatory or voluntary (subscription), depending on city policy. (2) In other communities, some households may subscribe to private recycling services independently, without a publicly negotiated contract. In 2015–2016, 19 percent of residences with curbside collection had subscription-based services, which could include franchise or individually contracted services (Sustainable Package Coalition, 2016).[4] Walls and colleagues (2005) found that residential private collection arrangements become much less common as population density rises.

BOX 4-3
U.S. Environmental Protection Agency Grant Programs

In 2020, Congress appropriated to the U.S. Environmental Protection Agency (EPA) $350 million for fiscal years 2022–2026 to improve waste management infrastructure and recycling (in the authorized Save Our Seas 2.0 Act of 2020 [Pub. L. 116-224] and the Infrastructure Investment and Jobs Act [i.e., Bipartisan Infrastructure Law, Pub. L. 117-58]). These funds were administered in two grant programs and one program to improve safe handling of used batteries (EPA, 2022):

- Solid Waste Infrastructure for Recycling (SWIFR) Grant Program ($275 million)
- Recycling Education and Outreach (REO) Grant Program ($75 million)
- Battery Collection Best Practices and Voluntary Battery Labeling Guidelines ($25 million)

The SWIFR grant program entails separate funding opportunities for states and territories; communities; and tribes and intertribal consortia. These grants are intended to support implementation of the National Recycling Strategy "to improve post-consumer materials management and infrastructure; support improvements to local post-consumer materials management and recycling programs; and assist local waste management authorities in making improvements to local waste management systems" (EPA, 2025b, para. 4). Awards to tribes and intertribal consortia can improve municipal solid waste management, including recycling management, and have mostly been used for materials recovery facility infrastructure (EPA, 2025a). Examples of infrastructure projects funded through SWIFR include $4 million awarded to Baltimore, Maryland, for a solar-powered compost facility and $3.3 million to Durham Country, North Carolina, for redesign of a drop-off station. States and territories receive funding to develop or improve on data collection or plan management rather than for specific investments or operational expenses.

The two other programs address social and behavioral considerations. The REO program focuses on reduction of food waste; expansion of the market for compost; and education and outreach to households. The program addressing collection of batteries primarily entails working sessions to develop best practices and guidelines (EPA, 2024).

Private collection arrangements are much more common for commercial establishments than for residences. In 1995, 47 percent of communities relied on private recycling arrangements for commercial establishments and another 6 percent had franchise arrangements (Walls et al., 2003). Commercial waste

---

[4] 73 percent of residences had curbside collection and 14 percent overall had subscription services.

makes up an estimated 55 percent of generated municipal solid waste (MSW) (EPA, 2013). Thus, private payment for recycling as a service probably contributes well over one-third of revenue for collection and processing of MSW recycling in the United States.

4.3 EVALUATING TRADITIONAL FINANCING FOR RECYCLING

The mix of financing for collection and processing of recycling (local government and private expenditures by households and firms) emerged as an extension of traditional garbage collection rather than as an active policy choice. Policymakers who seek improvements to the recycling system need to consider the strengths and weaknesses of this financial approach. This section discusses four considerations: incentives for recycling, cost control, risk management, and the distribution of financial burdens.

4.3.1 Incentives for Recycling

The current system for financing recycling both aligns and misaligns incentives with the social and environmental values of recycling. Incentives are aligned in that many of the benefits and costs of recycling vary by geographic location (see Chapter 3). Thus, if a local government receives greater benefits from recycling, it may be willing to bear higher financial costs. In contrast, financing recycling at a more centralized level, such as through the alternative financing approaches discussed later in this chapter, can reduce local control over this decision and reduce variability by community.

For example, a major benefit of recycling is avoiding the cost of disposal by landfill and incineration. Communities across the country face vastly different costs for disposal. The Northeast region had the highest waste disposal costs, with average tipping fees of $84 per ton in 2023, whereas the regional average in the Southeast was only $42 per ton (EREF, 2024). Thus, a government in the Northeast might be willing to spend twice as much on recycling (to avoid disposal costs) as would a government in the Southeast. In addition, communities vary in the value they attach to recycling (see Chapter 6); with locally organized and financed recycling, local governments can respond to the strength of local demand for recycling service.

On the other hand, many other benefits of recycling fall outside the horizon of a local government. These benefits include any gains from avoiding external costs, such as virgin material mining and production and global and regional pollution (e.g., methane from landfilling and air emissions of dioxins from incineration) (see Chapter 3). Local governments have limited direct interest in avoiding external costs because their populations are a small share of those affected. Thus, reducing external costs may not motivate local governments to offer recycling or to make it easy or extensive.

The concerns about alignment of incentives are even stronger for private financing of recycling. A commercial establishment that is paying privately for waste collection has only the incentive to sign up for recycling to reduce its own costs. Some public policy would be needed to give both local governments and private parties incentives to consider broader environmental benefits of recycling. Alternative financing approaches such as deposit-return systems, as discussed later in this chapter, can help fill in some of these missing incentives.

4.3.2 Cost Control

The current financing system provides opportunities for control of recycling costs. Almost half of local governments provide recycling collection services themselves, but 52 percent of local governments in 2017 contracted with private firms to provide recycling collection (ICMA, 2019). Because they pay for the recycling services either way, local governments have strong incentives to choose the lowest-cost options available for their circumstances. To meet local needs, they can make trade-offs regarding other features of the contract on which the local government is well informed (e.g., reliability, frequency of collection, materials collected). Private payers, such as households and commercial establishments, also apply pressure to keep costs down.

4.3.3 Management of Risk

The volatility of market prices for recyclables presents challenges for all approaches to funding recycling. MRF and other recycling processing facilities sell sorted recyclables for prices that vary dramatically over time. As discussed above, sometimes the recovered material revenue may cover MRF processing costs; at other times, it falls short.

Sometimes MRFs share this risk with municipalities. Contracts between municipalities and MRFs include provisions for sharing of revenues from the sorted materials (The Recycling Partnership, 2020b). A revenue-sharing component is typically combined with base processing fees or may enter the contract through a processing fee with a sliding scale that depends on prices (Anshassi and Townsend, 2024). Local governments typically enter into contracts of 3, 5, 7, or even 10 years with MRF owners to process and market recyclables. These contracts vary in scope and requirements but generally specify contamination rate limits, cost allocation for rejected loads, processing fees paid by the government, and share of revenue from recyclables paid to local governments after processing fees are paid. In Florida, for example, the current trend is toward higher processing fees paid by local governments and lower revenue shares to local governments (Anshassi and Townsend, 2024). Historically, MRFs charged about $50–$60 per ton to local governments, although sometimes no fee was assessed. Now, fees can reach up to $210 per ton, with an average of $107 per ton for Florida communities (Anshassi and Townsend, 2024).

Key parameters in these contracts include contamination levels and market values of materials (Anshassi and Townsend, 2024). Some contracts include minimum values for per-ton market revenue; if revenue from end markets falls below the minimum, the local government may make supplemental payments to the MRF. Lower fees are typically found when the materials are processed by a municipally owned MRF instead of a privately owned facility. In Florida, for example, MRF contracts rarely use a fixed processing fee; instead, fees (and sometimes revenue shares) are adjusted based on the fluctuating value of materials (Anshassi and Townsend, 2024). Outside of Florida, new MRF contracts sometimes have a flat tipping fee (that includes profit and annual escalators) that ensures that MRFs cover their cost and/or turn a profit. Then, in a separate agreement, the MRF might share with the municipality a portion of the end-market revenue from sale of materials ("revenue-sharing").

When a municipality agrees to greater revenue-sharing, it accepts more of the price risk and thus may reduce average overall costs (fees less revenues). Large municipalities may be able to manage these risks by dipping into contingency funds or by reordering immediate spending. They may also have good access to credit because of favorable tax treatment of municipal bonds, which eases the difficulty of covering short-term shortfalls. When states allow municipalities to engage in short-term borrowing, residents are less likely to have to pay privately for recycling (Walls et al., 2005). However, even large governments may sacrifice other priorities when recycling markets are unfavorable, and smaller municipalities have even more trouble managing this variability without cuts to services. Thus, imposing these risks on cities and towns may be a substantial disadvantage of the U.S. traditional system for funding recycling.

These risks probably raise the prices paid by customers for private recycling services and may reduce reliability. Residential subscription services pricing does not depend on commodity revenue, so private parties that collect and process the recyclables must bear the risk of commodity price movement. They likely charge higher prices on average to compensate for the risks, raising costs for households relative to a situation in which the government accepts some of the risk. Private companies offering subscription services may also be vulnerable to closure when markets are unfavorable and may be difficult to restart once closed.

4.3.4 Distribution of Financial Burdens

Funding approaches share the costs of recycling in different ways. In particular, public financing can reduce costs for vulnerable households. When a government uses general funds for recycling, it can take advantage of revenue sources such as property taxes that impose more burden on higher-income households. Alternatively, with subscription services or an enterprise fund, every household typically pays the same amount for the service, so low-income households pay a larger share of their income for recycling than higher-income households. Public payment for services may thus be more equitable than private funding.

However, reliance on local government financing limits the ability of the system to distribute costs across regions, because local governments can only seek revenue within their boundaries. Recycling programs may be especially expensive in rural areas, because of transportation costs and lack of access to end markets. Local governments in rural areas may also have few high-income households or large firms to tax. This combination of high costs and limited revenue opportunities may make financial burdens of recycling much higher in these areas. Relying more on state or national funding or on alternative models (e.g., extended producer responsibility, deposit-return systems) might help alleviate some of these disparities.

4.4 ALTERNATIVE FINANCING: RECYCLING PROGRAMS

In the early 1990s, European policymakers actively explored strategies for promoting environmentally responsible product design. These initiatives sought to reduce packaging, simplify recycling, and minimize waste directed to landfills. A central concept emerging from this period was extended producer responsibility (EPR), introduced by Thomas Lindhqvist (1990) in reports to the Swedish government. EPR asserts that producers should be accountable for the entire life cycle of their products and packaging. The approach aims to incorporate environmental costs into product pricing and shift the responsibility for end-of-life product management from municipalities to producers and consumers. EPR principle underpins Germany's "Green Dot" program, launched in 1991, as well as the European Union's 1994 Packaging Directive. The principle gained further recognition in a 2001 report by the Organization for Economic Co-operation and Development 2001), which described it as a policy framework that assigns end-of-life responsibility to producers. This approach is intended to encourage manufacturers to consider environmental impacts during the design phase (p. 9).

The concept of EPR is most directly embodied in individual—or "take-back"—policies, where manufacturers are required to reclaim product packaging, and eventually the product itself, once it has reached the end of its useful life. By imposing responsibility for disposal on the producer, these policies incentivize firms to design products and packaging with recycling and disposal costs in mind. Fullerton and Wu's (1998) economic model captures these incentives by demonstrating how market equilibrium—achieved when firms' production choices align with consumers' purchasing and disposal decisions—can drive optimal product design, output, and packaging choices, accounting for external disposal costs.

The model suggests three policies that can achieve socially optimal outcomes:

- The traditional economic solution places a tax per unit of consumer disposal at a rate equal to the marginal external damage (MED). If consumers have to pay both the private marginal cost (PMC) of disposal and the MED, they will choose to buy products with the optimal designs (less packaging and easier recycling).
- A general deposit-return system applies to bottles and lead-acid batteries, which are the most commonly collected materials in these systems. Instead it would collect a tax or deposit upon sale of every product, at a rate equal to the social marginal cost (SMC) of landfill disposal, and it would provide a refund upon proper recycling. To avoid paying a higher deposit, consumers would want goods with less packaging; to receive more refund, they would want to buy products that can be recycled more easily.
- An EPR policy requires the individual firm to take back the packaging and product and pay the full social marginal cost of disposal (equal to the PMC plus MED). This policy leaves the consumer with no direct disposal costs, but they must pay a higher equilibrium break-even price for the product, because firms must cover all production costs and disposal costs.

The outcomes are equivalent: either the consumers must pay social costs of disposal, and so they demand products with less packaging and with product designs that minimize disposal costs (see Chapter 6), or the firms must pay social costs of disposal and thus design products to require less packaging and easier recycling. Thus, a "perfect" individual EPR policy would achieve the socially optimal decisions about product designs, packaging, and consumer recycling decisions.

From the very beginning of EPR rules in the 1990s, however, European countries quickly realized that the perfect individual EPR policy would not be practical. It would require each firm to collect its own packaging and keep track of each product for years until the final consumer is done with it. Having many individual firms collect their waste in disparate locations is unnecessarily expensive relative to having one hauler collect waste at every house along each street. To address these collective EPR policies take advantage of these economies of scale in collection by having a "producer responsibility organization" (PRO) can collect all packaging and used products for disposal and then charge each firm for the weight of its waste collected.

A collective EPR policy, however, is very different from the ideal individual EPR. The collective EPR forces each producer to raise their sales price to cover the cost of disposal, so it effectively becomes a different policy called an "advance disposal fee," a charge for the purchase of a product that will ultimately require recycling or disposal. But it does not encourage the firm to increase recyclability, nor the consumer to recycle. The fee collects significant revenue, which replaces local government financing of recycling collection.

Thus, a collective EPR suffers from two significant problems not found in the "perfect" individual EPR policy. First, private firms pay the PRO's marginal costs of collection but not the external damages from waste disposal. Thus, incentives do not necessarily align the full costs of recycling with its benefits. For example, the extent of recycling is not sensitive to whether full direct and indirect (external) local disposal costs are high (i.e., where recycling is beneficial). Second, private firms pay the cost of collecting their own total waste (i.e., average cost per ton), but this payment is based only on weight and not on recyclability. The firm therefore has some incentive to reduce the total weight but not to design for recyclability. In only rare cases does the PRO's fee vary based on the recyclability of the materials it collects (i.e., "eco-modulation).

The economic impact of EPR varies depending on the scope of the law. For example, Canada's first packaging EPR policy was implemented in the province of Ontario and was established as a "shared responsibility" law, where producers and municipalities share the cost of recycling programs. However, the law resulted in financing disputes and lawsuits, with the system ultimately transitioning to a packaging EPR policy in 2019 with full responsibility on producers (with full implementation by 2025).

EPR policies may have both advantages and disadvantages in the way they redistribute the burden of paying for recycling. On the one hand, the need to pay producer responsibility organization fees will likely raise certain product prices, although prices may not rise by the full amount of PRO fees. These price increase pass along the costs of recycling to consumers and perhaps discourage consumption of these goods (source reduction of MSW that might be desirable). On the other hand, low-income households spend a larger share of their income on goods and thus may end up shouldering more of the burdens of recycling than they would if recycling were financed by local property tax, which can be designed to fall more on higher income households. Thus, the fact that an EPR follows a "polluter pays" principle does not necessarily make it fairer, because lower-income households may end up more burdened than they did under local government finance.

4.4.1 EPR Policies Outside the United States

Since the 1991 German "Green Dot" program, use of EPR policies has extended throughout the globe. EPR as a policy lever has grown, from voluntary programs and lightly enforced laws in Southeast Asian countries, to shared responsibility programs in the Canadian provinces of Manitoba and Saskatchewan, and to far-reaching producer responsibility programs in British Columbia and Quebec.

Empirical research supports the concern that collective EPR policies do not create the desired effect on product design. Joltreau (2022) studied the effects of European EPR policies, using data on the annual cost to comply with EPR from the organization PRO Europe (Packaging Recovery Organisation Europe) for 25 European countries and four packaging materials over 18 years (1998–2015). EPR rules vary in their stringency about the percentage of each firm's total cost that must be paid and what materials can be recycled. Joltreau (2022) argues that compliance cost is an effective overall measure of stringency and scope and observes the causal effects of EPR stringency on the amount of packaging and on substitution between packaging materials. Joltreau (2022) found that European EPR rules have a statistically significant but very small effect on the weight of packaging and no effect on substitution toward a type of material that could more easily be recycled.

What can be done to reestablish the original goal of using EPR policies to create incentives for recyclability? A recent innovation of EPR legislation in the European Union and North America is the development of "eco-modulation" of fees on producers. Eco-modulated fees vary by product characteristics, and they reward firms that design products to be more recyclable or to be more environmentally friendly. Lifset and colleagues (2023) discuss ideas for restoring the incentives for eco-design by using eco-modulation within EPR systems; they considered choices around products to be included in the fee structure, objectives to be pursued, criteria to be employed, differentiation within the structure of fees, and the amount of each fee to be charged. These choices all depend on the goal of the EPR system, such as (1) to achieve selected policy targets for eco-design, for recycling, or for all product-life management, including product durability and the ability to repair or to reuse it; (2) to account for the presence and toxicity of hazardous substances; (3) to charge for external costs of disposal and thus achieve socially optimal disposal; or, more comprehensively, (4) to cover external costs of production (Lifset et al., 2023).

Ultimately, Lifset and colleagues (2023) propose a fee structure based on each product's impact on the environment as measured through a life cycle assessment. They also recommended requiring better data from PROs about product characteristics and better use of those data for ex post policy evaluation. However, even eco-modulated charges do not vary with all the dimensions of heterogeneity in cost of recycling by location and over time, so they fall short of the ideal of imposing a tax that reflects the external costs of disposal for specific material types—a tax that would provide incentives for consumers to demand products with optimal durability, eco-design for reusability, and recyclability.

4.4.2 EPR for Packaging in the United States

For over 2 decades, many states have used EPR policies to fund and manage "hard to handle" materials. According to the Product Stewardship Institute (2025), 33 states and the District of Columbia have enacted 141 laws across 20 product categories. These policies have mostly addressed electronics, mercury thermostats, batteries, pharmaceuticals, paint, fluorescent lighting, and mattresses. What is more, the U.S. Environmental Protection Agency (2024) has published guidance on EPR policies for preventing plastic pollution (see Box 4-4).

Responsibility for waste and recycling management is decentralized in the United States, with authority delegated to individual states. This structure complicates efforts to implement federal recycling legislation, resulting in a patchwork of state and local recycling policies. Maine led the United States in passing a packaging EPR policy in 2021, followed by Oregon in 2021, Colorado and California in 2022, and Minnesota in 2024. Each has different financing and regulatory requirements (see Box 4-5). These states have used the policy as an opportunity to address other related packaging and recycling goals, as the new laws included regulatory requirements that are not part of the EPR policies themselves. In each of these five states, the enacted EPR legislation for packaging reflects the varying recycling gaps and perceived needs for that state. California and Oregon's bills were born of frustration about the growing use of plastic packaging and its low recycling rates, while Colorado's legislation (see Box 4-6) was driven by the high cost of recycling and low recycling rate. The need for economic support is also clear in the structure of Maine's law, while Minnesota is looking for future increases in its recycling program investments.

BOX 4-4
EPA National Strategy to Prevent Plastic Pollution: EPR Framework

In November 2024, the U.S. Environmental Protection Agency (EPA) (2024) finalized its National Strategy to Prevent Plastic Pollution, a comprehensive plan aimed at eliminating plastic waste from the environment by 2040. A cornerstone of the strategy is the proposed development of a national extended producer responsibility (EPR) framework, which would assign greater accountability for managing plastic waste to producers and manufacturers rather than municipalities. This policy aligns with international practices that require producers to take responsibility for the end-of-life management of their products, fostering a more circular economy. The strategy is organized around six objectives: reducing pollution from plastic production, encouraging innovation in sustainable material design, minimizing waste generation, enhancing waste management systems, improving plastic capture and removal, and reducing impacts on waterways and oceans (EPA, 2024).

The EPR framework is currently in its conceptual stage, with significant stakeholder engagement and policy development needed to translate it into a functional national program. Existing EPR programs in states such as Oregon, Maine, and California provide valuable insights and precedents, but aligning these diverse approaches into a cohesive federal framework remains a challenge. A national EPR system could standardize recycling and reuse targets, streamline efforts, and reduce inefficiencies. However, concerns have been raised regarding the potential economic impacts of implementing such a framework. These include increased operational costs for producers and the possibility of manufacturing activities relocating overseas. Successfully addressing these challenges may require collaboration among producers, policymakers, and environmental organizations to design a system that balances economic viability with the overarching goals of reducing plastic waste, improving resource efficiency, and advancing sustainability.

SOURCE: EPA, 2024.

BOX 4-5
Case Studies: State-Level Packaging and Postconsumer Recycled Content Legislation

Several states address packaging waste using extended producer responsibility (EPR) and postconsumer recycled content policies. These policies can improve recycling systems, reduce environmental impacts, and advance sustainable materials management through regulatory and market-driven mechanisms.

Maine (LD 1541)

In 2021, Maine became the first state to enact an extended producer responsibility (EPR) law for packaging. The legislation requires producers to fund recycling programs through producer responsibility organizations (PROs), which oversee the collection and recycling of packaging materials. This law primarily focuses on financing recycling systems, transferring cost burdens from municipalities to producers.

Oregon (Senate Bill 582)

Oregon's EPR law establishes a comprehensive framework for managing single-stream recyclables. It includes eco-modulated producer fees, life cycle assessment requirements, and recycling capture rate targets. The state's Department of Environmental Quality plays an integral regulatory role, particularly in addressing greenhouse gas impacts of recycling. Implementation is scheduled for 2025, excluding curbside collection costs already covered in urban and suburban areas.

Colorado (HB 1355)

Colorado's EPR law, passed in 2022, is one of the most comprehensive in the United States, requiring producers to cover all recycling system costs, including collection, processing, and market risk. The legislation mandates postconsumer content standards and provides funding for both traditional and compostable packaging management. The program will be operational by 2029, with a PRO responsible for implementation and performance monitoring.

continued

> **BOX 4-5** *continued*
>
> **California (SB 54)**
>
> California's SB 54 requires substantial packaging reductions and recycling improvements, including a 25 percent reduction in single-use plastics and a 65 percent recycling rate by 2032. Noncompliant packaging will be prohibited from sale in the state. The law also includes robust enforcement mechanisms, annual payments to environmental programs, and detailed reporting requirements for producers.
>
> **Minnesota (HF 3911)**
>
> Minnesota's EPR law incorporates a shared responsibility model, where producers reimburse municipalities for recycling costs, gradually increasing their financial contributions to 90 percent by 2031. Municipalities retaining operational control will receive funding to offset program expenses, while a PRO oversees operations in areas without municipal programs.

U.S. producers have lobbied for shared responsibility programs in several states. The result is a partial funding system in Oregon[5] and a phased-in fee approach in Minnesota.[6] Colorado's program[7] is a full producer responsibility program, where producers cover 100 percent of the costs and have broad program oversight. Full funding requirements come with the expectation of control by the producers, which is contentious in the United States because of existing public and privately funded systems.

EPR policies often have a basic fee structure, with specific discounts or penalties for products with specific characteristics. Eco-modulated fees have been built into programs in France and Canada, as well as some state programs in the United States. For example, California requires that producer responsibility organizations differentiate fees to reflect the presence of recycled content (CalRecycle, 2021). The State of Oregon has one of the more advanced eco-modulation systems, as described in Box 4-7.

With the exception of Maine's law, which serves primarily as a funding mechanism for recycling, the legislation includes specific requirements for producers to use postconsumer content in packaging, to achieve recyclability requirements, and to reduce waste. These requirements are all in addition to the producers' financial obligations to fund varying aspects of the state's recycling programs. Since producers must achieve the specified regulatory goals in order to sell their products in each state, the linkage between the financing requirements and the additional regulatory requirements creates an enhanced element of pressure on producers.

The Circular Action Alliance (CAA) is a 501(c)(3) nonprofit entity governed by a board of directors initially comprised of leaders from 20 corporations. The organization has established a national "umbrella" organization to respond to state EPR laws in the U.S. CAA works with producers that pay fees based on the specific requirements of each state's EPR program. CAA then establishes state-specific nonprofit organizations to manage the individual state EPR programs.

In the European Union, recycling packaging requirements are separate laws as part of the EU Waste Directive and are not built into the EPR funding laws. Conversely, U.S. state packaging requirements are built into the EPR policies, so they raise the stakes for packaging performance tied directly to providing the service in each state. In the United States, PROs manage compliance with state-specific EPR laws, coordinating services such as the collection, processing, and recycling of packaging materials, often in collaboration with municipalities or private service providers. EPR is being layered onto existing waste management and recycling programs in the United States (rather than creating new programs as in other countries). Thus, some companies within the U.S. recycling industry view EPR as a significant risk to their business invest-

---

[5] Oregon State Legislature. Senate Bill 582: Plastic Pollution and Recycling Modernization Act. 2021 Regular Session. https://olis.oregonlegislature.gov/liz/2021R1/Measures/Overview/SB582.
[6] Minnesota Statutes §§ 115A.144–115A.1463 (2024).
[7] Colorado General Assembly. House Bill 22-1355: Producer Responsibility Program for Recycling. 2022. https://leg.colorado.gov/bills/hb22-1355.

ments. To mitigate this concern, legislative negotiations in states with EPR policies have included requirements to prioritize the use of existing recycling facilities and to allow cities to continue to provide their own service or to contract with private service providers. Additionally, needs assessments can help states create effective EPR policies (Box 4-8).

BOX 4-6
Case Study: Colorado's Extended Producer Responsibility Policy

The State of Colorado is in the process of implementing an extended producer responsibility law requiring the producers to cover all of the costs of the State's residential recycling programs, as well as recycling programs for small and hospitality commercial business, and government entities.

As part of that process, the producer responsibility organization for the state, Circular Action Alliance, conducted a needs assessment of services available—and gaps that remain to evaluate the revenue it will need to charge to producers to achieve the requirements of the statute (see Table 4-2).

TABLE 4-2 Colorado Needs Assessment – Estimated Recycling Outcomes per Scenario

		2022 (Baseline)	2030	2035
Low	Recycling Rate (%)	22% - 28%	35% - 41%	48% - 54%
	Recycling Tonnage (k tons)	~310	~490	~670
Medium	Recycling Rate (%)	22% - 28%	38% - 44%	52% - 58%
	Recycling Tonnage (k tons)	~310	~530	~720
High	Recycling Rate (%)	22% - 28%	39% - 45%	54% - 60%
	Recycling Tonnage (k tons)	~310	~540	~750

SOURCE: Adapted from tables provided by CAA, 2024.

The state approved the medium scenario, requiring significant investments in infrastructure throughout the state, resulting in the estimated costs shown in Table 4-3, ranging from the baseline costs (2022) to the cost of the program when it is fully implemented in 2035.

TABLE 4-3 Colorado Needs Assessment – Estimated Costs per Scenario

		Baseline (2022) Lower	Baseline (2022) Upper	2030 Lower	2030 Upper	2035 Lower	2035 Upper
Low	Total Annualized Cost ($ millions)	80	140	130	210	160	260
	Cost Per Household ($)	60	90	60	100	70	120
	Cost Per Household ($)	260	430	270	440	240	380
Medium	Total Annualized Cost ($ millions)	80	140	160	260	190	310
	Cost Per Household ($)	60	90	70	120	90	140
	Cost Per Ton Recycled ($)	260	430	300	490	270	430
High	Total Annualized Cost ($ millions)	80	140	160	260	210	340
	Cost Per Household ($)	60	90	70	120	100	150
	Cost Per Ton Recycled ($)	260	430	300	480	280	450

SOURCE: Adapted from tables provided by CAA, 2024.

The medium cost is estimated to be $90/household/per year in 2022, growing to $140/household/year in 2035 when the program is fully implemented and all residents in the state will be able to recycle a long list of materials. The projected future costs in Colorado track more closely to the costs in Washington and California, since they reflect higher levels of service and greater participation (more trucks on the street and more material to manage as recyclable).

SOURCE: CAA, 2024.

BOX 4-7
Case Study: Oregon's EPR Policy

The State of Oregon's extended producer responsibility (EPR) policy mandates that producer responsibility organizations (PROs) adjust fees to provide incentives for actions that reduce the environmental and human health impacts of covered products (e.g., changes in the design, production, and distribution of products (ORS 459A.884[4]). The PRO must propose criteria for adjusting fees and for the magnitude of the adjustments. The Oregon Department of Environmental Quality (ODEQ) plays an important role in developing the fees by providing formulas for the PRO to use in evaluating fee-setting for eco-modulation.

Impacts of concern related to packaging include climate change, toxicity, and microplastic pollution. These factors contribute to two of six "planetary boundaries" for climate and novel entities (i.e., toxic and long-lived substances released to the environment).

In 2024, ODEQ (n.d.) recommended that PROs develop eco-modulation formulas that:

  i. Incorporate ODEQ's rules for life cycle evaluation. The approach should verifiably deliver environmental benefits based on the normalized and weighted results calculated following ODEQ's rules of life cycle evaluation.
  ii. Grant at least as many malus fees (penalties) as bonus fees, rather than emphasizing bonuses over maluses, to communicate adequate urgency for system change.
  iii. Increase the magnitude of fee adjustments over time to maximize their effect.

While some producers view EPR as a regulatory burden, others see it as an opportunity to gain greater control over packaging standards and materials, aligning compliance efforts with their sustainability goals. To achieve high recycling rates, producers need households to recycle more, resulting in more recyclable feedstock.

4.4.3 Deposit-Return Systems

In a deposit-return (or deposit-refund) systems (DRSs), consumers pay a small amount when buying a product—often beverages in single-use cans and bottles. This amount is reimbursed to them when they bring the empty beverage container to a collection point with a vending machine or manual handheld scanning readers.

BOX 4-8
Case Studies: State-Specific Needs Assessment Laws

Several states have enacted laws to evaluate existing recycling systems and identify gaps in infrastructure, capacity, and costs. These needs assessment laws are foundational steps toward the development of comprehensive recycling and EPR programs, offering insights to guide future legislative and programmatic actions.

Illinois (Public Act 103-0383): This law directs the state to conduct a comprehensive statewide recycling needs assessment to identify service gaps, evaluate current infrastructure capacity, and analyze costs associated with packaging and paper recycling.

Maryland (SB 222): Maryland's legislation requires an assessment of both traditional recycling and organics processing capacity. The findings will inform the establishment of a producer responsibility program for packaging materials, with recommendations developed by a designated advisory council.

A DRS replaces the local government as the party responsible for funding and managing the recycling of the items it covers. The collection or redemption center manages their collection and is usually funded from the DRS. This system can encourage consumers to recycle containers rather than dispose of them in the regular waste stream, and it penalizes the person who buys the container if they do not recycle it. As discussed in Box 3-3 in Chapter 3, an idealized version of a DRS could fully internalize the external costs of waste disposal without creating incentives for illegal disposal (Fullerton and Kinnaman, 1995). Moreover, a DRS can be generalized to apply not just to containers or batteries, but to help reduce pollution from any kind of material (Fullerton and Wolverton, 2000).

In the United States, ten states have deposit-return policies requiring fees on the sale of beverage containers. The process typically involves a network of collection points where consumers can return their containers; these may be located in supermarkets, recycling centers, or stand-alone automated machines. Once the containers are returned, they are sorted, cleaned, and prepared for recycling. This system is also commonly used for hazardous household wastes, because it can promote recycling and remove these materials from the ordinary solid waste stream. For example, DRSs in U.S. states and several other countries include collections for lead-acid batteries, motor oil, and tires (Sigman 2020).

NCSL (2020) also summarized DRS legislation in these ten states (see Table 4-4).[8] Notably, not much meaningful policy variation has occurred in the last 15 years. Exceptions include Oregon, which added products to the program as recently as 2018, and California, which expanded its DRS in 2024. See Box 4-9 for a case study on California's DRS.

TABLE 4-4 Deposit-Return Programs in the United States

State	Statute	Year	Summary Deposit Amount	Summary Beverages Covered
California	Cal. Public Resources Code §§14501–14599	1986	5¢ (<24 oz.) 10¢ (≥24 oz.)	Beer, malt, wine, and distilled spirit coolers; all non-alcoholic beverages, except milk
Connecticut	Conn. Gen. Stat. §§22a-243–22a-246	1978	5¢	Beer, malt, carbonated soft drinks (including mineral and soda water and any type of other flavored water), bottled water (bottled water covered starting in 2009)
Hawaii	Hawaii Rev. Stat. §§342G-101–342G-122	2002	5¢	Beer, malt, mixed spirits and wine; all non-alcoholic drinks
Iowa	Iowa Code §455C.1–455C.17	1978	5¢	Beer, wine coolers, wine, liquor, carbonated soft drinks, mineral water
Maine	Me. Rev. Stat. Ann. tit. 38, §§3101–3118	1976	15¢ (wine/liquor) 5¢ (all others)	All beverages
Massachusetts	Mass. Gen. Laws Ann. ch. 94, §§321–327	1981	5¢	Beer, malt, carbonated soft drinks, mineral water
Michigan	Mich. Comp. Laws §§445.571–445.576	1976	10¢	Beer, wine coolers, canned cocktails, soft drinks, carbonated and mineral water
New York	N.Y. Environmental Conservation Law §§27-1001–27-1019 (Amended 2013 SB 2608)	1982	5¢	Beer, malt, wine coolers, carbonated soft drinks, soda water, and water not containing sugar
Oregon	Or. Rev. Stat. §§459A.700–459A.740	1971	10¢ 2¢ (standard refillable)	Beer, malt, carbonated soft drinks, bottled water (will cover all beverages except wine, distilled liquor, milk, milk substitutes and infant formula by 2018)
Vermont	Vt. Stat. Ann. tit. 10, §§1521–1529	1972	15¢ (liquor) 5¢ (all others)	Beer, malt, mixed wine, liquor, carbonated soft drinks
Guam	Guam Code tit. 10, §§44101–44119	2010	5¢	Beer, ale, malt, mixed spirits, mixed wine, and all non-alcoholic beverages

SOURCE: NCSL, 2020.

---

[8] See https://www.ncsl.org/environment-and-natural-resources/state-beverage-container-deposit-laws.

> **BOX 4-9**
> **California's DRS Programs: A Redemption Value Case Study**
>
> California established a deposit-refund system (DRS) in the AB2020 Act, which requires consumers to pay a redemption value for each eligible beverage container. The retailer charges consumers the California Redemption Value (CRV) at the point of purchase. When the container is empty, the consumer can take it to a certified drop-off recycling center and receive a refund of the same CRV amount.
>
> **Heterogeneous Responses to the Refund Program**
>
> Berck and colleagues (2021) focused on the demand side of the recycling system and summarized some of the evidence around the efficacy of the California Department of Recycling (CalRecycle) deposit-refund recycling program in conjunction with other recycling alternatives, such as curbside recycling pick-up—in providing convenient recycling options to consumers.
>
> Using a representative sample of 1,000 Californian adults, Berck and colleagues (2021) found that 23 percent of respondents used drop-off recycling centers. Respondents seem to find recycling generally convenient and worth their time. About 32 percent of respondents reported using curbside recycling, and 5 percent reported recycling at a business or place of worship outside their home. Notably, those who reported using the latter two options were wealthier and more educated than those who report using drop-off recycling centers. Less affluent households had lower opportunity costs of time and hence were more likely to find it worth their time and energy to redeem their CRV at the drop-off recycling centers instead of using curbside recycling. Black and Hispanic respondents were less likely to use curbside programs than others, and more likely to return material through drop-off recycling centers (Berck et al., 2021).
>
> Berck and colleagues (2021) asked survey respondents whether a change in the CRV would encourage them to return containers to drop-off recycling centers. The number of people who said they would redeem their containers at a drop-off recycling center increased with the CRV. Of the people who said that they were currently throwing their beverage containers in the trash, it took an increase in the CRV from 5 cents to 15 cents before more than half said they would start taking containers to a drop-off recycling center. Among those currently using trash for disposal, only 11 percent said they would redeem at a drop-off center if the rate were increased to 7 cents. The survey respondents who use curbside were much more responsive to a potential increase in CRV than those who threw out their containers. The same increase of CRV to 7 cents would lead to a 34 percent reallocation of recycling from curbside to drop-off recycling centers (Berck et al., 2021).
>
> **Counterfactual Policy Effects**
>
> In a follow-up study, Berck and colleagues (2024) simulated whether an increase in the CRV amount would (1) increase the overall recycling rate, (2) simply induce a switch in recycling methods, or (3) have no effect at all. The answer was not clear ex-ante; the authors found that those who currently recycled but did not redeem their CRV (e.g., through curbside recycling) and those who use trash disposal may or may not be sensitive to small changes in the CRV amount. Berck and colleagues (2024) found that doubling the CRV amount would induce only a modest increase in overall recycling (and the benefits of this policy would mostly accrue to wealthier individuals). Hence, an increase in the CRV amount may not be the optimal policy to increase recycling.
>
> Berck and colleagues (2024) consider the effect of reducing or eliminating the state's subsidy for handling fee centers. A 2008 policy change that reduced handling fee payments for some centers caused many of these centers to close. In addition, many drop-off recycling centers throughout California have closed in response to China's 2017 decision to restrict imports of recyclable materials (i.e., the National Sword Policy), further limiting drop-off recycling center options for consumers. This development affected all drop-off recycling centers, not just handling fee centers, but it has further reduced the ability of handling fee centers to operate without a state subsidy. Berck and colleagues (2024) use the model estimates to predict recycling options under counterfactual hypothetical closures of nearby convenient handling fee centers. They find that handling fee center users would generally just switch to using processing fee centers. Hence, the fear of consumers not having convenient recycling options without handling fee centers is likely unfounded.

4.4.4 Benefits of Deposit-Return Systems

A unique feature of DRSs is that they provide incentives directly to consumer and commercial establishments to return their containers for recycling. Not surprisingly, recycling rates of these containers are much higher with a DRS than without one. For example, states that implemented a DRS for polyethylene terephthalate (PET) bottles saw a 56 percent recycling rate, compared with 18 percent in states with no such system (Container Recycling Institute, 2024). Globally, recycling rates exhibit a strong positive association with the deposit fee and refund amounts, (see Figure 4-4).

To determine the causal effect of DRSs and to address place-based selection bias, Viscusi and colleagues (2012) conducted a difference-in-difference analysis, comparing the change over time in two states that extended their DRS to include plastic bottles with changes in states that did not. After the extension, the number of households that recycled increased by a statistically significant 8.5 percent. Likewise, Ashenmiller (2009) assessed the material brought in for rebates to California recycling centers and concluded that 36–51 percent of the material would not otherwise have been collected. In 2017, Oregon increased its deposit from 5 cents to 10 cents, because of a provision that increases the deposit when the recycling rate falls below a certain threshold, and saw an immediate increase in the recycling rate of the bottles covered in its program from 64 percent to 81 percent. Their recycling rate for covered bottles is estimated at 90.5 percent for 2023, according to the Oregon Beverage Recycling Cooperative (2023).

DRSs may have other benefits as well. First, given the high degree of sorting involved in returning beverage containers to recycling systems, a DRS can provide a cleaner recycling supply stream with less contamination relative to material collected at the curbside. This source-separated waste has higher value for end uses and is less likely to be diverted to landfill.

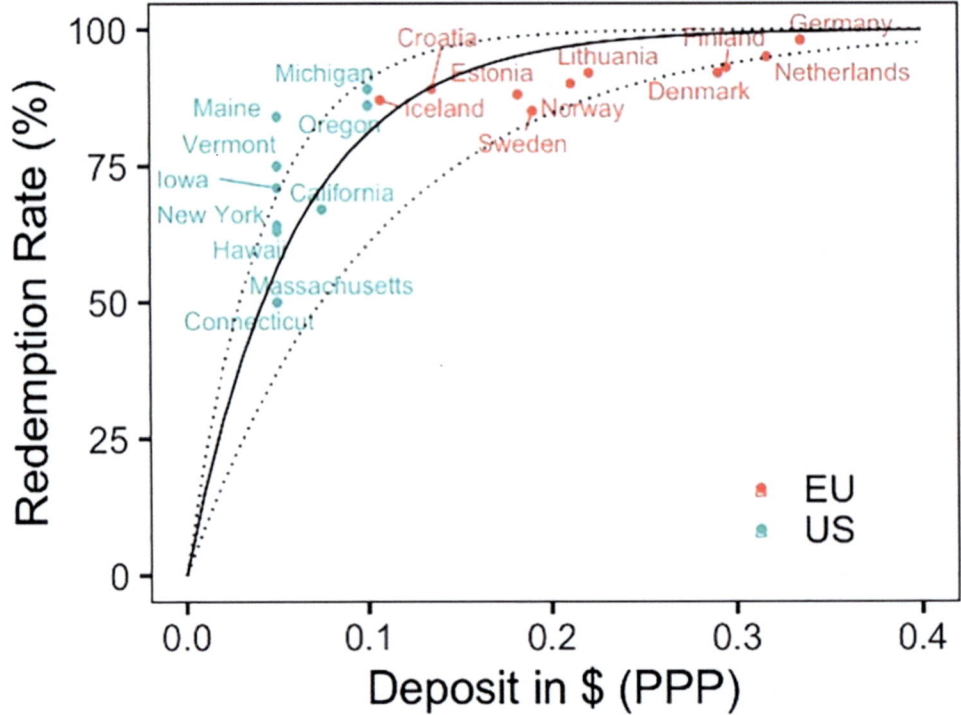

FIGURE 4-4 Comparison of redemption rates for deposit-return systems by deposit level.
NOTE: PPP = purchasing power parity.
SOURCE: Basuhi et al., 2024. CC-BY-NC-ND.

Second, the initial justification for using these systems for beverage containers included litter reduction; empirical research finds that they have been successful in this area. Levitt and Leventhal (1986) observed that New York's bottle bill reduced litter near highway exits and railways by a statistically significant 44 percent. And, using data from the United States and Australia, Schlyer and colleagues (2018) found that containers constitute a 40 percent lower share of coastal debris surveys in states with container DRSs. Similarly, Critchell and colleagues (2023) found that adopting a DRS significantly reduced plastic marine debris in Australia.

Third, these policies may have broader social benefits. Ashenmiller (2009) found that the beverage container DRS provide a significant source of income for a group of very-low-income "professional" recyclers. Extending this logic, she found that adopting beverage container DRS reduces petty crime by 11 percent (Ashenmiller, 2010).

4.4.5 Disadvantages of Deposit-Return Systems

Despite their many demonstrated advantages, deposit-return systems may have important disadvantages as well. First, while these programs increase the recycling of beverage containers through monetary incentives, they also remove high-value materials such as aluminum and PET bottles from the curbside stream. Studies have shown that aluminum and PET bottles are a small percentage of the recycling stream by weight but contribute disproportionately to MRF revenue because of their higher market value (Container Recycling Institute, 2022). Diverting these materials from the curbside stream reduces revenue for MRFs and could negatively impact the financial sustainability of curbside programs (unless MRFs were to receive some of the net revenue from the DRS). In fact, curbside recycling programs sometimes need to enforce against theft of containers from curbside bins (Lange, 2012).

Removing these materials from the curbside stream under a national DRS could reduce sales revenue to MRFs by more than the cost-saving from handling less curbside material. In addition, states with a DRS have struggled to invest in sophisticated MRF technology without the revenue associated with aluminum cans and PET bottles (Basuhi et al., 2024). Basuhi and colleagues (2024) estimate that implementing a national DRS would require increases in MRF processing fees net of commodity revenues by 13 percent and that $373 million more would be needed to offset these losses. Using unclaimed deposits to help fund MRFs, as California does, might help preserve this system. To date, DRSs have not confronted these issues because most were in place before curbside recycling programs were implemented.

A second potential drawback of DRSs is their high cost to redemption centers and consumers. Redemption centers, often in retail stores, must allocate space for collection and storage for the materials. Manual return systems rely on workers, and even automatic systems such as "reverse vending machines" require labor to maintain and manage, in addition to the capital costs of the machines. Estimates of handling costs vary. In Connecticut in 2018, handling costs were estimated at $0.0284 per container, which is within the range of handling fees paid by states that offer explicit reimbursement to redemption centers (Reloop, 2021). Handling fees offered by states range from 1 cent in Iowa to 3.5 cents for commingled containers in New York, Maine, and Vermont. An earlier estimate pegged the costs of California's system at 0.2 cents per container (Ackerman et al., 1995), which is an inflation-adjusted 0.41 cents per container today. About 50 percent of redemption centers in California, where the handling fee was 0.86, closed in the last decade, implying that centers may experience higher costs in many locations.

These estimates suggest relatively high costs relative to curbside recycling programs. For a PET bottle that weighs 9.25 grams (Recycling Today, 2015), a range of cost from $0.0041 to $0.0284 per bottle collected for refund implies a cost range from $400 to $2800 per ton of PET bottles collected, which may typically be higher than average costs for curbside collection and MRF processing. Adding a DRS to an area with curbside collection for non-deposit materials such as paper probably does not lower the costs for curbside collection dramatically. Thus, the combined costs of the two systems may be high.

Third, these redemption center costs exclude the substantial time and energy that households spend to sort, transport, and return materials for redemption. Household costs are likely lower for curbside recycling. Thus, the social costs of having a DRS may exceed curbside recycling costs substantially, suggesting that they are suited to collecting materials for which legal and illegal disposal is most harmful.

Finally, as with an EPR program, collecting materials through a DRS does not ensure socially valuable uses for the collected materials. The problem may be particularly acute for plastics. Virgin plastic resin is very inexpensive; manufacturers have little incentive to pay for U.S. Food and Drug Association–approved postconsumer resin without a policy that provides either a strong incentive or a requirement to do so. As a result, recycled plastic is more commonly used in nonfood applications (PET to textiles, high-density polyethylene [HDPE] to piping, low-density polyethylene [LDPE] to decking, polypropylene to paint cans). Even in these nonfood applications, however, demand may be low. A DRS may need to be coupled with policies to drive demand for the materials collected (e.g., a recycled content standard).[9]

4.4.6 Role of Unredeemed Deposits

Imperfect return rates in DRSs can generate net revenue via unredeemed deposits (i.e., when a customer forfeits deposit value by not returning the PET bottle). Handling of unredeemed deposits varies widely in the United States:

- Connecticut, Massachusetts, and Maine use it for general funds.
- New York gives producers a fraction to cover system costs.
- Iowa allows producers to keep it and operate the DRS.
- Michigan uses it to support retailers that handle deposits.
- California and Vermont earmark it for specific purposes such as a beverage container recycling fund and clean water programs.

If a 10-cent deposit fee were implemented nationwide, close to $1 billion in unredeemed deposits could be available. Policy design is needed for effective use of this revenue without introducing counterproductive or perverse incentives. For example, although perhaps politically expedient, allowing producers or retailers to keep forfeited deposits may undercut their incentives to encourage recycling (Calcott and Walls, 2005). Policymakers designing a DRS need to carefully consider the degree to which unredeemed funds are allocated back to DRS operations.

4.4.7 Considerations for Policy Implementation

In addition to unredeemed deposits, other considerations for implementing a DRS include the systems' limited scope and issues of end-market demand.

DRSs invite consumers to recycle, so expanding such a system nationwide could increase the amount of material collected for recycling. But it is important to note that the materials collected in these systems would account for only about 20 percent of the volume of recyclables in the United States. New deposit-return policies could be coupled with EPR policies to address the remaining 80 percent.

Additionally, increasing the supply of plastic and aluminum recyclables may have unintended market consequences, especially because demand for recyclables, particularly plastics, is always uncertain. A demand-side postconsumer recycling policy could balance the additional supply. Policies that address gaps in financing of plastic recycling are particularly important because the cost of virgin plastic is less than the cost of processing postconsumer plastic for recycling (see Chapter 5 for relevant recommendations).

---

[9] Another alternative would be to implement a variant of the system that applied refunds at the secondary producer, not the consumer, level (Palmer et al., 1995). In this policy, the refund would be paid as subsidy for use of the recycled material. This upstream subsidy would still increase material recovery because it raises prices for the secondary material and thus encourages its collection.

4.4.8 Deposit Legislation

Deposit-return legislation is sometimes thought to be a form of EPR, based on the role that producers play in the recycling process. However, EPR policies tend to require "invisible" fees built into the cost of the product, while DRSs require a visible fee to be paid by the purchaser (to be refunded only when that product is redeemed). Ten states in the United States have DRSs—also known as "bottle bills" (Figure 4-5; Container Recycling Institute, n.d.). These states tend to have high recycling rates for materials covered under the DRS; for example, bottle recycling returns are as high as 80 percent in Oregon.

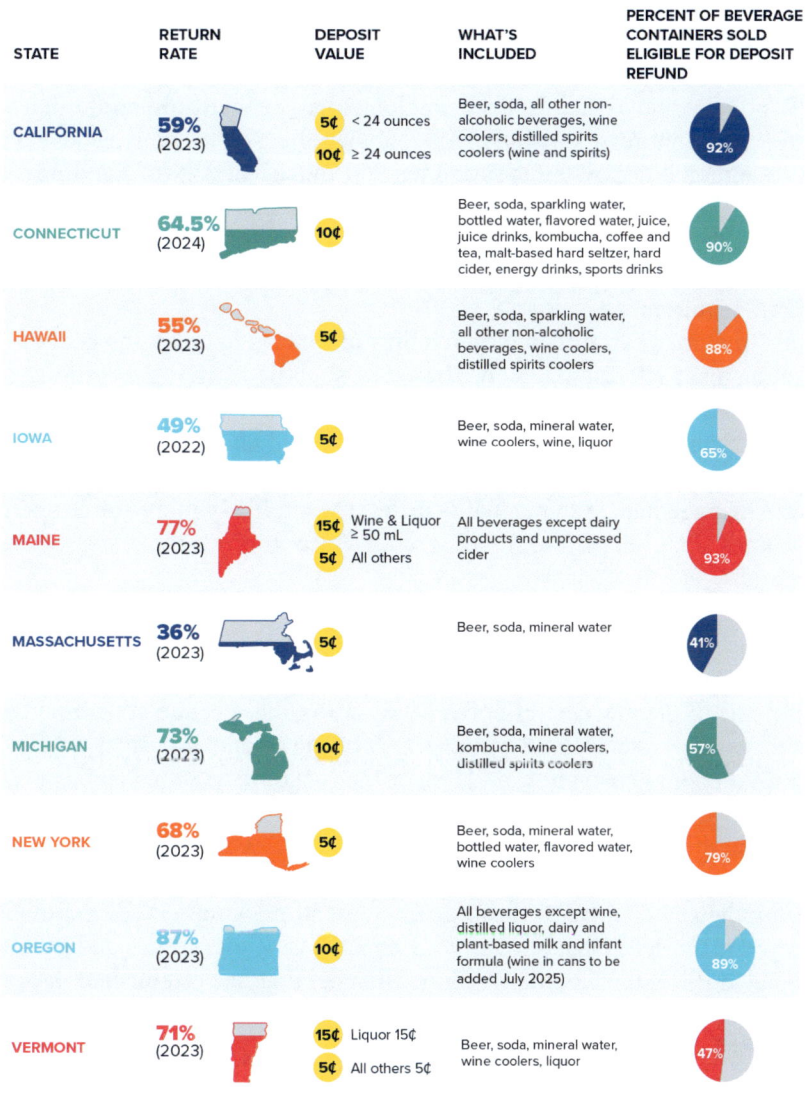

FIGURE 4-5 Bottle bill fact sheet.
SOURCE: TOMRA, 2023.

Except for Hawaii's program, all deposit-return programs in the United States were enacted before curbside recycling programs were implemented. Deposit-return policies include aluminum cans and PET bottles that could instead have been collected and sold as part of curbside recycling, reducing revenue for these programs unless the deposit-return policy provides compensation for them. Without compensation, the foregone revenue may make curbside recycling less affordable for local governments.

4.5 KEY POLICY OPTIONS AND RECOMMENDATIONS

The following section outlines key policy options and recommendations for improving recycling systems, focusing on strategies centered around the costs and approaches of recycling systems.

4.5.1 Extended Producer Responsibility

As discussed extensively in this chapter, EPR can be a policy strategy for decreasing the total environmental impact from a product by making the manufacturer of the product responsible for its entire life cycle and especially for the take-back, recycling, and final disposal of the product (Lindhqvist, 2000).

While EPR policies have already been implemented for hard-to-handle materials in the United States (e.g., e-waste, tires, mattresses, paint), EPR policies focused on plastics and other packaging and recyclables are less common but growing in prominence (France Ministry of Economy, 2025). As of 2024, five U.S. states have passed EPR policies for packaging.[10] While each state law is quite different, with language that reflects each state's unique recycling needs, four of the policies (all but Maine's) incorporate requirements for packaging reduction, recycling, use of postconsumer content, packaging design, and "responsible end-market" requirements.

The general structure of an EPR policy requires producers to cover the cost of the entire life cycle of packaging, including disposal, and to pay fees based on specified criteria. Additions or subtractions from the producer's disposal fee are intended to encourage changes in covered materials that would increase their recyclability or reduce their environmental harm. The fees are then used to pay for the cost of collection, processing, and recycling of the covered materials. In this way, EPR policies also establish a financing mechanism for waste management systems that is different, and likely more stable and reliable, than traditional financing schemes. This alternative funding scheme can improve recycling access and increase supply of recycled materials.

EPR policies for packaging typically allow for a nonprofit PRO to manage the program efficiently on behalf of the producers whose products are covered in the regulation. The organization collects fees from the producers and uses those funds to support and manage the recycling programs on their behalf to achieve the requirements of the law cost-effectively. By using the revenue generated by those EPR fees to support waste and recycling systems, EPR policies reduce the financial burden on local governments and recyclers, facilitate investment in recycling, and reduce the cost of recycling. Producer responsibility toward recycling programs increases access to recycling, ultimately increasing the supply of recycled materials. In addition, as discussed in this chapter, fee requirements under an eco-modulation scheme in the legislation can provide incentives for packaging design choices to help achieve the EPR policy goals, and disincentives for packaging design choices counter to those goals. Cost implications of EPR policies vary, but initial ranges can be observed in the states that have already enacted EPR for plastics and packaging. For example, Oregon has a designated PRO, the Circular Action Alliance, that has established packaging fees in a proposed program plan, ranging from 1 to 273 cents per pound depending on the packaging type (ODEQ, 2024).

EPR policies have trade-offs as well. First, at least a portion of the EPR fees imposed on manufacturers would be passed along to consumers in the cost of goods (e.g., groceries) and thus could be considered a regressive tax, as these goods make up a larger-than-average percentage of spending by low-income households. Second, EPR policies have not historically been designed to address end-market demand for recyclables. Adding supply of material without addressing demand has the potential to erode market prices,

---

[10] Maine, Oregon, Colorado, California, and Minnesota.

which would likely yield negative impacts on various actors in the waste management system, including collectors, haulers, sorters, and processors. This potential drawback of EPR policy could be addressed by a fee per ton of virgin plastic resin purchased by manufacturers and a reward per ton of recycled plastic resin purchased by those same manufacturers. Both the fee and the reward are intended to increase demand for recycled plastic resin and thus complement EPR policies that increase its supply.

States have passed EPR legislation with variations that reflect their specific recycling needs. While federal legislation could provide more consistent implementation and widespread adoption, state-by-state heterogeneity still needs to be considered. Instead of a nationwide EPR law, developing a federal EPR framework that recommends key elements for states to include in their EPR legislation could help facilitate effective use of this policy at the state level, while promoting consistency to the extent practicable. This federal framework would include recommending that states conduct a needs assessment to identify gaps in current services and programs, materials to be covered in the law ("covered materials"), eco-modulation concepts, and program implementation guidelines. Furthermore, Congress could encourage state adoption of these federal EPR guidelines by offering federal support for state recycling infrastructure that depends on adoption of federal EPR guidelines and standards.

Once passed, EPR policies for plastic packaging and other recyclables require multiple years to implement. Reasonable estimates based on states that have already passed EPR policies range from 3 to 5 years for implementation. After enactment, the results of these policies can be assessed via annual producer reporting to the regulatory agencies (including state or federal agencies). Reporting requirements are typically embedded within EPR policies and include data such as tons collected, tons recycled into new products, tons of packaging reduced, and carbon emission reductions associated with packaging reduced and recycled. Consistent with the committee's recommendation below regarding data needs, these data need to be collected and analyzed to ensure the EPR policy is working as intended and accomplishing appropriate recycling goals.

In summary, the committee offers the following conclusions, recommendation, and policy options regarding EPR policies.

Conclusion 4-1: EPR policies can achieve multiple policy objectives for recycling, including providing stable financing for recycling programs, decreasing contamination, increasing the efficiency of recycling systems, increasing demand for recycled material, and enhancing environmental and social benefits of recycling.

Recommendation 4-1: The United States should increase reliance on extended producer responsibility (EPR), which should cover packaging and expand to other materials as appropriate. EPR policies should include eco-modulation to create economic incentives for manufacturers to design for recyclability, and funding streams for recycling systems and infrastructures. State government should enact EPR policies to account for regional heterogeneity but should be supported and informed by a national framework with guidelines.

The committee developed key policy options to accomplish this recommendation, targeted at different levels of government.

Key Policy Option 4-1: The U.S. Environmental Protection Agency (EPA), with appropriate funding and authority from Congress, could develop and facilitate a national extended producer responsibility (EPR) framework, as outlined in its 2024 report *National Strategy to Prevent Plastic Pollution*. If it pursues this framework, EPA should consult with state, local, and tribal governments; nongovernmental organizations; industry; and other relevant partners. This framework should provide guidelines on key elements of state-level EPR policies and recommend minimum state-level standards and best practices. A national framework should provide as much consistency across states as possible and support multistate efforts, while allowing for state-level variation in targets, fees, covered materials, and methods to reflect heterogeneity in costs and benefits across states.

Key Policy Option 4-2: State governments could enact extended producer responsibility (EPR) policies, informed by any minimum standards provided by the U.S. Environmental Protection Agency. State-level needs assessments should identify gaps in current services and programs and serve as a basis for setting EPR fees. Within an EPR framework, state governments could consider policies, such as recycled content standards, to enhance end markets for recyclable materials.

4.6.2 State-Based Landfill Surcharges for Recycling

MSW landfills charge a "tipping fee" for the waste that they accept. This fee is typically charged as dollars per ton of solid waste (or in some cases, dollars per cubic yard). The tipping fee comprises (1) a baseline fee paid to the owners of the facility to cover their capital and operating costs and (2) a surcharge to address the social and environmental impacts associated with landfilling (i.e., external costs relative to other means of MSW management). A landfill tipping fee raises the cost of disposal and reduces its cost advantage over recycling. Furthermore, the funds collected can be used to support statewide or local recycling programs, further shifting MSW management from landfills toward recycling. Along with these incentives, reducing landfill use confers benefits by reducing the social and environmental externalities that they generate (see Chapter 6). Costs are also reduced by lengthening the lifetime of available landfill capacity.

The landfill tipping fee surcharges is an example of a Pigovian tax or fee intended to internalize the social and environmental external costs of a management option (as discussed in Chapter 3). Under ideal competitive market assumptions, a Pigovian tax induces reduction of costly polluting behavior in an economically efficient manner, by setting the tax rate equal to the marginal external cost (Arrow, 1970; Burrows, 1979; Cremer et al., 1998). The process for determining an appropriate externality fee involves two major steps: first, environmental impact studies such as life cycle analyses calculate the social or environmental damage expected from each disposal option (discussed below); second, economic studies are necessary to value this impact. These economic studies can employ behavioral (revealed preference), or attitudinal (stated preference) methods (see Chapter 6; see also Mendelsohn and Olmstead, 2009). Examples of social and environmental costs that have been explored for use in cost-benefit studies include Matthews and Lave (2000), Sovacool and colleagues (2021), and Comineti and colleagues (2024). Applications of externality pricing in cost-benefit analysis for solid waste management are demonstrated by Goddard (1995), Fullerton and Kinnaman (1995, 1996), Nahman (2011), Paes and colleagues (2020), Medina-Mijangos and colleagues (2021), Matheson (2022), and Massarutto (2024).

Landfills are regulated and permitted by state governments; however, their baseline tipping fees are set by the landfill owners based on market conditions (private landfills) or costs (public landfills). These fees can be paid at the landfill or at a transfer station that feeds into a particular landfill. At present, many but not all states have a landfill tipping fee surcharge (essentially, a tax), which is collected by landfill operators and paid by the individual or entity disposing of MSW at that landfill. In urban areas, this entity is often a hauler of MSW, but in rural areas the fee may apply to others transporting their own household waste to a landfill. Regardless of the entity directly paying this surcharge at the landfill, the surcharge would mostly be passed along to the original waste generator, raising the cost of waste disposal relative to other management options (e.g., recycling). State governments can also tailor fee levels to their state's conditions and can choose the appropriate local governments and programs to receive the funds raised.

Current Surcharge Rates Across the United States

By 2009, 30 states had imposed tipping fee surcharges, with an average fee of about $2 per ton (Jenkins and Maguire, 2012). In response to a survey from the National Conference of State Legislatures (2021), 12 states reported surcharges as of 2021, as shown in Box 4-10. These states have tipping fee surcharge rates that range from $0.25 to $6.00 per ton, with a mean of $2.40 per ton, a median of $2.00 per ton, and a standard deviation of $1.70 per ton.

| \multicolumn{4}{c}{**BOX 4-10**} |
| \multicolumn{4}{c}{**States with Landfill Tipping Surcharges**} |

State	Landfill Surcharge	State	Landfill Surcharge
Arizona	$0.25	Maine	$2.00
Arkansas	$1.50	Michigan	$6.00
California	$1.34	Nebraska	$1.25
Georgia	$0.75	Ohio	$2.00
Illinois	$2.00	Pennsylvania	$4.00
Iowa	$4.25	South Dakota	$3.00

NOTE: Rates are listed for the 12 states that provided them for the study.
SOURCE: NCSL, 2021.

The Environmental Research & Education Foundation (2024) reported total tipping fees in 2023 had a mean of $58 per ton, with regional averages ranging from $43 to $83 per ton, the $2 per ton surcharge represents 3.4 percent of the average total tipping fee (and 2.4–4.7 percent of the low and high regional averages). With a proposed additional surcharge of $2 per ton, the decrease in the financial advantage of landfilling relative to recycling is likely to be minor. However, the very presence of a landfill surcharge and its use to support recycling awareness and infrastructure may be enough to motivate public and stakeholder awareness, support, and action to begin and to continue the process of transition from landfill disposal toward recycling. While the most appropriate landfill surcharge rate for a state will be determined by its legislature with input from its citizens, industry, and environmental and economic advisors, the results of our estimates and analysis suggest that a $2 per ton surcharge can serve as a moderate and representative point of departure to initiate and support state-specific recycling legislation.

Appropriate (Moderate) Rates for a Tipping Fee Surcharge?

While states may select tipping fee surcharge rates based on the environmental and social externalities associated with disposal at a landfill, external damages are difficult to calculate in a consistent manner across sites and are subject to high measurement uncertainty. Furthermore, calculated rates may exhibit wide variation across a state, due to both physical and demographic heterogeneity and the use of differing methods and modeling assumptions. Guidance is needed for states to identify a range of tipping fees to be deemed as moderate and appropriate. For a total tipping fee to be considered moderate, it should encourage systems to begin recycling, due either to the modified cost differential or the new funding for recycling efforts generated by the fees. However, if they are too high, they present a threat to an MSW system's financial viability while it undergoes the transition from landfill dominated MSW service toward more recycling, nor to penalize communities that lack the necessary conditions and resources for this transition. A first test for a state or system is to ask, "How does a proposed surcharge compare with tipping and surcharge rates at other landfills, nationally or in their state or region, especially those with similar waste, economic, and environmental conditions?"

While surcharges affect costs only minimally, it is reasonable to assume they are passed along to the original waste generator, thereby raising the cost of waste disposal relative to other management options, including recycling and composting. This may help address difference in cost between landfill disposal (which are usually lower) versus recycling and composting (which are usually higher). Addressing this difference is especially important when revenues from the sale of recovered recyclables are in a market downswing. Increasing surcharges for landfill disposal could increase economic pressures and incentives for recycling and composting.

Surcharges can further reduce the price differential between landfilling and recycling if the revenues generated are allocated by state governments to support local recycling efforts (or other MSW treatment, including composting or other treatments of organics). These funds could offset the cost of local recycling operations by being allocated to recycling enterprise funds, or could support grants for local recycling infrastructure, fund local social modeling programs (i.e., locally organized programs that promote recycling norms and behaviors [see Section 6.3 in Chapter 6]). The funds could even support research efforts that inform recycling systems and promote their cost efficiency.

The economic pressure and incentive to recycle from landfill tipping fee surcharges would be felt by consumers differentially. For example, businesses and rural households that pay for private waste hauling (and rural households that transport their own waste for disposal) would see the direct effects of this policy and have greater incentives to substitute recycling for disposal, when possible, as well as reducing their overall waste generation. Local governments that pay for MSW collection will likewise have stronger economic incentives to provide curbside and recycling drop-off programs and municipal composting programs. Households served by these government-funded systems may see an increase in taxes or waste management fees, but not a change in costs that would motivate changes in their waste management behavior. Exceptions would include households covered by pay-per-bag or subscription programs that are sensitive to the waste quantities. Thus, a limitation of this policy is that it may not encourage appropriate or desired changes in behavior for these households, which make up most of the population.

Additional disadvantages of this policy may arise but can be managed by keeping landfill tipping fee surcharges at an appropriate level. For waste generators who experience greater direct costs of disposal, surcharges may encourage not only reuse and recycling, but also illegal dumping of waste or greater contamination of recycled materials. The unintended adverse effect of a landfill surcharge on illegal dumping is likely to vary by location and the associated demographic characteristics of that locality. Residents of rural areas may have easy access to unmonitored locations for dumping waste illegally; urban locations could also see illegal dumping in vacant lots or in commercial dumpsters, if those locations are not monitored. Revenues raised from surcharges could be spent on efforts to monitor and/or discourage illegal disposal and "wish-cycling" behavior (i.e., the act of placing non-recyclable items in recycling bins under the mistaken assumption that they can be processed, leading to contamination and inefficiencies in recycling systems).

The distribution of burdens of landfill tipping fee surcharges may also be a concern. As described above, direct costs of waste management will rise for households that transport waste to landfills themselves or pay for private waste hauling. It will also raise costs for local governments who will see an increase in their waste management fees. When local governments use enterprise funding for MSW and impose uniform charges across all households, these increased costs will place a higher relative burden on low-income families than on high-income families. Some of these effects might be mitigated by focusing the redistribution of surcharge revenue to address recycling costs for the households most burdened by the surcharges. Lastly, for local government-owned and -operated landfills, landfill tipping fees are often used to support general funds for other government-provided public goods. If the tipping fee surcharges work as intended and decrease the amount of recyclable material entering a landfill, these landfills may see a decline in revenue that reduces funding for local governments.

Landfill tipping fee surcharges to encourage and support recycling need to be mandated at the state rather than local level. Some local governments own and operate their own landfill and therefore have different incentives for the volume of waste that is landfilled. Furthermore, residents or business could bypass local surcharges could be bypassed by transporting waste to landfills in localities without surcharges. State governments can tailor surcharge levels to their appropriate state-level conditions and can choose the appropriate local governments and program to receive the funds raised.

Especially as compared with other policy options for recycling, the time frame necessary to implement surcharges on landfill tipping fees could be short. If a state legislature adopts or increases a landfill surcharges, changes can likely begin within 1–2 years. Funding for recycling programs arising from these surcharges can thus likely be available within 2–3 years (i.e., 1 fiscal year after the initiation of the surcharge). To measure its effectiveness as a policy, states could monitor both funds raised and MSW diversion rates. However, states also need to consider evidence of increased illegal disposal or recycling contamination from segments of the MSW system that see direct effects of the surcharges.

Conclusion 4-2: State-based landfill tipping fee surcharges can provide a dedicated revenue source to support recycling programs and can provide incentives for waste diversion from landfills (especially recyclable materials and organics) from landfills. As such, landfill tipping fee surcharges can offset some of the costs of recycling, enhance social and environmental benefits associated with recycling, and provide stable financing for recycling systems.

Key Policy Option 4-3: State governments could implement mandatory surcharges on landfill tipping fees to provide incentives for recycling, support recycling and composting efforts, and divert waste from landfills. Moderate surcharges would minimize harmful responses (e.g., illegal dumping, increased contamination of recycling streams). State governments could collect and redistribute the funds to various recycling activities based on state and local priorities. Local uses of these revenues may vary with needs but could include grants for recycling infrastructure, shoring up enterprise funds for recycling operations, and funding local social modeling programs.

4.6.3 Federal and State Government Funding for Tribal and Rural Recycling Programs

One of the primary objectives of MSW management programs is to make it easy and convenient for residents to participate in them. Achieving this objective is a key reason why curbside collection services are provided on a regular basis to single-family residences.

While this type of service can be provided cost-efficiently to residents in urban and suburban communities, curbside collection is cost-prohibitive for many tribal and rural communities because of their low population density and long distances between households. While 80 percent of the United States is rural land, only about 20 percent of U.S. households are located in these areas. Therefore, the primary method of providing recyclables and organics collection to rural communities is through drop-off recycling and composting centers.

While the economics of recycling in rural parts of the United States are different from suburban or urban settings, recycling can sometimes provide similar benefits. Recycling in these areas can reduce landfilled waste and ensure that landfills last longer. Recycling services can provide similar job availability as in urban settings and achieve positive environmental benefits over alternative means of disposal. However, overall life cycle impacts of materials are different in rural settings from urban settings; this heterogeneity is an important consideration for tribal and local governments in rural settings when determining which materials to accept in their recycling programs. As discussed in Chapter 2, local governments need to conduct assessments based on sustainable materials management and set goals to ensure their programs are both economically and environmentally sound.

Recycling programs in rural areas face other economic challenges besides low population density. In general, these programs have higher transportation costs than urban settings; they tend to have more difficulty in reaching end markets for recycled materials, and in reaching economies of scale. As mentioned in Chapter 3, a hub-and-spoke system may help address these challenges.

Rural recycling programs have other economic differences compared with urban settings. Local drop-off centers and facilities that process fewer amounts and different kinds of materials often have lower capital cost requirements than MRFs in urban settings. Operational costs (e.g., staffing) may also be lower, but access to capital and financing to cover those operational costs may be less reliable than in urban areas. As such, alternative financing models may be needed to design cost-efficient and environmentally sound rural recycling programs. Alternative funding mechanisms for rural recycling areas include dedicated financing generated from statewide federal level grants such as the Solid Waste Infrastructure for Recycling programs, state-level EPR policies that promote recycling programs in rural areas, and landfill tipping fee surcharges.

While tribal and rural recycling often run into similar issues—such as lack of infrastructure; long distances to access waste disposal; recycling, or composting; and lack of funding—tribal recycling programs face many unique challenges. If a reservation-based tribe is not checkerboarded with non-tribal residents (i.e., if the land is solely tribal land), then they often are responsible for their own solid waste responsibilities as they are not tied to a municipality or county collection system. Funding is seriously lacking to address this problem, since tribal programs are not locally, state or federally funded. These locations may have reduced populations without the ability to collect taxes to cover costs, substantially raising the fees to address open dumping of solid waste. Cultural differences also arise between tribes and rural citizens. Tribal citizens who live on reservations solely inhabited by tribal members believe that the U.S. government will provide for them based upon treaty rights that are in place. Non-tribal citizens who live in rural areas may try to do the right thing but many simply do not want to pay to dispose of their waste and therefore dump it wherever they can—to avoid paying fees.

Unique challenges are also present for each tribe. Of the 574 sovereign tribal nations across the United States and Alaska, 229 are Alaskan Native Villages. Nearly half of those Alaskan Nations must send out their waste by barges or planes that may only be able to operate for 6 months of the year. For example, the Alakanuk Traditional Tribe of Alakanuk, Alaska, does not have good roadways in and out of the Village, so waste is addressed through barges and planes. Conversely, the Chickaloon Village of Palmer and Sutton, Alaska, relies on its roadways to dispose of waste. The Fort Belknap Indian Community, the fourth largest reservation in Montana is only 40 miles south of the Canadian border. This community have challenges with finding resources for waste disposal because of the distance from larger communities. The Mille Lacs Band of Ojibwe, located in East Central Minnesota approximately 2 hours north of Saint Paul, Minnesota, has developed a successful hub and spoke system but funding remains a challenge.

Conclusion 4-3: Relying on local government financing limits access to recycling programs, particularly for residents of rural areas, where recycling costs are high. It also limits access for many small business owners, who often face difficulties participating in municipal recycling programs. Alternative funding mechanisms, such as state or federal grants or EPR programs, would help distribute recycling costs across a broader population.

Key Policy Option 4-4: The U.S. Congress and state legislatures could authorize and appropriate funds for rural and tribal recycling. These funds could help communities overcome transportation distances and achieve economies of scale, through purchase of infrastructure such as trucks, drop-off and transfer facilities, and processing facilities. In parallel, the U.S. Environmental Protection Agency could continue to provide Solid Waste Infrastructure for Recycling grants for rural and tribal communities. State government funding could be derived from revenues generated from extended producer responsibility policies, landfilling tipping fee surcharges, or other state-based revenues.

REFERENCES

Ackerman, F.D., J. Stutz, and B. Zuckerman. 1995. Preliminary analysis: The costs and benefits of bottle bills. 94(229) Tellus Institute.

Anshassi, M., and T.G. Townsend. 2023. The hidden economic and environmental costs of eliminating kerbside recycling. *Nature Sustainability* 6(8):919–928.

Anshassi, M., and T.G. Townsend. 2024. Residential recycling in Florida: A case study on costs, environmental impacts, and improvement strategies. *Resources, Conservation and Recycling* 206. https://doi.org/10.1016/j.resconrec.2024.107627.

Arrow, K. 1970. Political and economic evaluation of social effects and externalities. In *The analysis of public output* (pp. 1–30). National Bureau of Economic Research. http://www.nber.org/chapters/c3349.

Ashenmiller, B. 2010. Externalities from recycling laws: Evidence from crime rates. *American Law and Economics Review* 12(1):245.

Ashenmiller, B. 2009. Cash recycling, waste disposal costs, and the incomes of the working poor: Evidence from California. *Land Economics* 85(3):539–551.

Basuhi, R., K. Bhuwalka, R. Roth, and E.A. Olivetti. 2024. Evaluating strategies to increase PET bottle recycling in the United States. *Journal of Industrial Ecology* 1–12. https://doi.org/10.1111/jiec.13496.

Berck, P., G. Englander, S. Gold, S. He, J. Horsager, S. Kaplan, M. Sears, A. Stevens, C. Trachtman, R. Taylor, and S.B. Villas-Boas. 2021. Recycling policies, behavior and convenience: Survey evidence from the CalRecycle program. *Applied Economic Perspectives and Policy* 43(2):641–658. https://doi.org/10.1002/aepp.13117.

Berck, P., M. Sears, R.L.C. Taylor, C. Trachtman, and S.B. Villas-Boas. 2024. Reduce, reuse, redeem: Deposit-refund recycling programs in the presence of alternatives. *Ecological Economics* 217:108080. https://doi.org/10.1016/j.ecolecon.2023.108080.

Burrows, P. 1979. Pigovian taxes, polluter subsidies, regulation, and the size of a polluting industry. *The Canadian Journal of Economics/Revue canadienne d'Economique* 12(3):494–501.

Calcott, P., and M. Walls. 2005. Waste, recycling, and "Design for Environment": Roles for markets and policy instruments. *Resource and Energy Economics* 27(4):287–305.

CalRecycle. 2021. *Carpet Stewardship Law*. https://www.calrecycle.ca.gov/carpet/law.

Circular Needs Alliance. 2024. *Colorado needs assessment*. Circular Action Alliance. https://circularactionalliance.org/co-needs-assessment.

Comineti, C.D.S.S., M.M. Schlindwein, and P.H. de Oliveira Hoeckel. 2024. Socio-environmental externalities of sewage waste management. *Science of The Total Environment* 945:174109.

Container Recycling Institute. 2024. https://www.container-recycling.org/index.php/key-facts.

Cooperman, A., A. Eberle and E. Lantz. 2021. Wind turbine blade material in the United States: Quantities, costs, and end-of-life options. *Resources Conservation and Recycling* 168:105439. https://doi.org/10.1016/j.resconrec.2021.105439.

Cremer, H., F. Gahvari, and N. Ladoux. 1998. Externalities and optimal taxation. *Journal of Public Economics* 70:343–364.

Critchell, K., L. Edge, and M.J. Traurig. 2023. Legacy community science data suggest reduced beached litter in response to a container deposit scheme at a local scale. *Marine Pollution Bulletin* 195(1):115471.

Crowe, LLP. 2024. *Study results: Processor commodity risk fee and contamination management fee*. https://www.oregon.gov/deq/recycling/Documents/croweCRPFfeesRep.pdf.

Duran, A.S., A. Atasu, and L.N. Van Wassenhove. 2021. Cleaning after solar panels: Applying a circular outlook to clean energy research. *International Journal of Production Research* 60(1):211–230. https://doi.org/10.1080/00207543.2021.1990434.

EPA (U.S. Environmental Protection Agency). 2013. *MSW Residential/Commercial Percentage Allocation–Data Availability*. Office of Resource Conservation and Recovery. https://www.epa.gov/sites/default/files/2016-01/documents/rev_10-24-14_msw_residential_commercial_memorandum_7-30-13_508 fnl.pdf.

EPA. 2022. *The Bipartisan Infrastructure Law: Transforming U.S. Recycling and Waste Management*. U.S. Environmental Protection Agency. https://www.epa.gov/system/files/documents/2022-02/orcr_bil_brochure.pdf.

EPA. 2024. *Consumer Recycling Education and Outreach (REO) Grant Program*. https://www.epa.gov/system/files/documents/2024-09/24-12.pdf.

EPA. 2025a. *Recycling Grant Selectees and Recipients*. January. https://www.epa.gov/infrastructure/recycling-grant-selectees-and-recipients.

EPA. 2025b. *Solid Waste Infrastructure for Recycling Grant Program*. February. https://www.epa.gov/infrastructure/solid-waste-infrastructure-recycling-grant-program#about.

EREF (Environmental Research & Education Foundation). 2024. *Analysis of MSW Landfill Tipping Fees—2023*. www.erefdn.org.

Fullerton, D., and T.C. Kinnaman. 1995. Garbage, recycling, and illicit burning or dumping. *Journal of Environmental Economics and Management* 29(1):78–91. https://doi.org/10.1006/jeem.1995.1032.

Fullerton, D., and T.C. Kinnaman. 1996. Household responses to pricing garbage by the bag. *American Economic Review* (September).

Fullerton, D., and A. Wolverton. 2000. Two generalizations of a deposit-refund systems. *American Economic Review* 90(2):238–242.

Fullerton, D., and W. Wu. 1998. Policies for green design. *Journal of Environmental Economics and Management* 36(2):131–148.

Goddard, H.C. 1995. The benefits and costs of alternative solid waste management policies. *Resources, Conservation and Recycling* 13(3–4):183–213.

ICMA (International City/County Management Association). 2019. *2017 Alternative Service Delivery Survey – Summary of Survey Results*. Washington, DC: ICMA. http://icma.org.

Jenkins, R.R. and K.B. Maguire. 2012. An examination of the correlation between race and state hazardous and solid waste taxes. In H.S. Banzhaf (ed.), *The political economy of environmental justice* (pp. 249–266). Stanford University Press.

Joltreau, E. 2022. Extended producer responsibility, packaging waste reduction and eco-design. *Environmental and Resource Economics* 83:527–578. https://doi.org/10.1007/s10640-022-00696-9.

Lange, R. 2012. Resource Recycling. https://www.resource-recycling.com/images/e-newsletterimages/Lange 0212rr.pdf.

Levitt, L., and G. Leventhal. 1986. Litter reduction: How effective is the New York State Bottle Bill? *Environment & Behavior* 18(4):467.

Lifset, R., H. Kalimo, A. Jukka, P. Kautto, and M. Miettinen. 2023. Restoring the incentives for eco-design in extended producer responsibility: The challenges for eco-modulation. *Waste Management* 168:189–201. https://doi.org/10.1016/j.wasman.2023.05.033.

Lindhqvist, T. 2000. Extended producer responsibility in cleaner production. IIIEE dissertation. https://www.iiiee.lu.se/thomas-lindhqvist/publication/e43c538b-edb3-4912-9f7a-0b241e84262f.

Lindhqvist, T., and K. Lidgren. 1992. Towards an extended producer responsibility-analysis of experiences and proposals. *Ministry of the Environment and Natural Resources*. Stockholm.

Lindhqvist, T., and K. Lidgren. 1990. "Modeller för förlängt producentansvar" ("Models for Extended Producer Responsibility," in Swedish). *Ministry of the Environment, From the Cradle to the Grave – Six Studies of the Environmental Impact of Goods* 7–44.

Massarutto, A. 2024. Monetary evaluation in LCA of WM: Everything engineers always wanted to know about it (but were afraid to ask). *Waste Management* 178:12-25.

Matheson, T. 2022. Disposal is not free: Fiscal instruments to internalize the environmental costs of solid waste. *International Tax and Public Finance* 29:1047–1073. https://doi.org/10.1007/s10797-022-09741-1.

Matthews, H.S., and L.B. Lave. 2000. Applications of environmental valuation for determining externality costs. *Environmental Science & Technology* 1390–1395.

Mazancourt, T., A. Thonier, P. Follenfant, M. Pascal, C. Ghesquieres, M. Outters-Perehinec, G. Mikowski, and A. Paugam. 2024. *Performance and governance of sectors with extended producer responsibility*. Ministry of Economy, Finance and Industrial and Digital Sovereignty. https://www.economie.gouv.fr/cge/filieres-rep.

Medina-Mijangos, R., A. De Andrés, H. Guerrero-Garcia-Rojas, and L. Seguí-Amórtegui. 2021. A methodology for the technical-economic analysis of municipal solid waste systems based on social cost-benefit analysis with a valuation of externalities. *Environmental Science and Pollution Research* 28(15):18807–18825.

Mendelsohn, R., and S. Olmstead. 2009. The economic valuation of environmental amenities and disamenities: Methods and applications. *Annual Review of Environment and Resources* 34(1):325–347.

Nahman, A. 2011. Pricing landfill externalities: Emissions and disamenity costs in Cape Town, South Africa. *Waste Management* 31(9–10):2046–2056.

NCSL (National Conference of State Legislatures). 2020. *State Beverage Container Deposit Laws*.

NCSL. 2021. *States with Landfill Tipping Charges*. https://cdn.ilsr.org/wp-content/uploads/2022/02/State-landfill-tipping-surcharges.2021.pdf.

ODEQ (Oregon Department of Environmental Quality). 2012. *Materials Management in Oregon: 2050 Vision and Framework for Action*. State of Oregon Department of Environmental Quality. https://www.oregon.gov/deq/filterdocs/mmanagementor.pdf.

ODEQ. 2023. *Overview of Scenario Modeling: Oregon Plastic Pollution and Recycling Modernization Act.* State of Oregon Department of Environmental Quality. https://www.oregon.gov/deq/recycling/Documents/RMAModeling-ExecSum.pdf.

OECD (Organisation for Economic Co-operation and Development). 2001. *Extended producer responsibility: A guidance manual for governments.* OECD Publishing, Paris. https://www.oecd-ilibrary.org/environment/extended-producer-responsibility_9789264189867-en.

Oregon Recycling Beverage Cooperative. 2023. *Delivering Results for Oregon: 2023 Annual Report.* https://obrc.com/wp-content/uploads/2024/03/About_Reports_2023_Annual.pdf.

Paes, M.X., G.A. de Medeiros, S.D. Mancini, A.P. Bortoleto, J.A.P. de Oliveira, and L.A. Kulay. 2020. Municipal solid waste management: Integrated analysis of environmental and economic indicators based on life cycle assessment. *Journal of Cleaner Production* 254:119848.

Pressley, P.N., J.W. Levis, A. Damgaard, M.A. Barlaz, and J.F. DeCarolis. 2015. Analysis of material recovery facilities for use in life-cycle assessment. *Waste Management* 35:307–317. https://doi.org/10.1016/j.wasman.2014.09.012.

Product Stewardship Institute, Inc. 2025. https://productstewardship.us.

Recycling Today. 2015. *Weight of water bottles decreases, while recycled content increases.* https://www.recyclingtoday.com/news/water-bottle-weight-decreases-recycled-content-increases.

Reelop. 2021. *Handling fees in deposit return systems.* https://www.reloopplatform.org/wp-content/uploads/2021/05/DRS-Fact-Sheet-Handling-Fees-4May2021.pdf.

Resource Recycling. 2022. *From green fence to red alert: A China timeline.* https://resource-recycling.com/recycling/2018/02/13/green-fence-red-alert-china-timeline.

Sheahan, J. 2024. Comments to the Committee on Costs and Approaches for Municipal Solid Waste Recycling Programs. National Academies of Sciences, Engineering, and Medicine.

Sigman, H. 2020. Deposit refunds. In K. Richards and J. van Zeben (eds.), *Policy instruments in environmental law.* Cheltenham, UK: Edward Elgar (pp. 377–386). https://doi.org/10.4337/9781785369520.

Smith, B.L., A. Sekar, H. Mirletz, G. Heath, and R. Margolis. 2024. *An Updated Life Cycle Assessment of Utility-Scale Solar Photovoltaic Systems Installed in the United States.* Golden, CO: National Renewable Energy Laboratory. NREL/TP-7A40-87372. https://www.nrel.gov/docs/fy24osti/87372.pdf.

Sovacool, B.K., J. Kim, and M. Yang. 2021. The hidden costs of energy and mobility: A global meta-analysis and research synthesis of electricity and transport externalities. *Energy Research & Social Science* 72:101885.

Sustainable Packaging Coalition. 2016. *2015–16 Centralized Study on Availability of Recycling.*

The Recycling Partnership. 2020a. *2020 State of Curbside Recycling Report.* https://recyclingpartnership.org/wp-content/uploads/dlm_uploads/2020/02/2020-State-of-Curbside-Recycling.pdf.

The Recycling Partnership 2020b. *2020 Guide to Community Material Recovery Facility Contracts.* https://recyclingpartnership.org/mrf-contracts.

The Recycling Partnership. 2024. *State of Recycling: The Past and Present of Residential Recycling in the US, 2024.* https://recyclingpartnership.org/wpcontent/uploads/dlm_uploads/2024/01/Recycling-Partnership-State-of-Recycling-Report-1.12.24.pdf.

TOMRA. 2023. *Bottle bill states and how they work.* https://www.tomra.com/reverse-vending/media-center/feature-articles/bottle-bill-states-and-how-they-work.

TOMRA. 2025. *Rewarding recycling: Lessons from the world's highest performing deposit refund systems.* https://circular-economy.tomra.com/resources/drs-white-paper.

UNC (University of North Carolina) School of Government. 2024. *North Carolina Benchmarking Project Benchmarking 2.0.* https://benchmarking.sog.unc.edu.

U.S. Census. 2024. Annual Survey of State and Local Government Finances and Census of Governments downloaded through Tax Policy Center web tool. https://state-local-finance-data.taxpolicycenter.org.

Viscusi, K.W., J. Hueber, and J. Bell. 2012. Alternative policies to increase recycling of plastic water bottles in the United States. *Review of Environmental Economics and Policy* 6(2):190.

Walls, M., M.K. Macauley, and S.T. Anderson. 2003. *The Organization of Local Solid Waste and Recycling Markets: Public and Private Provision of Services.* Resources for the Future. https://doi.org/10.22004/ag.econ.10892.

Walls, M., M.K. Macauley, and S.T. Anderson. 2005. Private markets, contracts, and government provision. *Urban Affairs Review* 40(5):590–613.

5
Materials and Markets

Summary of Key Messages

- **Commonly recycled materials:** Most U.S. communities focus on recycling plastics, paper, cardboard, glass, and metals.
- **Managing food and yard waste:** Some states prohibit the landfilling of yard waste and have mandatory recycling policies for these materials. Municipal solid waste authorities have increasing interest in managing these materials through composting and anaerobic digestion. While these management strategies deliver high environmental benefits through the reduction of greenhouse gas emissions, separate collection present challenges.
- **Variability in recycling efficiency:** Recycling rates vary significantly across the United States, with some cities achieving higher efficiency due to mandates, specialized programs, effective education and enforcement, and focused public policies. These efforts demonstrate the potential for recycling efficiency to surpass national averages through targeted local actions.
- **Importance of reducing contamination:** Contamination in recycling streams hinders the economic and environmental effectiveness of recycling systems; reducing contamination is a critical focus for national recycling strategies.
- **Challenges in recycling plastics:** Recycling plastics is important because of their persistence in the environment, their generation from non-renewable sources, their contribution to litter problems, and more. However, their recycling rates are low, partly because only certain resin types are accepted for recycling (as influenced by market demand and technological limitations).
- **Role of end markets:** End markets play a critical role in sustaining recycling systems, with cardboard and high-value materials such as metals and HDPE and PET plastics contributing the most reliable revenues.
- **Global recycling markets:** Global market shifts, such as China's import bans on recyclables have reshaped global recycling markets, exposed the need for resilient domestic markets, and led to increased recycling capacity and market development in the United States.

This chapter explores the various materials commonly accepted for recycling that are within the committee's scope of task, and it identifies relevant issues that are unique or specific to individual materials. It also builds on previous chapters' discussions of end markets, which are critical to an effective recycling system, and outlines various public policies relevant to enhancing those markets.

5.1 RECYCLING RATES

In the United States, most communities focus on five material types collected curbside or at drop-off centers and processed at materials recovery facilities (MRFs): plastics, paper, cardboard, glass, and metals (ferrous and aluminum). Less commonly collected are food and yard wastes, which can be managed through composting or anaerobic digestion. Industry plays an important role by collecting and recycling some of its own materials. Understanding the materials accepted by MRFs across the country, and the rates at which those materials are collected, is complicated because of a lack of standardized reporting; see Box 5-1 for legislation proposed to address these issues.

> **BOX 5-1**
> **Recycling and Composting Accountability Act**
>
> The Recycling and Composting Accountability Act (S. 3743) is aimed at strengthening recycling and composting systems across the United States through enhanced data collection, standardized reporting, and actionable analysis for policymakers and stakeholders. Passed by the Senate in July 2022 with support from 11 cosponsors (5 Democratic, 5 Republican, and 1 Independent), the act directs the U.S. Environmental Protection Agency (EPA) to evaluate the nation's recycling and composting infrastructure, establish baseline metrics, and provide comprehensive reports on material recovery. These measures seek to address existing inefficiencies, improve transparency, and guide improvements in waste management practices.
>
> Central to the act is a focus on improving data reporting. It tasks EPA with creating detailed inventories of public and private materials recovery facilities, cataloging the types of materials—such as plastics, paper, metals, and glass—that each facility can process. This initiative provides a clearer picture of the national recycling landscape, highlighting gaps in infrastructure and opportunities for optimization. Furthermore, the act mandates the standardization of recycling and composting rate reporting across states, ensuring consistent and comparable data nationwide. These standardized metrics will help reduce confusion about recycling and composting capabilities and improve the ability of local and federal agencies to make data-driven decisions.
>
> The act also emphasizes the development of metrics for contamination and capture rates within recycling systems. By identifying inefficiencies in material processing and recycling outcomes, these metrics will guide targeted improvements in collection, sorting, and recycling practices. This data-centric approach will enable federal, state, local, and tribal governments to enhance the efficiency and effectiveness of recycling systems, ultimately reducing contamination and promoting a more sustainable use of resources. By prioritizing comprehensive and consistent data collection, the act lays the groundwork for informed policymaking that supports both environmental and economic goals.
>
> SOURCE: Recycling and Composting Accountability Act, S. 3743, 117th Congress (2022), see https://www.congress.gov/bill/117th-congress/senate-bill/3743/text.

Absent standardized reporting, data on national collection and processing of recyclables are updated rarely. The most recent data were published in 2020 and reflect estimates of diversion in 2018 (see Figure 5-1 and Table 5-1; EPA, 2020). Figure 5-1 shows the amount of material recycled rose rapidly in the 1990s and early 2000s but has risen much more slowly since.

In addition to total recycling material tonnages, efforts have been made to calculate recycling rates for different materials over time. As displayed in Figure 5-1B, Arbex and Mahone (2024) use data from the U.S. EPA, together with material flow accounts, which track economy-wide material use (Eurostat, 2018), to calculate an overall recycling rate for recycled materials from 1970-2015. While recycling rates for materials differ, all materials display an increase in recycling rate during this time. Recent changes in recycling may be explained by a combination of technical factors such as differences in the inherent flexibility of materials for sorting and reuse, and economic and regulatory factors affecting waste processor costs and consumer preferences (Bening et al., 2021; Butler and Hooper, 2005; Le Pera et al., 2023).

In Figure 5-1, the tonnages for all recycling categories increase over the entire 50-year period, but the largest tonnage in all those years is for "paper and paperboard." The next largest tonnage across this period is for "metals." Interestingly, Figure 5-1 shows that tonnages for the glass category remain small relative to the others, but Figure 5-2 shows that the recycling rate for glass has increased more than any other category. The recycling rate for plastics started small but has increased somewhat more than the increase for metals. These recycling rates are each an average across the country, so they understate rates for individual materials currently recycled in locations such as cities that invest heavily in recycling (see Box 5-2).

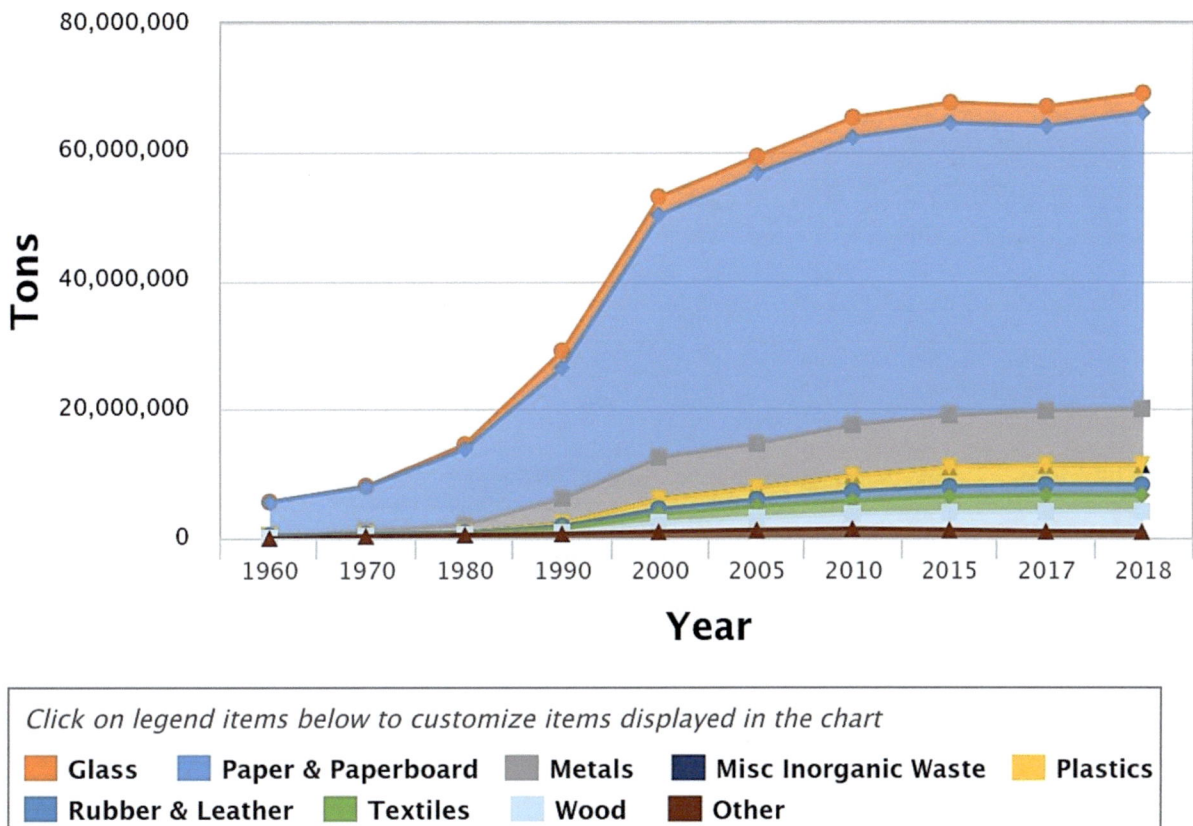

FIGURE 5-1 Recycling tonnages in the United States, 1960–2018.
SOURCE: EPA, 2020.

To help understand these trends in recycling tonnages and recycling rates, we might like to see trends in total recycling expenditures by government. However, most local governments report total MSW expenditures, and do not break down collection of garbage for the landfill and other collection for recycling. In general, recycling might be an increasing fraction of total MSW expenditures, but it is difficult to determine how much that share has increased. Thus, recycling expenditures cannot be shown for the several time-trends in Figure 5-3. However, the dark blue trend does show total U.S. MSW expenditures as a fraction of gross domestic product (GDP). That ratio doubles from 100 percent of its 1970 level to 200 percent of that level in 1992 and then levels off until 1995, before falling steadily thereafter. In other words, after 1995, any growth in MSW expenditures was exceeded by overall growth in GDP. This decline in relative MSW spending may or may not partially explain the slowdown in growth of the overall recycling rate (shown in the orange trend line for the aggregate of all materials in the previous figure).

Another influence on recycling tonnages or recycling rates over this period might be changes in prices, so Figure 5-3 also shows a time-trend for the prices paid by firms for recycled materials (in light blue) and for the price of virgin materials (in green). For average recycled materials prices, data are not available until 1990 and then unavailable again in 1997 and 1998. These prices are highly volatile, though they end in 2015 somewhat higher than in 1990. In contrast, the price for virgin materials bounces up and down across the entire period, but it falls by 2015 to about 70 percent of its 1970 level. That price decline is only partly explained by recent reductions in prices for petrochemicals used to make plastic.

TABLE 5-1 Recycling Tonnages in the United States, 1960–2018

Recycled	1960	1970	1980	1990	2000	2005	2010	2015	2017	2018
Paper and Paperboard	5,080	6,770	11,740	20,230	37,560	41,960	44,570	45,320	44,170	45,970
Glass	100	160	750	2,630	2,880	2,590	3,130	3,190	3,070	3,060
Metals										
Ferrous	50	150	370	2,230	4,680	5,020	5,800	6,070	6,170	6,360
Aluminum	Neg.	10	310	1,010	860	690	680	670	600	670
Other Nonferrous	Neg.	320	540	730	1,060	1,280	1,440	1,290	1,710	1,690
Total Metals	*50*	*480*	*1,220*	*3,970*	*6,600*	*6,990*	*7,920*	*8,030*	*8,480*	*8,720*
Plastics	Neg.	Neg.	20	370	1,480	1,780	2,500	3,120	3,000	3,020
Rubber and Leather	330	250	130	370	820	1,050	1,440	1,550	1,670	1,670
Textiles	50	60	160	660	1,320	1,830	2,050	2,460	2,570	2,510
Wood	Neg.	Neg.	Neg.	130	1,370	1,830	2,280	2,660	3,030	3,100
Other	Neg.	300	500	680	980	1,210	1,370	1,230	990	970
Total MSW Recycled	*5,610*	*8,020*	*14,520*	*29,040*	*53,010*	*59,240*	*65,260*	*67,560*	*66,980*	*69,020*

SOURCE: EPA, 2024a.

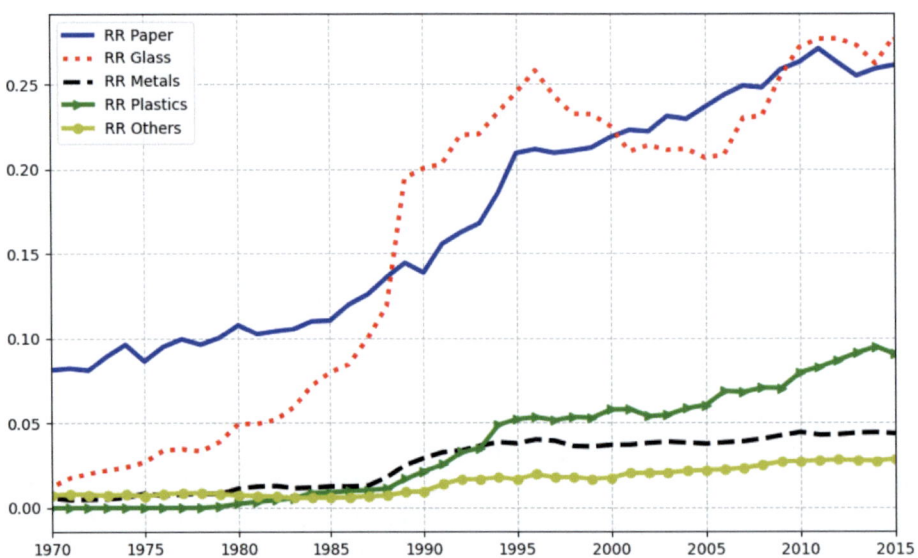

FIGURE 5-2 Recycling rates for five categories of materials in the United States, 1970–2015.
SOURCE: Arbex and Mahone, 2024.

As it turns out, the trends in recycling tonnages and rates cannot be explained either by changes in government MSW expenditures nor by these two price trends. This is because the time-trends are determined simultaneously by the same complicated economic influences. For example, an increase in economic growth could explain increased demand for virgin materials, but any resulting increase in price of virgin materials could shift some of that demand to recycled materials. That shift might increase recycling rates, with or without more MSW expenditures. The trends cannot readily be disentangled.

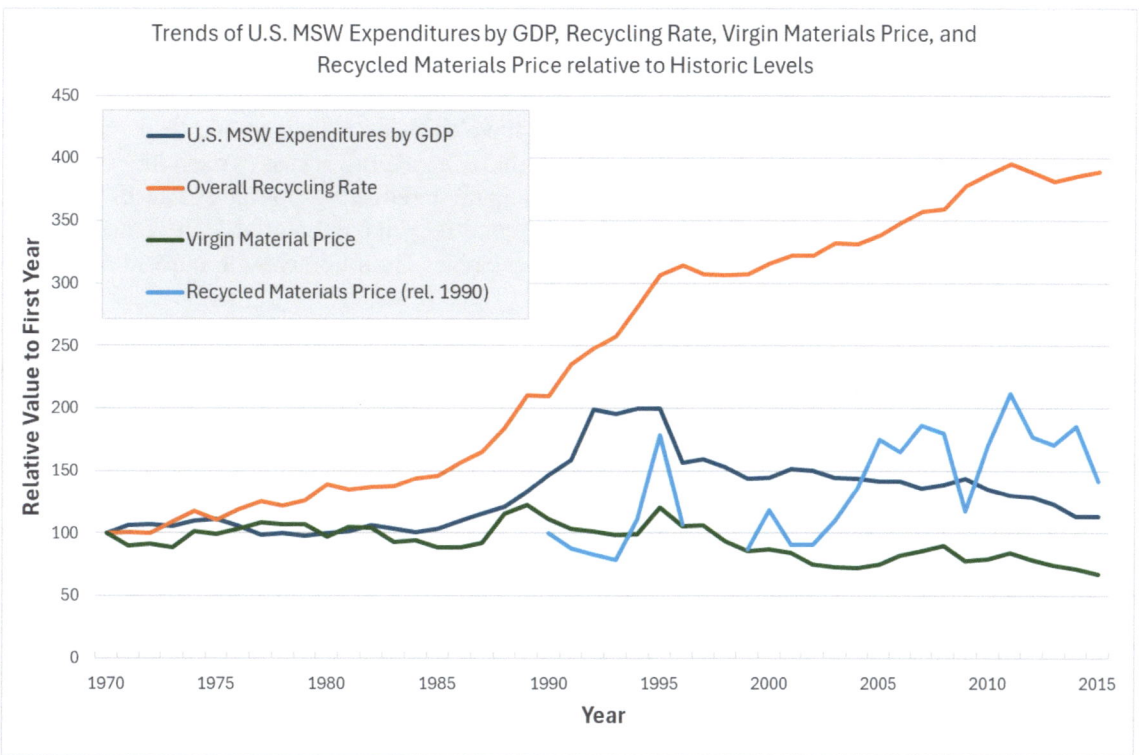

FIGURE 5-3 Trends of U.S. municipal solid waste (MSW) expenditures by gross domestic product (GDP), 1970–2015, each relative to its 1970 level.
NOTES: The virgin material and recycled material prices have been corrected for inflation. Data on recycling materials price are not available prior to 1990, so subsequent years are shown relative to 1990.
SOURCE: Derived from Arbex and Mahone, 2024. MSW and recycling data primarily from EPA, and virgin material price primarily from Federal Reserve Economic Data.

BOX 5-2
Case Study: Recycling Excellence in the United States

Three U.S. cities report efficiency rates that far exceed national averages (see Table 5-2). These rates differ in part because some cities likely include a broader range of materials, including organic waste, particularly in cities such as Seattle, Washington, and Portland, Oregon. For example, Seattle's 60 percent recycling rate reflects contributions from organics, as the city has local mandates for food waste separation. This material differs from the traditional recyclables highlighted in subsequent figures, which may not capture the same breadth of materials.

TABLE 5-2 Examples of Recycling Excellence in the United States

Community	Reported Overall Recycling Rate	Tactics
Boulder, Colorado	38%	100% curbside collection, special event recycling, occupational trash tax, commercial recyclable separation
Seattle, Washington	60%	Local food waste separation mandate, lower collection charges, variable garbage rates, recycling requirements
Portland, Oregon	39% (statewide)	State mandates, supervised drop-off centers, recycling requirements, every-other-week garbage collection

SOURCES: Generated by the committee with data from Brewer, 2025; ODEQ, 2024.

EPA (2020) estimated the financial benefits of recycling in terms of jobs (681,000 in 2012), wages ($37.8 billion), and tax revenues ($5.5 billion). Yet, according to The Recycling Partnership (2024), over 76 percent of residential recyclables are landfilled or incinerated. Part of the loss in recyclables can be attributed to the fact that only 85 percent of communities have curbside collection and 63 percent of multi-family communities have no access to recyclables collection for an overall access to recycling of 73 percent. Unfortunately, over 57 percent of households do not participate in recycling (for lack of access or interest), and the average MRF captures only 87 percent of its accepted recyclable materials. Therefore, The Recycling Project (2024) estimates that over 50 percent of recyclables from homes with curbside collection is landfilled or incinerated.

Efficient recycling has technical, behavioral, and economic barriers, and multiple challenges that include contamination of recycling streams, single-stream recyclable collection, limited access to recycling collection, failure to properly separate recyclables, and inefficient processing of recyclables at MRFs.

Contamination of materials in recycling bins and drop-off facilities reduces the economic benefits of recycling. EPA (2021) identified contamination reduction as a key element of its National Recycling Strategy. The Recycling Partnership (2020) estimates that nationally the contamination rate is 17 percent, but in some cases it may be as high as 50 percent (Runewese et al., 2020). Contamination can occur because of poor communication between municipalities and participants, confusion over what can be recycled, indifference to the environment, and "wish-cycling" (placing materials in bins that participants hope can be recycled).

The following sections discuss the collection and use of the five main materials collected for recycling, as well as food and yard waste.

5.1.1 Plastics

Plastic is a ubiquitous component because of its strength, low cost, durability, environmental benefits, and wide range of properties (Avery et al., 2025). Dozens of types of plastic resins are or have been in use (see Table 5-3). Most have single or short-term use and therefore routinely and quickly find their way into municipal solid waste (MSW) streams. The production of plastics consumes 6 percent of the world's refined crude oil (Dal et al., 2022).

Although these resins are all types of plastic, each resin has different characteristics that are beneficial for different types of manufacturing and product uses. Different challenges also relate to recycling for each resin type. Thus, to be useful for manufacturing, plastic recyclables must be separated by resin type.

Although plastics make up only 12 percent of MSW generated in the United States, many people see it as the most problematic type for several reasons. Plastic has low density, so it represents a greater share of waste by volume than by mass. Plastic waste has especially high environmental costs because it is long lasting, highly visible in the environment as litter, and has recently been identified as the source of micro and nano plastic particles in the environment. Approximately 5 percent of plastic waste is managed inappropriately, harming marine and terrestrial life when animals become entangled in it or ingest it (NASEM, 2023). Micro and nano plastics have been found in every environment on earth and may harm human health.

Some types of plastics are collected by almost every curbside recycling program. EPA (2024a) estimated that, in total, approximately 9 percent of plastic waste was recycled in 2018. Recycling of plastics avoids the consumption of energy and fossil fuels to produce virgin plastics. Figure 5-4 shows the breakdown of plastic waste by resin and product type as determined by Milbrandt (2022), who reported that 5 percent of plastics were recycled globally in 2019. Although most thermoplastics (plastics that can be remelted and molded into new plastics) can be recyclable, only resin numbers 1 and 2 are commonly collected in household curbside collection programs or at drop-off centers (see Table 5-3). Types 3–7 are considered hard to recycle, although interest is increasing among developing programs to manage these waste types. Film, wrap, and bags made of low-density polyethylene (LDPE) dominate plastic wastes, representing 5.8 percent of MSW by weight (Milbrandt, 2022; see Figure 5-2). LDPE is still poorly recycled because of the high cost of processing and difficulty of processing for MRFs. However, the industry is investing in new sorting and recycling technologies that may help address these challenges.

TABLE 5-3 Resin Identification Codes

Resin Identification Code (RIC)	Polymer Name	Common Products
1 PETE	Polyethylene terephthalate (also PET)	Bottle, jars, containers, trays, carpet
2 HDPE	High-density polyethylene	Bottles, milk jugs, bags,[a] containers, toys
3 V	Polyvinyl chloride	Pipes, siding, pool liners, bags, shoes, tile
4 LDPE	Low-density polyethylene (another form is linear low-density polyethylene, or LLDPE)	Bags, wrap, squeezable bottles, flexible container lids, agricultural film, cable coating
5 PP	Polypropylene	Tupperware plastics, yogurt tubs, hangers, diapers, straws
6 PS	Polystyrene or expanded polystyrene (PS/EPS)	Disposable cups and plates, take-out container, packing peanuts
7 OTHER	Polycarbonate, nylon, acrylonitrile, butadiene styrene (ABS), acrylic, polylactide (PLA)	CDs, safety glasses, medical storage, baby bottles

[a] Plastics bags are manufactured with HDPE, LDPE, and LLDPE. While most shopping bags (e.g., grocery store bags) are manufactured with HDPE, LDPE is typically used for tear-away dry cleaner bags, and LLDPE is used for heavier and thicker bags, for example those used by clothing stores.
SOURCES: ASTM, 2022; NASEM, 2023.

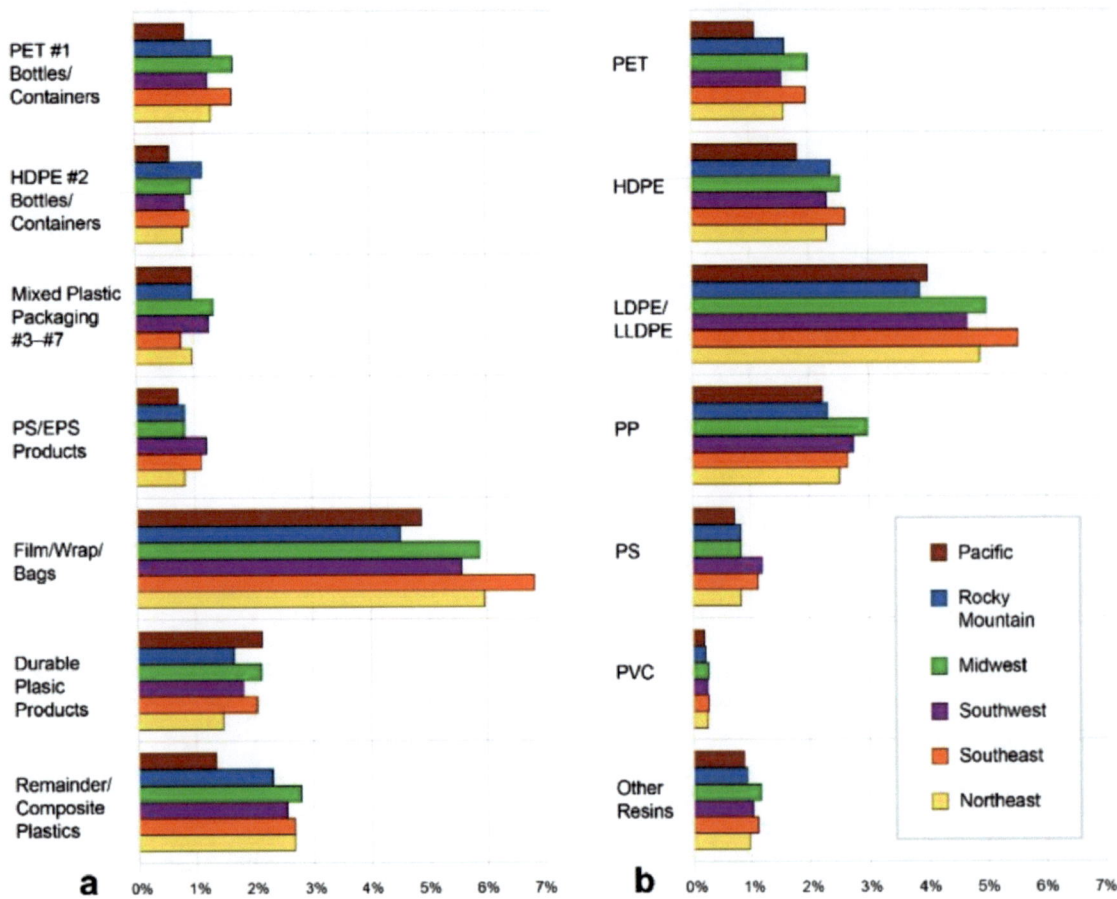

FIGURE 5-4 Plastics waste composition in the United States (percent by weight), for each material type and resin code.
NOTE: EPS = expanded polystyrene (Styrofoam); HDPE = high-density polyethylene; LDPE = low-density polyethylene; LLDPE = linear low-density polyethylene; PET = polyethylene terephthalate; PS = polystyrene; PVC = polyvinyl chloride.
SOURCE: Milbrandt et al., 2022.

Milbrandt and colleagues (2022) also reported that about 86 percent of plastics was landfilled and 9 percent was combusted in the United States in 2019. Based on these data, the authors calculated the total mass of plastics landfilled as 44 million tons, representing $2.3 billion in value and 3.4 Ej (exajoule) of embodied energy (or 12 percent of the energy consumed by the industrial sector).

Since 2018, when EPA published its latest dataset regarding national waste generation and management, several confounding events have occurred. First, China banned imports of recyclables with its National Sword policy in 2018 (discussed further in later sections of this chapter). Nearly 3 million metric tons of U.S. recyclables (largely plastics and mixed paper) that would have been exported to China diverted elsewhere, often to landfills (Sigman and Strow, 2024) or to other countries, where they may have been mismanaged (Taylor et al., 2024).

Additionally, beginning in early 2020, the COVID-19 pandemic had multiple effects on MSW generation and recycling. As many people worked from home, waste from commercial businesses decreased, while household waste increased markedly. Plastic wastes increased because personal protection equipment (predominately gloves and masks) was discarded in large numbers. Take-out packaging also increased, as fewer families ate at restaurants (Olawade et al., 2024). Finally, online shopping grew to over 13 percent of U.S. commerce in 2020. E-commerce packaging can use up to 4.2 times the material of that used for brick-and-mortar packaging (Kim et al., 2022), including paper bills, envelopes, cardboard, plastic bags, tape, protective bubble wrap, and Styrofoam. These factors have not been captured in national data yet. Estimates suggest increasing tonnages of plastic and reduced recycling since the start of the pandemic (OECD, 2022). See Box 5-3 for innovative packaging choices meant to increase recyclability.

BOX 5-3
Case Study: Navigating Recycling Challenges for Toothpaste Containers

Many toothpaste tubes are made from multilayered materials, including plastic and aluminum, which can complicate the recycling process. Some manufacturers, including Colgate, have introduced recyclable tubes made from high-density polyethylene. However, variability in municipal solid waste (MSW) recycling programs and unclear labeling can make it difficult for consumers to determine whether and how these items can be recycled.

Research has shown that uncertainty about recycling guidelines often leads consumers to discard recyclable materials in general waste (Schneller et al., 2023). Without consistent instructions or assurances that local facilities can process such items, consumers may dispose of the entire toothpaste container rather than attempt to separate recyclable components. Colgate's "Recycle Me!" campaign aims to improve awareness about recyclable packaging (Colgate-Palmolive, 2022).

In addition to consumer education on proper disposal practices, the successful diversion of toothpaste containers from landfills depends on more uniform recycling systems.

SOURCES: Colgate-Palmolive, 2022; Schneller et al., 2023.

5.1.2 Paper and Cardboard

Keeping paper and cardboard out of landfills reduces greenhouse gas emissions. Paper and cardboard waste is a general category that captures multiple types of wood-derived fibrous materials, including newspapers, mechanical papers, directories, inserts, direct mail printing, and others. Almost every curbside recycling program collect materials in this category.

Recycling paper and cardboard reduces the demand for tree harvesting and requires less water and up to 40 percent less energy than production from virgin materials (Kumar, 2017). On the other hand, producing paper and cardboard from virgin pulp allows manufacturers to make nonfossil biofuels for heat and electricity, using pulp byproducts (bark, branches, and leaves).

Certain types of paper cannot currently be recycled, including napkins, tissues, paper towels, toilet paper, waxed paper, receipts and sticky paper/notes (note, however, that some can be composted). EPA (2024a) reported that more than 67.4 million tons of paper and paperboard (cardboard) were generated in 2018, and 68.2 percent was recycled, the highest product recycling rate measured. Estimates of more recent recycling rates from the American Forest & Paper Association (2025) are similar, with recycling of 65 to 69 percent of paper and 71 to 76 percent of cardboard. However, using a bottom-up rather than a top-down methodology, Milbrandt and colleagues (2024) found that 56 percent of paper and cardboard waste in the United States was landfilled, 6 percent combusted, and only 38 percent recycled. The difference in estimates derives from a larger estimate of paper and cardboard generation in the study by Milbrandt and colleagues than by EPA. Milbrandt and colleagues (2024) estimated that landfilled paper and paperboard could have generated $4 billion in end markets and embodied 2,158 joules of energy.

5.1.3 Metals

Metals in MSW are valuable commodities that are generally collected through curbside programs and received at drop-off or buy back centers. Most scrap metal, however, is not part of MSW but instead is generated from end-of-life vehicles and in various manufacturing facilities. EPA (2024a) reported that 25.6

million tons of metal waste (8.76 percent of MSW) were generated in 2018, of which 34 percent was recycled. Metal waste collected from households is primarily ferrous (steel, cast iron, wrought iron, carbon steel) (e.g., some food cans and small metal objects) and nonferrous (aluminum) (e.g., some beverage and food cans). Approximately 50 percent of aluminum cans was recycled in 2022, although this number has been higher in the past (The Aluminum Association, 2024). Metal can be recycled many times, which makes it particularly suitable for the circular economy.[1] Environmental benefits of recycling metals are numerous, including reduced energy requirements compared with metal ore processing and extraction. Environmental benefits associated with avoided mining include reduced greenhouse gas, water, and air emissions. Therefore, low recycling rates mean valuable materials are ending up in the landfill.

5.1.4 Glass

Glass waste is commonly collected curbside and at drop-off centers. In 2018, 12.2 million tons of glass waste were generated in the United States, representing 4.2 percent of MSW (EPA, 2024a). Approximately 3.1 million tons were recycled, representing 25 percent of generated glass waste; (a percentage that has persisted for about 25 years) (EPA, 2020). Glass recycling benefits include savings on raw materials (sand, soda ash, limestone, and feldspar), reduced energy use, and greenhouse gases production (because of lower furnace temperature); additionally, in most cases, glass is 100 percent recyclable (Glass Packaging Institute, 2024). Glass recycling is particularly successful in states with deposit-return systems (see Chapter 4).

Despite these many benefits, glass presents challenges in single-stream collection programs. Glass breaks during collection, and its inherent abrasive characteristics become problematic for other commodities and for equipment (trucks and MRF conveyers). Broken glass is a significant problem when collected alongside paper, adding cost to the paper recycling process. As the MRFs sort the glass pieces (called "fines") from the other recyclables, the result is a highly contaminated glass stream with rocks, lids, and other small contaminants that are removed from the stream along with the broken glass. This ultimately creates a low-value product requiring an expensive process before it can be remanufactured into new glass bottles. The weight of glass compared with its value also limits transporting over long distances to manufacturing facilities.

Consequently, according to a 2023 survey, 43 percent of curbside collection programs no longer accept glass (Glass Recycling Coalition, 2024). One area of increased use for glass waste is in construction, including concrete aggregates, geotechnical applications, tile and brick manufacturing, and water filtration (Kazmi et al., 2019); this may be especially relevant for projects located near MRFs that process glass. In many cases, glass waste leaving MRFs is too contaminated to reuse and is diverted to landfills, where it is used as "alternative daily cover" (i.e., non-earthen material placed on the surface of an active MSW landfill at the end of the day) (NERC, 2023).

5.1.5 Food Waste and Yard Waste

EPA (2024a) reported that in 2018, 21.6 percent of MSW (63.1 million tons) was food waste and 12.1 percent (35.4 million tons) was yard trimmings. Approximately 4.1 percent of food waste was composted (2.6 million tons), down from 6.3 percent in 2017. EPA (2024a) analyzed information from state composting programs to calculate the composting of yard trimmings. This analysis resulted in an estimate of 22.3 million tons of yard trimmings composted or wood waste mulched in 2018 with a 63 percent composting rate. Badgett and Milbrandt (2021), however, calculated that 75 percent of food and yard waste was landfilled and 18 percent combusted.

---

[1] Note that for aluminum recycling, each cycle may result in relatively small net losses of material induced through mill operations (as residual slag) and oxidation during the recycling process. See https://www.aluminum.org/Recycling (accessed May 25, 2025).

Food waste has the highest methane generation potential of all wastes landfilled. This fact, coupled with the high degradation rate of food waste and the typical delay in formal gas collection from landfills, means that most, if not all, of the methane potential in food waste is generated prior to collection of landfill gas (Amini and Reinhart, 2011).

Beyond the impact landfilled food waste has on the environment, the social implications of food wasted (at the retail level) and lost (during harvesting, storage, or transport) on a global scale mean that 24 percent of the world's calories go uneaten (Goodwin, 2023). Also lost are the land, fertilizer, water, and energy used to generate uneaten food. Regulations at the federal, state, and local levels have tried to prevent food waste through more accurate date labeling; rescue of food for the hungry through food safety and liability protection; and diversion from landfills through animal feeding, organic waste bans, and mandatory recycling policies.

Food and yard wastes can be composted, digested, or converted into biofuel. Kiran and colleagues (2014) estimated that 1.32×10^9 m^3/year of methane could be produced globally from food waste, which would generate some 2.6×10^7 joules of energy. Today, composting is the most common alternative to landfilling for food and yard waste, but interest is growing in anaerobic digestion of organic waste, which produces biogas and digestate that can be used as a soil amendment. Badgett and Milbrandt (2021) reported that 283 anaerobic digestion facilities, or digesters, in the United States are processing food waste, including 82 located at wastewater treatment plants and 68 at livestock facilities.

One of the impediments to recovery of food wastes is the difficulty and cost associated with its collection. Many communities that have banned food waste from landfills have created drop-off locations (e.g., New York City) or added collection routes for separated food waste, or allow food waste to be collected along with other green waste. Additional collection routes increase the number of trucks on the road, along with tailpipe emissions, increased accident risks, and cost. However, these disadvantages are offset by reduced landfill gas emissions and potential sale of biogas generated by food waste treatment. A recent study of the impact of adding a food waste collection line (beyond garbage, recyclables, and recyclables) found that food waste diversion reduces the global warming potential of waste collection significantly at a relatively low cost (Reinhart et al., 2023). Furthermore, separation of food waste may decrease the contamination of recyclables, which can have significant advantages for reducing cost and global warming potentials.

5.2 DEMAND FOR RECYCLABLE MATERIALS: END MARKETS

Sales of recyclable commodities in end markets provide revenues that reduce the expense of recycling for local governments and private parties. Commodity revenues can cover a substantial share of the costs of processing recyclables (see Chapter 4). Ideally, recycled commodity prices would be high and steady enough to contribute reliable revenue toward supporting the recycling system. However, the effect of end markets is not exclusively financial, because end uses determine much of the environmental benefit of recycling. Recycling is more likely to improve environmental quality when secondary materials successfully substitute for extraction or production of environmentally damaging primary materials and when these recycled materials can be incorporated into new products without themselves requiring resource-intensive processing. Thus, end markets and programs to support them should be assessed not just in terms of revenue but also in terms of environmental attributes.

Table 5-4 provides summary information about common, current end uses for materials in the United States and Canada. The prices represent the average sales prices of each material after it has been sorted at a MRF in the U.S. Northeast and Canadian Maritime provinces. The price ranges reflect variation in values of the average prices by week for the first half of 2024. Only for some materials do these end markets meet the goals for supporting the recycling system: high prices, steady prices, and environmentally favorable replacements for extraction of primary materials.

TABLE 5-4 End Uses and Prices for Recycled Commodities

Waste	Price per Ton[a] (Jan–July 2024)	Typical End Uses[b,c]
Steel cans	$155–$210	Many steel products
Aluminum cans	$1,210–$1,550	Aluminum cans, industrial aluminum sheet
Glass bottles (color separated)	$10–$58	Glass bottles
Glass bottles (mixed)		Fiberglass, asphalt, other aggregates, landfill cover
Corrugated cardboard	$93–$110	Corrugated cardboard
Other cardboard		Cardboard
Mixed paper	$48–$68	Cardboard, some other paper grades
Plastics:		
PET bottles	$250–$360	Fiber (e.g., carpet, clothing), bottles (31%)
PET clamshells		Fiber
Uncolored HDPE bottles	$610–$770	Nonfood-contact HDPE bottles
Colored HDPE	$370–$540	Nonpackaging plastics (e.g., pipes)
Polypropylene packaging		Durable plastics
Plastic film		Plastic lumber
Flexible plastic packaging		None

NOTES: HDPE = high-density polyethylene; PET = polyethylene terephthalate. Prices ranges are for a recent 6-month period. A longer time series is shown in Figure 5-3 below.
[a] Data from wv.gov based on recycle.net data.
[b] Miller, 2024.
[c] PET bottle reuse: Brian Staley in presentation to committee, June 11, 2024.

As Table 5-4 reports, prices per ton vary greatly across recycled commodities. Materials that are good substitutes for valuable primary materials, such as metals and corrugated cardboard, typically maintain high prices. However, many materials do not easily substitute for primary materials. For example, dyes are often added to color high-density polyethylene (HDPE) and glass bottles, making the postconsumer material suitable only for uses that are not sensitive to the color of the material. And, for another example, reprocessing strips some plastics and paper of desirable chemical or physical properties, restricting the number of times they can be recycled or requiring they be reused only when mixed with primary materials (Basushi et al., 2023). Thus, improved markets for many materials will require either invention of new technologies or vastly improved sorting processes. Federal research funding could help address this important weakness in the recycling system by supporting development of these new technologies. Pointing to the importance of end markets, Bradshaw and colleagues (2025) conclude that "large MRFs also demonstrate success finding markets for a variety of plastics and gain significant revenue from plastics compared to smaller MRFs" (p. 326).

Competition with low-price virgin materials in end markets creates a challenge for recyclables. Virgin materials prices are often lower than the full social costs of producing these materials—both because of the environmental impacts of mining and other raw material production, and because of subsidies for energy production that lower the cost of feedstocks for virgin plastic manufacture. With low prices for virgin materials, manufacturers have limited incentives to use recycled materials in their production. Virgin plastic has become especially cheap in recent years because of fracking in the production of petrochemicals, and because of low-cost imports from China. For 2023–2024, Figure 5-3 shows prices for food-grade recycled polyethylene terephthalate (PET) and a similar grade of virgin PET. It demonstrates that the recycled resin price is consistently higher than the virgin resin price over time, which limits demand for recycled products.[2]

---

[2] That these price lines track each other is not a coincidence. If the virgin material price were to fall, for example, then demand would shift toward virgin material and away from recycled material, which would reduce the price of recycled material along with the reduced price of virgin material.

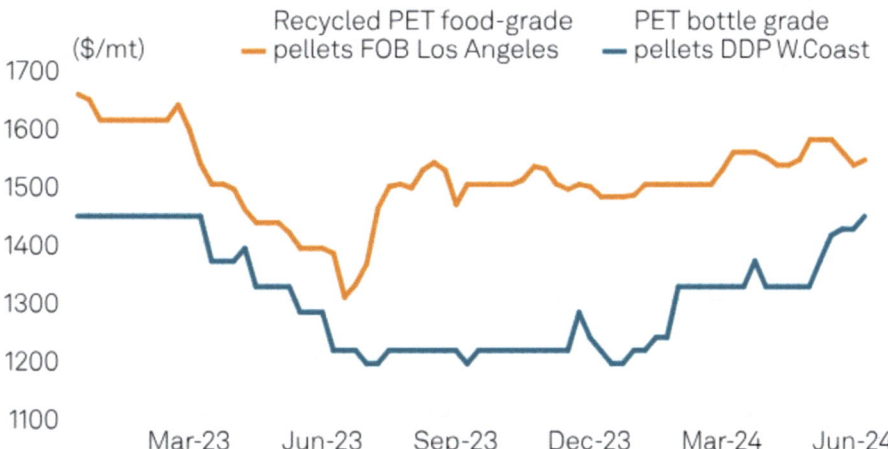

FIGURE 5-5 Virgin and recycled plastic prices in 2023 and 2024.
NOTES: The recycled price is free on board or freight on board (FOB), meaning that the buyer may be responsible for delivery and related costs. The virgin price is delivered duty paid (DDP), meaning that the seller pays those costs. Because of this distinction, the figure may understate the true price gap. Mt = metric ton; PET = polyethylene terephthalate.
SOURCE: Platts, part of S&P Global Commodity Insights, 2024.

Volatile prices present another challenge. Price volatility creates stress on local governments that finance most recycling, and it creates fragility in the recycling system because investments are risky. In the history of prices for several commodities, shown in Figure 5-6, prices commonly double or triple and then fall back again on a year-to-year basis. Recycled commodity prices are substantially more volatile than other commodity prices (Moore et al., 2022; Timpane, 2024).

Several features of markets for recycled commodities may explain this price volatility. First, supply of recycled materials is insensitive to price ("price inelastic"). Households do not receive any compensation for recycled materials, so the amount they send to MRFs and other processors is unrelated to the value of the material. MRFs may be contractually obligated to process recyclable material and attempt to sell it regardless of whether they expect to cover their costs in doing so. With this price-inelastic supply, fluctuations in demand for the material translate into large price swings. In addition, in many recycling markets, demand for the materials is driven largely by the prices of the primary commodities for which they substitute, and these primary materials prices themselves are volatile. In the United States, plastics derive from natural gas and from feedstocks from natural gas processing and crude oil refining, resulting in prices that vary with energy market conditions. They are also subject to other global production shocks, such as a recent surge in low-cost imports of virgin PET from China (Staub, 2024). Finally, volatility in these prices relates to the fact that prices are low, because small absolute changes in value can create large percentage price swings. Many of these prices can be very low: mixed paper prices were below zero in 2019 but rebounded dramatically by 2021 with the development of new domestic paper mills (NERC, 2021). These factors are fundamental features of these markets; recycling policies must manage volatility but cannot eliminate it.

5.2.1 Evaluating End Uses

When a recycled commodity is a close substitute for the primary material, their prices are similar, and the recycled commodity's price reflects the private costs avoided when 1 ton of the recycled material replaces 1 ton of the primary material. These savings may include savings in labor and energy inputs in addition to mineral or other raw material values. They also include costs of environmental controls on production of the primary material, so tightening environmental regulations on primary production will be reflected in higher prices for the recycled material.

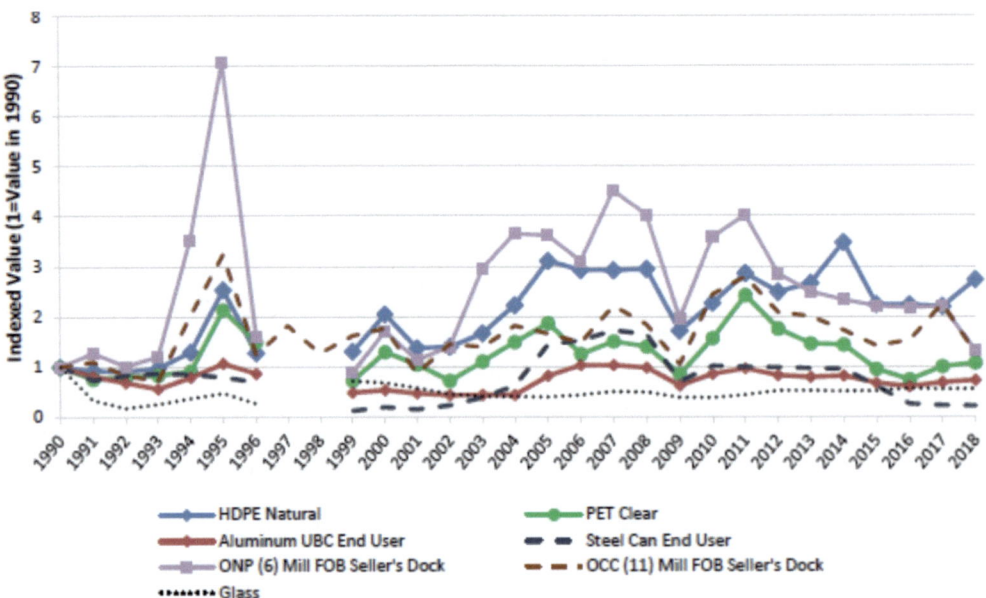

FIGURE 5-6 Historical recycled commodity prices.
NOTES: Prices adjusted by the Consumer Price Index. OCC = old corrugated cardboard; ONP = old newspaper; UBC = used beverage container.
SOURCE: EPA, 2020.

For some materials, the private sector saves a lot less using recycled commodities. For example, the mixed-color glass cullet that results from MRF-sorted curbside glass recycling often contains a high level of contamination (bottle caps, rocks, and other small "fines").[3] It is expensive for glass processors to make into usable feedstocks for manufacturing. When it is spread over landfills as alternative daily cover, it only avoids the use of some other low-cost material, such as soil, or even another form of waste, such as construction debris (NERC, 2023).

However, end uses may have social benefits not captured in these commodity prices. Primary resource extraction and production generates pollution and other environmental degradation. For example, additional recycling of paper can reduce timber harvesting and thus promote forest carbon sequestration (Lorang et al., 2023). Thus, policymakers also need to consider non-monetized effects of substitution for primary materials.

Recyclable material types differ radically in their value when analyses include both the private values reflected by their prices and the avoided environmental damages. For example, recycling aluminum saves valuable energy, and material resources and thus has social benefits that greatly exceed the benefits of avoiding the use of landfill space. However, given current market opportunities and technologies, recycling glass provides much more limited benefits. Thus, focusing on the weight-based average recycling rates as a policy goal is insufficient and does not reflect the heterogeneity in benefits of recycling for different materials.

In addition, "closed-loop" recycling, or returning a material to its original use, is not always the best target. When recycled materials are used to create another product, or "open-loop" recycling, the result

---

[3] Because the value of glass is low, it is rarely worthwhile for MRFs to invest in glass-sorting technology. Secondary-glass processing facilities have optical sorters to separate the colors of glass, but the cost of extracting the rocks and bottle caps is very high, especially considering the low end-market value of glass.

is sometimes called "down-cycling." However, neither open-loop nor down-cycling are always inferior to closed-loop recycling in terms of either the avoided use of primary material or the environmental consequences. Geyer and colleagues (2016) demonstrate that closed-loop recycling may increase rather than decrease materials use when the virgin and recycled materials are not good substitutes for each other. Virgin and recycled fiber may instead be complements in paper production, increasing the recyclability of the resulting product (Mondi, 2024).

Allaway (2024) provided examples of how closed-loop recycling can raise resources costs relative to open-loop processes in the context of glass and plastic recycling. First, grinding recycled glass bottles into a fine powder and using it as pozzolan (open-loop recycling), has far less environmental impact than using the glass in new bottles (closed-loop recycling). In the closed-loop process, the recycled glass displaces virgin silica, which reduces energy needs and does have some environmental benefit relative to use of virgin material. However, in the open-loop process, the recycled material (pozzolan) substitutes for cement and thus reduces the substantial greenhouse gas emissions from cement production, thereby providing a greater environmental benefit. In addition, no heating is required to melt the recycled glass cullet when it is used in pozzolan, making the open-loop process more energy efficient than the closed-loop process. A second example concerns plastics recycling. Recycling HDPE packaging into packaging (closed-loop) or pipes (open-loop) both offset the need for virgin HDPE (Allaway, 2024). But preparing scrap for reuse in packaging requires more processing to meet safety standards and thus has higher market and environmental costs.

Additionally, closed-loop recycling does not always offset an equivalent amount of virgin resin. For engineering reasons, some products require more plastic if they are made from postconsumer plastic than from virgin plastic. In those cases, the increased material use diminishes the environmental benefits of using recycled resin. For more on closed- and open-loop recycling, see Box 5-4.

5.2.2 New Technologies

Technological change in end uses may present new opportunities and change future priorities. Figure 5-7 reports the counts of patents for recycling of various materials (the count of patents is commonly used as a metric for innovative activity). These patent counts are for all countries in the Organisation for Economic Co-operation and Development (OECD), but data for the United States alone show the same patterns. Plastic recycling has the most innovative activity and has experienced a boom starting in 2018; a similar, and perhaps related, increase appears in patenting for recycling and reuse of packaging beginning in 2017. The increased patenting activity may have been driven in part by market pressures: the decline in exports markets for postconsumer plastics after the Chinese import ban in 2017 may have spurred developed countries to focus their efforts on creating end uses for these materials. However, the timing of the patenting increases also coincides with greater attention to global harms from plastics, including from microplastics, marine debris, and exposure in lower-income countries.

After plastics and packaging, the next most active area for patenting is paper recycling, but it has not experienced a rapid recent increase. Notwithstanding occasional progress (e.g., Andini et al., 2024), textile recycling has received little innovative attention, and the current deficit of opportunities for recycling textiles seems likely to continue. However, the wave of innovation for plastic and packaging recycling visible in the figure may bring expanded end-use opportunities within a few years. The next section discusses advanced methods for plastics recycling, which may account for some of the surge in innovative activity seen in the patent data.

5.2.3 Mechanical and Chemical Recycling Technologies

Recycling technologies are generally categorized by mechanical or chemical processes. Mechanical recycling processes are used in virtually all types of materials collected for recycling, whereas chemical recycling is associated only with plastics. It is a particular focus as a potential solution for flexible film plastics and post-use plastic food packaging, which are difficult to recycle into closed-loop products (e.g., food packaging).

BOX 5-4
Closed- and Open-Loop Recycling

Closed-loop recycling is a manufacturing process that leverages the recycling and reuse of postconsumer products to supply the material used to create a new version of the same product.

In a closed-loop recycling system, products are designed in a way that benefits the overall supply chain, emphasizing universal collection and recovery, ease of remanufacturing, and economic feasibility.

Prime examples of closed-loop recycling products include glass used for bottles and jars, aluminum and steel used for cans and other metal products, and a very limited number of plastics. Glass[a] and aluminum[b] are near infinitely recyclable with no degradation of quality, making them valuable to the loop (Deer, 2021; Roadrunner, n.d.). In fact, about 75 percent[c] of all aluminum ever produced is still in use today. Unfortunately, only 2%[d] of global plastic production is reused for the same or similar products (TOMRA, n.d.).

Closed-loop differs from the far more widespread and achievable concept of *open-loop recycling*, which does not depend on the output of the process. In this system, the end of the product life cycle can take multiple routes: recycled as raw material for new, yet inferior, products or rejected as waste. For an example of policymakers prioritizing closed loop systems, see National League of Cities (2021).

In a sense, an open-loop system can be semi-circular[e] or nearly linear depending on how much is recycled (Bell, 2020). When products are recycled in an open-loop system, the new items produced are of a lesser value than the original product—a process known as *down-cycling*.

Cardboard and paper (which can be recycled five to seven times before unrecoverable loss in quality), many plastics, and in some cases, food waste lack the structural advantages of aluminum and cardboard, making them better suited for open-loop recycling. Examples are countless: polyethylene terephthalate (PET) plastics become fleeces and carpets, and cardboard returns as cereal boxes and paper towels.

[a] See https://www.roadrunnerwm.com/blog/why-is-glass-recycling-going-away.
[b] See https://www.roadrunnerwm.com/commingled-recycling.
[c] See https://www.aluminum.org/Recycling.
[d] See https://www.tomra.com.
[e] See https://www.roadrunnerwm.com/blog/circular-economy.

Differences Between Mechanical and Chemical Recycling

The primary way chemical recycling differs from mechanical recycling is that mechanical recycling preserves the original polymer structure, while chemical recycling breaks polymers down into their molecular constituents (however, these processes can be resin specific). Chemical recycling is more expensive than mechanical recycling processes, operates at higher temperatures, is less energy efficient, and is available at a smaller scale today. Because mechanical recycling is better established and requires fewer steps, it tends to be cheaper and more energy efficient. Thus, most recyclables will be processed using mechanical recyclers where possible, with the more expensive chemical recycling use for those types of feedstocks that cannot be recycled via mechanical technologies (e.g., flexible film plastic).

Chemical recycling technologies can create building blocks for new resins that will have the quality of virgin resin. While studies have shown chemical recycling to have a lower carbon footprint than virgin processes, they are more energy intensive and have a higher carbon footprint than mechanical processes.

Conversion processes often require the use of mass balance accounting because not all output is reprocessed into polymers. Some outputs may be used in other applications, such as fuels or base chemicals. Under certain certification programs and policy frameworks, these processes may still qualify as recycling when properly documented and linked to plastic waste feedstocks.

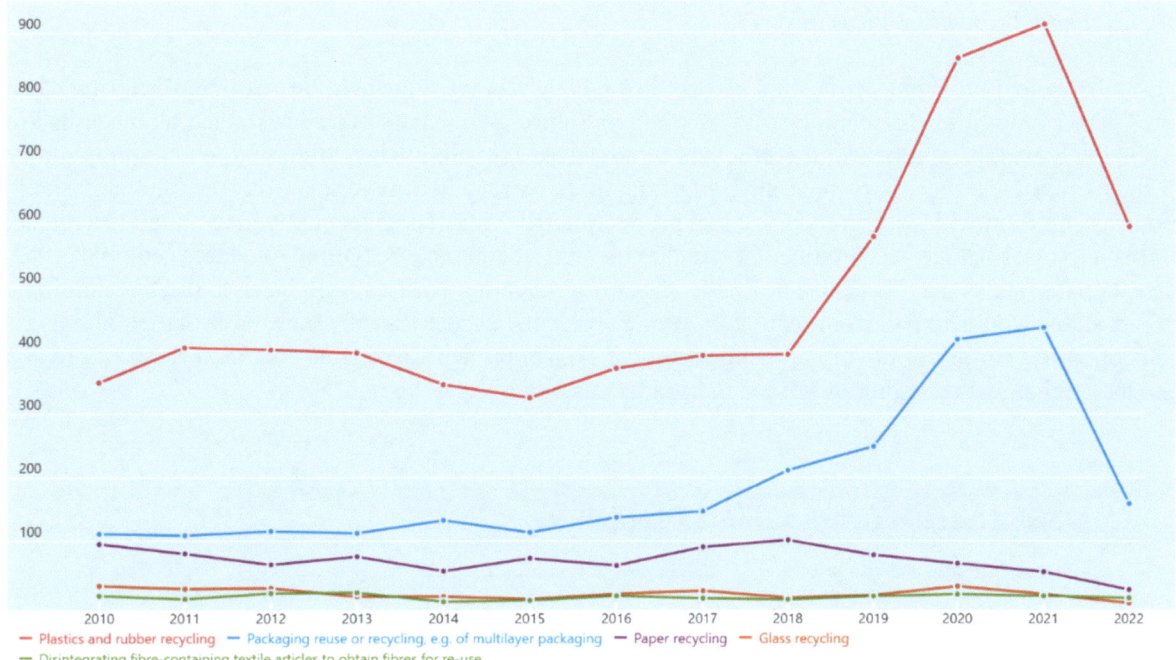

FIGURE 5-7 Patents for recycling by material in countries in the Organisation for Economic Co-operation and Development, 2000–2022.
SOURCE: Data from OECD, 2025, patent statistics on environment-related technologies. See https://data-explorer.oecd.org.

TABLE 5-5 Key Definitions of Recycling Technologies

Recycling Technology	Definition	Key Examples
Mechanical Recycling	Physically reprocessing materials without altering their chemical structure, enabling repeated recycling within a closed-loop system.	Paper mills that use water and heat to make new paper products; aluminum smelters and steel plants that use shredders and high heat to recycle metals; glass manufacturing plants that crush and melt glass into liquid form for new bottles; plastics facilities that melt recyclable feedstock for manufacture into new products.
Chemical Recycling	Breaking down plastic polymers into their original monomers, which can then be repolymerized into new plastic products. A common way of grouping chemical recycling technologies is purification, depolymerization, and conversion.	Purification: molding recycled polypropylene into food containers and automotive parts. Depolymerization: converting polyethylene terephthalate (PET) into monomers used to create new PET bottles and textiles, matching the quality of virgin materials. Conversion: transforming mixed plastic into synthetic fuels and naphtha, which serve as chemical feedstocks for new plastic production or fuel alternatives.

Policies Around Chemical Recycling

Chemical recycling legislation reclassifies advanced recycling facilities as manufacturing plants, rather than as facilities that handle solid waste (see Figure 5-8). Many state laws, such as those in West Virginia and Mississippi, permit the products of advanced recycling to be "returned to economic utility," but they often lack specificity regarding what qualifies as recycled material. Conversely, states such as Arkansas and Kentucky do not recognize fuel production from plastics as recycling, necessitating a nuanced approach to mass balance accounting. Given the potential of chemical recycling to address complex plastic waste streams, states may foreclose desirable uses from recycling with outright bans on these technologies. Instead, states would benefit from regulatory frameworks that ensure transparency, environmental integrity, and economic viability of novel recycling methods. This approach can enable chemical recycling to complement mechanical recycling in achieving broader sustainability goals.

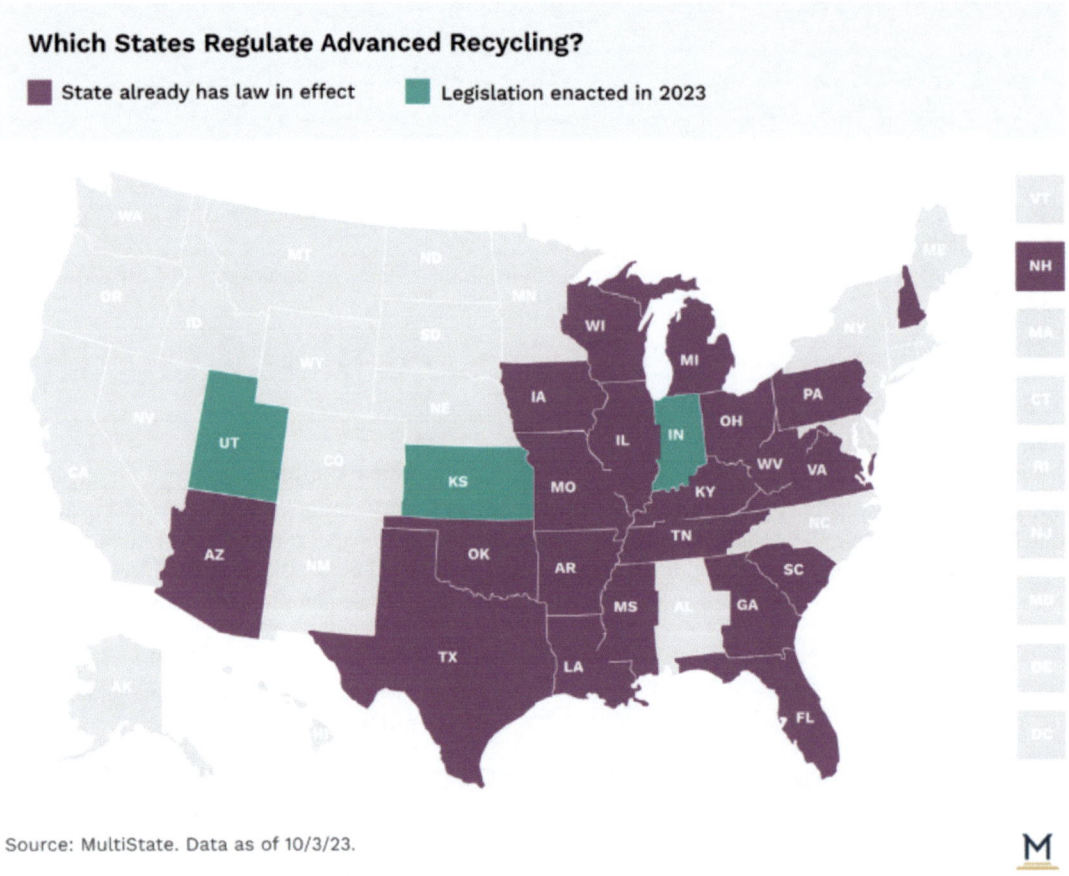

FIGURE 5-8 Legislation related to chemical recycling.
SOURCE: Crawley, 2023, see https://www.multistate.us/insider/2023/11/1/state-legislators-work-to-address-plastics-via-advanced-recycling.

5.3 GLOBAL, REGIONAL AND LOCAL RECYCLING MARKETS

Recyclables are commodities that flow to markets with the highest value. U.S. trade laws generally allow shipment of commodities across the globe to these markets based on the economics associated with these transactions.

5.3.1 Global Markets for Recyclables

While the United States does not limit the export and import of nonhazardous waste, trade is subject to the applicable laws and regulations in the country or counties that control the receipt of waste and recyclables, as well as applicable international agreements. In 2021, most plastic scrap was added to the Basel Convention, which regulates international movement of hazardous wastes (EPA, 2021).

Packaging is needed where products are produced. From the 1980s through the early 2010s, China's economy grew and became a primary source of products in the global economy. As its finished-product exports grew, China imported more recyclable materials to use in the production of their products and packages.

China's appetite for feedstock to make new products and packages grew during a time when the United States was collecting more material for recycling. As China's production of low-cost goods increased, U.S. production shifted to a reliance on products made in China and other developing countries. This development reduced demand for packaging, and the increasing costs of environmental controls and labor resulted in increased costs for U.S. domestic manufacturing companies and a corresponding increase in exports of recyclables to China.

In the 1990s and early 2000s, China purchased increasing volumes of recycled paper and plastic from across the globe to feed its growing economy. Its low-cost labor and mill production (with few environmental controls) led to a massive shift in the production of goods and packaging worldwide. In China, imported recycled plastics were turned into flakes or pellets to be repurposed by plastic manufacturers. This practice was made profitable by favorable shipping rates—in cargo vessels that carried Chinese consumer goods abroad and would otherwise return to China empty—coupled with the country's low labor costs and high demand for recycled materials.

Operation Green Fence

The reliance on China and its insatiable appetite for recycled feedstock began to change in 2010 with China's preparation for the 2012 Olympics and with the growth of internet communication about the harms of China's pollution. Photos and videos were commonly shared showing plastic-filled rivers and ditches, and neighborhoods filled with smoke from plastic and paper mills. China's air and water pollution became global news at a time when it hoped to attract positive global attention.

As domestic press and global visibility into China's "pollution problem" grew, Chinese authorities began their first of several constrictions on imports—called "Operation Green Fence"—starting with plastics in 2013. These restrictions banned many types of plastics from being imported into the country. Operation Green Fence was a wake-up call for the U.S. recycling industry. As demand for recycled materials continued to increase in the United States, demand for recycled plastics previously shipped to China came to a halt. Some plastic in the United States was warehoused until end markets could be found, but anecdotal evidence indicates that most low-quality plastic (resin codes 3–7) was ultimately landfilled.

The change in Chinese policy was a precursor to subsequent restrictions and may have begun a shift to more investments in domestic U.S. solutions for materials previously exported across the globe. Research suggests that the policy's effect on overall waste trade was limited at first, causing a temporary decline in exports to China but not otherwise altering patterns on international trade in waste (Balkevicius et al., 2020).

National Sword Policy

In July 2017, China informed the World Trade Organization of its intent to ban imports of all recycled plastic and mixed paper, effective January 2018. The policy, part of a broader crackdown on imports called the "National Sword," was an effort to halt a deluge of soiled and contaminated materials that was overwhelming Chinese processing facilities and leaving the country with yet another environmental problem (Katz, 2019).

Because China was handling nearly half of the world's recyclables at this point (primarily paper and plastic), the Chinese ban had a dramatic effect on global markets. Before the ban, 75 percent of U.S. recycled plastics was exported, with 40 percent going to China (Sigman and Strow, 2024). Europe exported 46 percent of its recycled plastic, even as the phase-in of the Chinese ban began in 2017 (Bishop, 2020). U.S. exports to China in the waste categories covered by the ban fell from 2.9 million metric tons in 2016 to only 135,000 metric tons in 2018. Some exports diverted to other international buyers, such as Vietnam and India, but these countries also began to implement their own restrictions on imports through their ports (Resource Recycling, 2022).

The loss of the Chinese markets had profound effects on U.S. waste management and on the environmental effects of trade in recyclables. Landfilled waste rose in the states most affected by the ban; the ban also seemed to undermine local recycling systems and led to reductions in employment in MRFs (Sigman and Strow, 2024). Taylor and colleagues (2024) found that other countries experiencing the largest increases in recycled plastic imports following the Chinese ban also experienced increases in plastic bottle litter.

Basel Convention and International Trade in Recyclable Scrap and Waste

The Basel Convention on the Control of Transboundary Movements of Hazardous Wastes and their Disposal (Basel Convention) was adopted in 1989 and came into force in 1992. It is the most significant multilateral treaty governing the movement of hazardous waste, solid waste, and scrap for recovery and disposal internationally (UNEP, 2011). Currently, 187 countries are ratified members of the Basel Convention. The United States was an original signatory to the Basel Convention in 1989 but never ratified the agreement and is not a current member of the Convention.

The Basel Convention establishes rules and regulations for international shipments of all hazardous waste and solid waste for reuse, recovery, and disposal. It thus has important impacts on international markets for recyclable materials members and nonmembers alike. A notable feature of the Basel Convention is that member nations are prohibited from trading scrap and waste with nonmember nations if it is classified as having hazardous characteristics (UNEP, 2011). This rule has the potential to severely limit the markets available for the United States to export recyclable materials that are classified in categories deemed to have hazardous characteristics.

A significant example involves trade in plastic waste. Beginning in January 2021, provisions in the Basel Convention reclassified plastic scrap and waste from a single category that was previously freely traded to two new categories that are now subject to Basel Convention trade restrictions (even plastics that are not hazardous). The first new category covers mixed-plastic scrap and waste, plastic waste mixed with other forms of scrap, and plastic waste that is contaminated (e.g., with food or other nonhazardous substances). The second new category covers plastic scrap and waste that the Basel Convention classifies as hazardous waste. Both categories now require prior notice and consent, which means an importing country must accept in writing any shipment before it leaves the port of the exporting country. The Basel Convention plastics amendments significantly reduced international trade in plastics. For example, Ishimura and colleagues (2024) found that the Convention reduced plastic waste trade volume from developed to developing countries by 64 percent.

Importantly, the Basel Convention contains a provision that its members are not permitted to trade Basel-controlled plastic scrap and waste with nonmember countries except under the terms of an agreement or arrangement provided for by Article 11 of the Convention (Department of State, 2025). Thus, for the 187 Basel member countries, plastic scrap and waste that was uncontrolled prior to January 2021 cannot be traded with the United States unless a separate agreement is negotiated. The United States has negotiated separate agreements for importing and exporting hazardous waste with five countries: Canada, Mexico, Costa Rica, Malaysia, and the Philippines.

The United States and Canada first established a bilateral agreement in 1986 to address the movement of hazardous and municipal waste for recycling and disposal between the two countries. In 2020, they developed an additional agreement regarding the transboundary shipments of nonhazardous scrap and waste products, including nonhazardous plastic scrap and waste. These shipments are not subject to prior notice-and-consent rules as long as the movement of the material is only between the two countries and is being sent to locations following environmentally sound practices (EPA, 2018). The U.S. agreement with Mexico was also established in 1986 and allows for the movement of hazardous waste from Mexico to the United States for the purposes of recycling *or* disposal, but it only allows for hazardous wastes to be shipped from the United States to Mexico for the purposes of recycling. The United States also has bilateral agreements with Costa Rica, the Philippines, and Malaysia stipulating that wastes may be exported from those countries to the United States for recycling or disposal, but the United States may not export waste to any of them.

In addition to those five bilateral agreements, the United States is a member of OECD and is thus subject to the Council on the Control of Transboundary Movements of Wastes Destined for Recovery Operations, as well as OECD's plastic waste amendments effective January 2021. These amendments grant OECD member countries the right to govern nonhazardous plastic waste trade according to their own national and international laws. Given that most OECD member countries are also members of the Basel Convention, most have adopted the new Basel Convention restrictions on plastic waste.[4]

U.S. Trade in Recyclable Waste and Scrap Material

Recyclable scrap and waste materials operate in interconnected global commodity markets. From 2012 to 2023, the world traded more than 776 million metric tons of scrap and waste glass, paper, plastic, and aluminum across international borders.[5] Of the four commodities traded internationally in Figure 5-9, scrap and waste from paper and paperboard products are the largest by weight. Prior to 2018, plastic scrap and waste was the second-largest recyclable commodity of the four (UN Comtrade, n.d.). But after China's National Sword policy was implemented in 2018 and plastics were reclassified under the Basel Convention in 2021, plastic scrap and waste trade has fallen off dramatically, from 15 million metric tons in 2012 to 4 million metric tons in 2023 (UN Comtrade, n.d.). This trend could continue with more countries restricting imports of plastic scrap and waste. In January of 2025, Thailand will completely ban the imports of plastic scrap and waste (Igini, 2023).

Trade in scrap and waste glass and aluminum has remained fairly steady over this period, with minor increases from 2012 to 2023 (UN Comtrade, n.d.). The United States has been, and remains, a large player in all four scrap and waste commodity markets internationally. Of total world trade in 2023, the United States was either an importer or exporter of 4 percent of glass traded, 50 percent of paper and paperboard traded, 27 percent of aluminum traded, and 22 percent of plastics traded (UN Comtrade, n.d.). However, as demonstrated in Figures 5-10 through 5-12, the United States has been a net exporter of scrap paper and aluminum for the past decade but a net importer of glass from 2012 to 2023.

The biggest changes have been observed in plastic scrap and waste trade. Figure 5-12 shows that, from 2012 to 2023, the United States has gone from being a net exporter (1.69 million metric tons per year) to being a net importer (15,000 metric tons per year) of plastic scrap and waste (UN Comtrade, n.d.). This dramatic decline follows the general worldwide trend for scrap and waste plastics, shown in Figure 5-12.

Importantly, much of the reversal in the U.S. trade surplus in plastic scrap has been driven by changes in trade with non-OECD nations. Figure 5-13 shows that U.S. exports of plastic scrap to OECD countries remained relatively stable, but exports to non-OECD countries, such as China, Thailand, and

---

[4] For specific rules for each OECD member country, see https://www.oecd.org/en/data/tools/transboundary-movements-of-waste.html.

[5] Data are from the United Nations Comtrade database at https://comtradeplus.un.org. These data are for the 4-digit harmonized system (HS) classifications of waste, pairings, and scrap for plastics (HS 3915); scrap and waste of paper and paperboard (HS 4707); glass cullet and other scrap and waste of glass (HS 7001); and aluminum scrap and waste (HS 7602).

Malaysia, began to decline rapidly around 2017 (UN Comtrade, n.d.). During the same time, imports of plastic scrap from non-OECD countries (Figure 5-14) has been increasing, even though imports from OECD nations have remained fairly stable. U.S. imports of plastic scrap and waste from non-OECD countries increased 235 percent over 2012–2023, from 40.5 thousand metric tons in 2012 to 1.35 million metric tons in 2023 (UN Comtrade, n.d.).

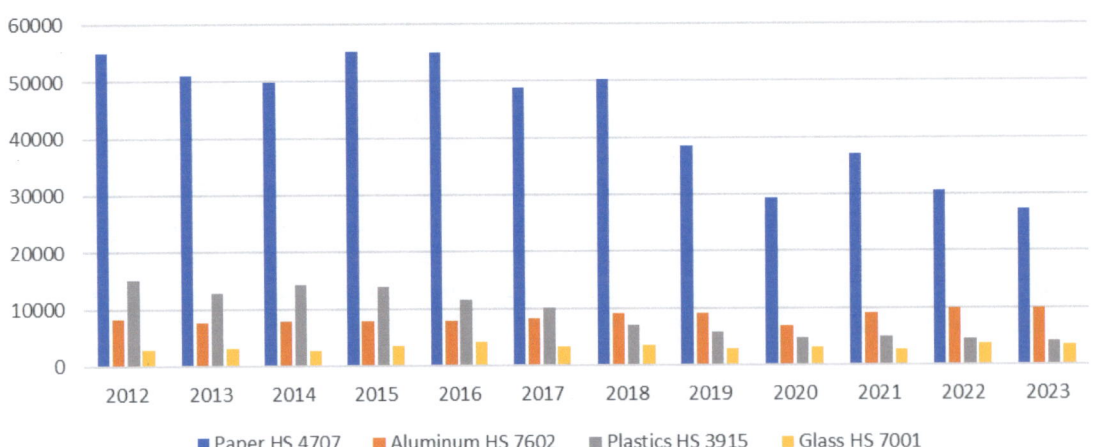

FIGURE 5-9 Recyclable waste and scrap materials (four types) traded worldwide, in thousands of metric tons, 2012–2023.
NOTE: HS = harmonized system.
SOURCE: Data from UN Comtrade database at https://comtradeplus.un.org.

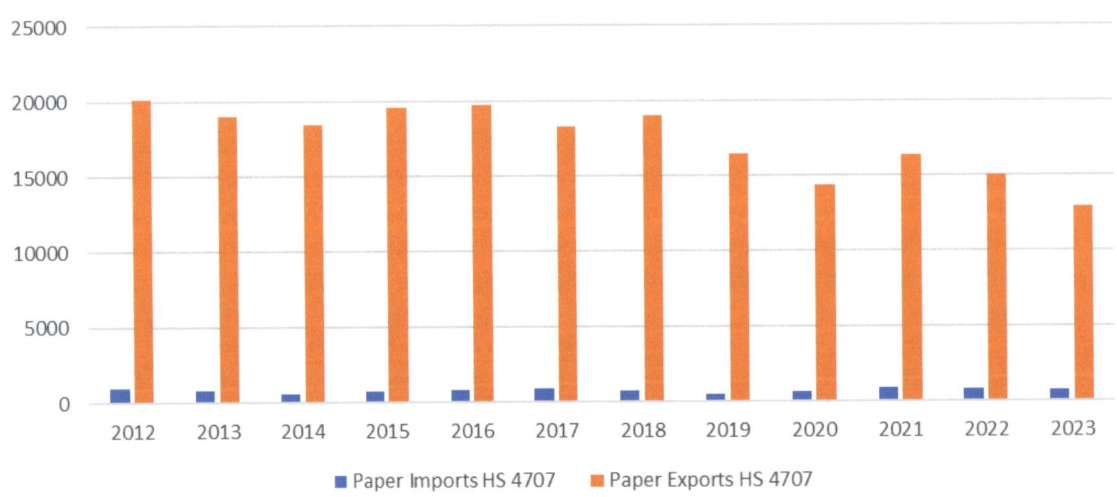

FIGURE 5-10 U.S. waste and scrap paperboard imports and exports, in thousands of metric tons, 2012–2023.
NOTE: HS = harmonized system.
SOURCE: Data from UN Comtrade database at https://comtradeplus.un.org.

Shifting Trade in a World after China Sword and Basel Convention

As a result of both the Basel Convention and China's National Sword policy, the United States and other countries have increased their focus on developing domestic markets for plastics collected for recycling. Ultimately, China's policy spurred increases in recycled paper capacity in North America, with over

25 new recycled paper mill projects completed since the National Sword policy was implemented (NERC, 2024). After decades of shuttering paper mills as production moved overseas to developing countries, starting in 2018, the United States saw commitments for new domestic paper mill capacity and plastic recycling infrastructure. In 2024, significant new capacity for paper mills producing cardboard from recycled feedstock has come online, and domestic markets for plastics have sufficiently replaced capacity to handle the quantity previously exported. These changes have contributed to a rebound in the price for postconsumer paper, which decreased in the immediate wake of the 2018 restrictions but has now surpassed pre-2018 prices.

FIGURE 5-11 U.S. waste and scrap aluminum imports and exports, in thousands of metric tons, 2012–2023.
NOTE: HS = harmonized system.
SOURCE: Data from UN Comtrade database at https://comtradeplus.un.org.

FIGURE 5-12 U.S. waste and scrap plastics imports and exports, in thousands of metric tons, 2012–2023.
NOTE: HS = harmonized system.
Data from UN Comtrade database at https://comtradeplus.un.org.

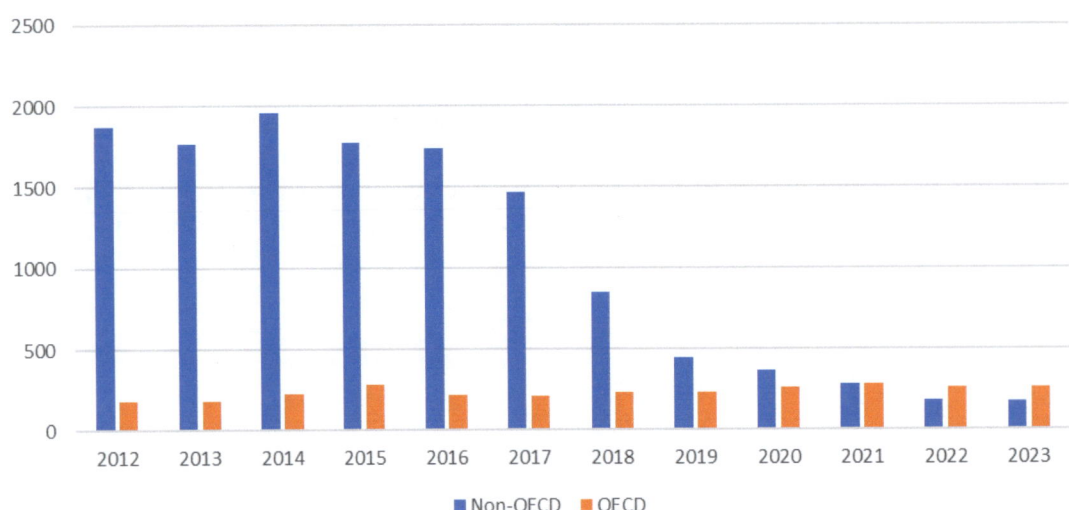

FIGURE 5-13 U.S. exports of plastic waste and scrap to OECD versus non-OECD countries, in thousands of metric tons, 2012–2023.
NOTE: OECD = Organisation for Economic Co-operation and Development.
SOURCE: Data from UN Comtrade database at https://comtradeplus.un.org.

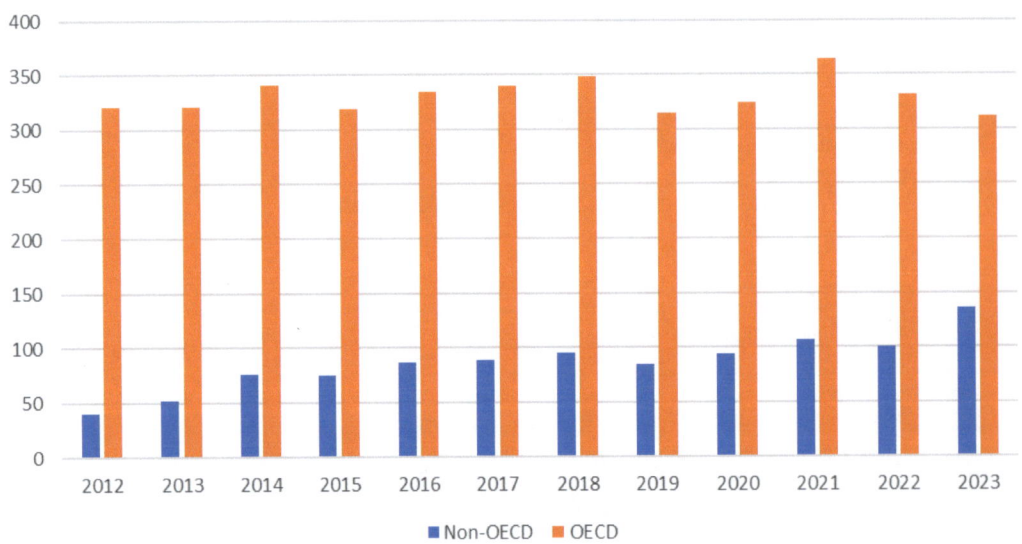

FIGURE 5-14 U.S. imports of plastic waste and scrap to OECD versus non-OECD countries, in thousands of metric tons, 2012–2023.
NOTE: OECD = Organisation for Economic Co-operation and Development.
SOURCE: Data from UN Comtrade database at https://comtradeplus.un.org.

The Role of the Global Marketplace

Although the amount of paper and plastic recycled domestically has grown since the 2010 peak in exports of recyclables, the global marketplace plays an important role in creating balance between supply and demand. When demand (and prices) are low because of seasonality or changing economics, the export market tends to help keep material moving. Even when domestic demand is high, the availability of international trade partners helps keep markets competitive, ensuring that domestic purchasers such as mills do not develop too much power over prices. These relationships can help stabilize prices for metals, paper, and some plastics and ensure that those materials obtain higher-valued uses than a domestic market alone might provide.

However, international trade in recyclable materials may raise environmental concerns that stem from the possibility of a "waste haven" effect, in which waste management and possibly recycling tend to occur in jurisdictions that provide fewer environmental protections (Kellenberg, 2012). Taylor and colleagues (2024) found that increases in a country's plastic imports increase plastic bottle litter, with the increase concentrated in countries that mismanage their plastic waste.

Virgin Plastic Imports

The production of virgin resin has surged globally in recent years, resulting in an oversupply that drives down the cost of virgin plastic on the market (Plastics Europe, 2021). This oversupply has led to substantial imports of low-cost virgin resin into the United States, creating significant challenges for recycled plastic markets. With virgin resin prices frequently undercutting the cost of producing postconsumer resin, the financial sustainability of recycling is at risk as those operations struggle to compete against inexpensive, newly manufactured plastic (Chowdhury et al., 2022).

This influx of low-priced virgin plastics not only affects the competitiveness of recycling operations but also has broad implications for sustainability goals. Manufacturers' preference for cheaper, imported virgin resin reduces demand for recycled materials, complicating efforts to close the recycling loop and establish a circular economy (Ellen MacArthur Foundation, 2020; UNEP, 2021). This trend underscores the need for supportive policies and industry incentives that prioritize recycled content (European Commission, 2020).

5.3.2 Regional and Local Markets

The desire to develop local and regional markets for recyclables dates back to early recycling programs in the United States. These programs arose from a growing recognition that effective management of sustainable recycling systems requires robust, consistent demand for recyclable materials within close proximity to collection centers. Developing regional markets can help minimize transportation costs, stimulate local economies, and provide greater resilience in managing material streams that can be vulnerable to fluctuations in international markets.

Early Market Development Efforts in the 1990s

In the 1990s, as local recycling programs grew in number, several states took it upon themselves to support efforts to develop local or regional markets for recyclables. Specific recycling market development programs were created in New York, New Jersey, North Carolina, and Washington State. North Carolina's Department of Environmental Quality (n.d.) issued grants through its Recycling Business Development Grant Program to support the state's recycling businesses and promote material recovery and end-use development. Washington established the Clean Washington Center, housed in its Department of Commerce.

EPA (1995) followed in states' footsteps, publishing *Recycling Means Business*, which outlined a market development strategy. The program included funding for state recycling and reuse business centers, economic development advocates, and the establishment of recyclable commodities trading systems.

Renewed Market Development Efforts in 2018

Efforts to maintain investments in domestic end markets continued but slowed during the 2000s because of financial constraints, political factors, and other limitations. However, in response to China's National Sword policy, state policy initiatives related to recycling market development regained momentum. As a result, several states (including Minnesota, South Carolina, Pennsylvania, and Michigan) that had previously established recycling market development programs saw those programs revitalized with a renewed sense of urgency, bolstered by increased financial and personnel resources. See Boxes 5-5 and 5-6 for examples of recent state policies and initiatives.

BOX 5-5
Case Studies: State Recycling Market Development Programs

Several states have initiated efforts to establish new recycling market development centers or programs since 2018:

- **Colorado** (SB20-055, 2020)[a] directed the state's Department of Public Health and Environment to create a plan for a new recycling market development center. Colorado (HB 1159, 2022)[b] also created a circular economy development center to grow existing markets; create new markets; and provide the necessary infrastructure, systems, and logistics to create a sustainable circular economy for recycled commodities in the state. One of the center's primary roles is connecting end markets with existing state grants and other incentives.
- **Maryland** (HB0164, 2021)[c] required the state's Office of Recycling to "promote the development of markets for recycled materials and recycled products" through efforts such as helping connect recycled materials, especially hard-to-recycle ones, with suitable end markets.
- **New Jersey** (S-3939, 2018)[d] called for the creation of a Recycling Market Development Council that would report on best practices to reduce the contamination of recyclables and recommend ways to stimulate demand for the materials.
- **Texas** (SB 649, 2019)[e] mandated the Texas Commission on Environmental Quality (2021) to produce a market development plan for using recyclable materials as feedstock in processing and manufacturing.

[a] See https://leg.colorado.gov/bills/sb20-055.
[b] See https://leg.colorado.gov/bills/hb22-1159.
[c] See https://mgaleg.maryland.gov/mgawebsite/Legislation/Details/HB0164?ys=2021RS.
[d] See https://legiscan.com/NJ/bill/S3939/2018.
[e] See https://legiscan.com/TX/bill/SB649/2019.

Regional and Local Market Development Focus by Material Type

Efforts to develop regional and local markets for recyclable materials have long centered on addressing the challenges presented by the heterogeneity of material types and variations in market demand, transportation costs, and end-use applications. Although each material has unique obstacles, sustained efforts have aimed at fostering viable domestic recycling solutions and reducing dependency on global export markets. The following sections summarize current challenges and initiatives for glass and plastics.

Glass Recycling and Market Development

Glass has consistently posed a challenge for regional recycling efforts because of its high weight, low market value, and high transportation costs that often exceed the material's economic value and thus limit its accessibility to end markets. The widespread use of imported-glass packaging in consumer products—such as wine, beer, and food items—has created an oversupply in many areas. Additionally, the quality of glass collected through curbside programs tends to be low, further complicating the ability to market this material effectively. As a result, glass has become a persistently difficult commodity to manage within recycling systems, particularly in areas with limited local end markets.

Various initiatives have tried address the need for new end markets for recycled glass. Programs such as those led by the Northeastern Recycling Council (NERC) and the Clean Washington Center have explored solutions to increase the value and usability of recycled glass. While large-scale solutions have remained elusive, niche markets for recycled glass have developed, including applications in abrasives, art, drinkware, and specialty products. Despite these efforts, comprehensive regional markets for glass recycling remain limited, so glass recycling continues to pose challenges for many local systems.

BOX 5-6
Case Study: Washington's NextCycle Program

Washington State (SB5545, 2019)[a] authorized the creation of a recycling development center to encourage new companies to process recyclable materials and develop more local end markets, guided by an advisory board that includes recycling industry representatives (Department of Ecology, n.d.).[b] The center, now called NextCycle (n.d.),[c] has evolved to include the City of Seattle and King County. NextCycle seeks to advance a circular economy by fostering public–private partnerships that support innovative recycling and material recovery solutions (Department of Ecology, 2023).

A core component of NextCycle is the Circular Accelerator, which uses a competitive model for identifying and supporting projects with high potential impact. In its first round, 14 teams proposed scalable end-market solutions targeting materials prevalent in the regional waste stream, such as glass, textiles, food waste, demolition debris, and used bedding (King County Solid Waste Division, 2023). These teams competed for $26,000 in funding to further develop initiatives aimed at enhancing local recycling infrastructure and reducing material sent to landfills.

The NextCycle Washington program highlights the importance of developing resilient end markets for recyclables, particularly as global recycling markets fluctuate. By facilitating regional market development and supporting viable recovery pathways for traditionally discarded materials, this initiative aligns with evidence-based goals for sustainable materials management (Washington State Recycling Association, 2022). In doing so, NextCycle Washington addresses both environmental and economic objectives, providing a replicable model for improving material recovery systems through integrated, locally focused approaches.

[a] See https://app.leg.wa.gov/billsummary?BillNumber=5545&Initiative=false&Year=2019.
[b] See https://ecology.wa.gov/Waste-Toxics/Reducing-recycling-waste/Strategic-policy-and-planning/Recycling-Development-Center.
[c] See https://www.nextcyclewashington.com/teams.

Development of Domestic Markets for Plastic

Growing restrictions on the export of plastics—intensified by the Basel Convention's regulations—have been accompanied by an increased focus on developing domestic end-market solutions for various plastic types. Notably, significant progress has been made in recycling PET and HDPE plastics to meet stringent standards for bottle-to-bottle recycling, thereby supporting sustainable domestic markets.

Historically, less than 25 percent of recycled PET bottles was reprocessed into new bottles, with the majority instead repurposed in the textile industry. However, in 2023, for the first time, the volume of PET recycled into bottles surpassed that used in textiles, reflecting advancements in processing technology and market demand for high-quality, recycled PET suitable for food-grade applications (NAPCOR, 2023).

And traditionally, HDPE recycling has focused on applications like piping. Recently, however, an increasing percentage of natural-colored HDPE bottles are being recycled to Food and Drug Administration standards, allowing these materials to be used to produce new milk jugs and other food-grade containers.

5.4 PUBLIC POLICIES FOR END MARKETS

Public policy interventions may improve and support end markets for collected materials by (1) increasing materials supply by encouraging households to recycle and (2) increasing demand for the secondary material. The latter can be achieved through postconsumer recycling (PCR) content standards and government procurement policies.

5.4.1 Postconsumer Recycling Content Policies

PCR content laws (PCR policies), also called minimum content legislation, can play a vital role in promoting recycling by requiring manufacturers to reach a minimum percentage of inputs that are recycled material. These policies apply to postconsumer content, as opposed to postindustrial content (a byproduct of manufacturing) or preconsumer content (a product that never reaches the consumer). Postconsumer content is sometimes less pure and harder to use compared with these other types, so legislation has focused on these materials.

PCR policies can increase regional and national demand for recycled materials, thereby supporting prices (Palmer and Walls, 1997). They may also help stabilize prices by providing a reliable source of demand, because manufactures cannot easily substitute from the recycled commodity when primary materials prices drop. Thus, these laws reduce the economic risk for investments in capital-intensive recycling infrastructure. By supporting the recycling industry, they help provide many of the benefits of recycling discussed above, such as reducing reliance on virgin materials and the adverse climate and environmental impacts associated with raw material extraction and production.

Experience with PCR Policies

Although no federal PCR policies have been passed to date, states have used them for more than 30 years as part of a demand-side market development strategy. The first wave began in the 1980s and focused on newsprint. By 1990, eight states enacted recycled content mandates: Arizona, California, Connecticut, Illinois, Maryland, Missouri, Oregon, and Wisconsin (National Solid Wastes Management Association, 1991). In the 1990s, California enacted PCR policies for fiberglass, glass containers, plastic containers, and plastic trash bags. In 1995, Oregon set a 25 percent PCR content standard for plastic beverage containers.

Although these early PCR policies played an important role in supporting the expansion of recycling, they were not adopted by other states. Most of the focus of recycling programs between 1995 and 2017 focused on collecting more material and processing it to grow supply for end markets. Few policies were designed for growth demand of recycled materials.

As discussed above, cheap virgin resin has undercut demand for postconsumer plastic. In response to pressure from recyclers, states have passed a new wave of minimum PCR content standards. These new initiatives primarily target packaging materials, beverage containers, and plastic bags. For example, California's AB 793 mandates that plastic beverage containers must meet specific recycled content thresholds, which are set to increase over time. New Jersey has phased in PCR content mandates across various packaging types, including rigid plastic and glass containers, plastic bags, and trash bags, with rates that will rise incrementally through 2027. Similarly, Washington State's SB 5022 sets minimum PCR content for plastic beverage bottles and other containers, aiming to drive the use of recycled materials across common consumer products (Retail Compliance Center, 2023).

Problems with PCR Policies

Although PCR content standards can have benefits, they also have drawbacks. First, these standards often require that materials are reused in consumer products, which may not always be the least resource-intensive use of recycled materials (as some indicated by some of the examples in the discussion of closed loop recycling above). The standards may thus raise the overall social costs of the recycling system. The Oregon Department of Environmental Quality's 2020 report, "Evaluating the Effects of PCR Content on Environmental Impacts" stated:

> The second finding is that recycled content by itself is not a good predictor of lower environmental outcomes when comparing functionally equivalent (substitutable) packaging made from different

materials. Just because a package contains higher levels of recycled content (on a mass or percentage basis) does not indicate that it has lower negative impacts because materials have substantially different production burdens. For example, a glass bottle may contain a higher percentage of recycled content than a lightweight flexible pouch or plastic bottle. Based on recycled content one may be inclined to prefer the glass container, but because of its higher overall weight glass may use more virgin material overall, and because of how it is made, may result in higher impacts such as emissions and resource depletion.[6]

Another drawback of PCR policies is that they may raise the prices or reduce the quality of products, such as packaging, that must have a minimum recycled content. To be effective, the policies must force manufacturers to use more recycled material inputs than they would choose without the policy; the necessary adjustments will either reduce the quality of the products made with these inputs or raise costs (and thus price). The resulting price increases may create desirable incentives for source reduction (Sigman, 1995). States have chosen to focus PCR content mandates on packaging, which may reflect a desire to discourage its consumption at the same time as increasing demand for recycled materials. However, restrictive content rules may raise final product costs more than necessary and discourage consumption of the products that have the best potential for closed-loop recycling.

Flexibility in the design of PCR content standards can reduce these adverse impacts. A form of flexibility used in practice is to allow averaging of recycled content across all covered products made by a company. These "portfolio standards" allow firms to choose the product lines in which recycled materials can most easily be integrated; they may also decide to concentrate their recycled materials in product lines that they sell at a discount or as a "green choice" that the market might not otherwise sustain. An even more flexible system would allow averaging across an industry with trading: producers who can more easily substitute recyclables will exceed the standards and receive payments from other producers who do not meet the standards.[7] Such flexibility might have been useful in the early days of recycled newsprint content standards. For example, Canadian producers relied mostly on virgin pulp from their extensive forests, and they were forced to import U.S. recycled pulp. Thus, they complained that these policies put them at a competitive disadvantage (McCarthy, 1994).

The effectiveness and costs of PCR policies are likely reduced by the fact that they have been implemented at the state level rather than the national level. Requirements that vary across states create complexity for producers and logistical issues that raise costs. And a single state can increase demand for only a small share of the overall product market. The resulting prices increases may be muted because recycled commodity prices are determined regionally or nationally. However, state policies may have spillover effects to other states, as producers find it easier to comply with the strictest state policy across their entire sales region, rather than tailor the manufacture of their products state by state. This phenomenon is known as the "California effect"; it may allow large states to assume an outsized role in national production patterns.

Finally, the most significant concern about PCR legislation is that, as demand-side policy, it does not ensure adequate supply of recycled materials. Although it will likely increase the prices of recycled materials and thus help MRFs support increased collection and processing of these materials, the extent of this indirect supply response is unclear. Any supply response might require local governments to adjust their policies, which they might not do in response to small changes in prices. If the supply does not react adequately, the policy may create excessive price increases. To ensure adequate supply to meet the demand that these PCR policies create, they need to be combined with public policy interventions to support supply—such as EPR or mandatory recycling policies.

---

[6] See https://www.oregon.gov/deq/FilterDocs/pcr.pdf.
[7] Standards that allow trading have not been used in practice for recycling but have been used successfully for other environmental objectives, in for example, state renewable energy portfolio standards and in the U.S. phase-out of lead in gasoline.

5.4.2 Government Procurement

Government procurement may be an important public policy in supporting recycling end-use market. The U.S. federal government is among the largest consumers of goods in the world and thus has considerable influence on many product markets. The 1976 Resource Conservation and Recovery Act required EPA to set requirements for recycled content in government purchases, resulting in the Comprehensive Procurement Guideline Program (EPA, 2024b). EPA set requirements for 62 products in eight categories that range from paper to construction materials and vehicular parts. Although difficult to assess empirically, the program's benefits may come not just from its direct effect on demand but also from the information it provides to other purchasers about which products contain or could contain recycled materials. States and local governments also have recycled content requirements for their purchases, but their reach is narrower than that of the federal program.

5.4.3 Taxes and Fees on Virgin Material

Another mechanism that governments use to stimulate end markets for recycled material is to impose taxes or fees on virgin (i.e., nonrecycled) material. This kind of policy places an economic disincentive on the use of virgin material while creating an economic incentive to use recycled material. A recent example of this kind of tax is from the United Kingdom (Box 5-7), which has a policy that focuses on plastic use in packaging.

The United States also imposes taxes on virgin plastic resin through the Superfund program. These fees, discussed in Box 5-8, are not designed for their effects on recycling but nonetheless slightly discourage virgin plastic use.

5.5 KEY POLICY OPTIONS

Policies that influence material use and recycling can help balance economic and environmental considerations while shaping market dynamics and consumer behavior. Effective approaches aim to address cost differences between virgin and recycled materials, encourage the use of recycled content, and support waste reduction efforts. Well-designed policies can also enhance market stability, promote innovation in recycling technologies, and improve the overall efficiency of material management systems.

BOX 5-7
Case Study: UK Plastic Packaging Tax

In April 2022, the United Kingdom implemented a tax on plastic packaging that does not contain at least 30 percent recycled material. The initial tax of £200 per metric ton of plastic packaging material rose to £217.85 per metric ton in April 2024. It applies to plastic packaging used in the supply chain and in single-use consumer packaging. Two parties are responsible for paying the tax: UK-based manufacturers of plastic packaging and importers of plastic packaging material. This tax was implemented to increase domestic demand for recycled plastic packaging and to reduce the amount of plastic waste ending up in landfills or in the environment, or being incinerated.[a]

[a] See https://www.gov.uk/government/publications/plastic-packaging-tax-rate-change-from-1-april-2023/rate-change-from-1-april-2023-for-plastic-packaging-tax.

5.5.1 Revenue-Neutral Fee-and-Reward Program

For two major reasons, the societal costs of virgin materials extraction exceeds the direct dollar costs to companies and their customers. First, for many years, various direct and indirect government subsidies have promoted oil and gas industries that provide petrochemicals used to make virgin plastic resins

(Metcalf, 2018). Second, as highlighted in this report, those industries do not pay all the substantial indirect environmental, social, and economic costs of extraction. For both reasons, the use of virgin materials exceeds the socially optimal amount, and its subsidized supply on the market outcompetes the demand for recycled materials.

BOX 5-8
Superfund Excise Tax on Hazardous Substances

The reinstatement of the Superfund excise tax through the Inflation Reduction Act of 2022 (Pub. L. 117-169) will affect incentives for use of virgin versus recycled materials. Effective mid-2022, the tax imposes levies on crude oil and petrochemical feedstocks products, which will be adjusted annually for inflation (IRS, 2023). The levy applies to several feedstocks for virgin plastics, including ethylene and propylene, at an initial rate of $9.74 per ton. By increasing the financial burden of using virgin petroleum-based materials, the tax may indirectly encourage substitution toward recycled plastic inputs, as well as away from plastics toward other materials. It may modestly increase the cost of manufacturing plastic from petrochemicals but is not designed to reduce reliance on single-use plastics (Mann, 2024). Revenues generated from this tax are directed to the Superfund Trust Fund, managed by the U.S. Environmental Protection Agency, which pays for cleanup of hazardous sites—especially when settlements have not been reached for private clean-up (GAO, 2023).

Meanwhile, many policies have successfully encouraged the supply of recycled materials from households and businesses. Without sufficient demand for those materials, recycled materials markets—particularly plastics markets—experience significant imbalance. As such, a simple fee by weight of virgin resin could address part of this imbalance, with various possibilities for the use of the revenues generated by that fee.

The costs of MSW recycling programs and curbside collection include the need to collect, sort, clean, crush, and bale the different materials for transport, and to process the material into recycled resin that can be used as a feedstock. These costs mean that the market price of recycled materials may exceed the often-subsidized price of virgin feedstock. For example, Figure 5-4 shows persistently higher prices for the recycled material than for the virgin material for PET that is used to make plastic packaging. As such, recycled materials may face challenges in competing against clean, homogenous virgin material. This economic challenge is particularly acute for plastic resins and for glass collected in curbside recycling programs, as explained earlier in this report. While glass markets are mature, large-scale virgin plastic resin production is increasing the amount of low-cost virgin plastic on the market. A subsidy per weight of recycled material would reduce its price as an input to manufacturing and address another part of that market imbalance. However, such a subsidy or reward for using recycled materials would require a source of funds.

For these reasons, an effective policy would combine both approaches: a revenue-neutral policy that applies a fee per ton of virgin resin used in domestic manufacturing of plastic packaging and single-use products, with all revenue used to fund a reward per ton of recycled resin used in those same processes. In parallel, a border adjustment fee on fully manufactured imported plastic packaging and single-use products would be needed to sufficiently discourage the relocation of manufacturing processes. Because it applies to national and international markets, it would need to be enacted at the federal level. Other policies encourage the supply of PCR content; firms would earn the reward when they buy postconsumer, not postindustrial or preconsumer recycled content.

This policy would help achieve several recycling objectives: (1) it makes the cost of recycled materials more competitive with virgin materials used as feedstock in manufacturing; (2) it increases demand for recycled materials, which raises the effectiveness of efforts to increase household participation and recycling supply; (3) it helps reduce environmental and social damages from landfills and from litter or dumping; and (4) it helps build end markets for recycled materials (the importance of which is discussed earlier in this chapter; see also Bradshaw et al., 2025).

The policy reduces environmental and social damages from disposal by incorporating some of the external costs of these materials into the prices that manufacturers and consumers pay for these materials.[8] By enhancing demand for recycled plastic resin and reducing demand for virgin material, this policy would reduce environmental costs from virgin materials extraction and landfilling of plastics. It would increase the effectiveness of MSW recycling programs and may have additional effects of decreasing plastic litter. Lastly, an additional strength of this policy approach is that it avoids mandated rules about company behavior and allowable materials, instead leveraging price incentives in market choices.

Trade-offs need to be acknowledged for this policy approach, however. The committee acknowledges the challenges in enacting and administering large-scale tax policy. Furthermore, while companies that can readily adjust to the new incentives for using more recycled inputs receive a cost advantage, this policy may place some cost disadvantages on companies less capable of readily adjusting their processes to these changes in relative material prices. Finally, while suppliers of recycling material would likely see increased profits on account of increased market demand for their product, producers of virgin plastic resins would likely see declines in their profits on account of increased market competition and decreased demand for their products.

In general, this policy would require a multiyear process to establish the appropriate fee and reward levels, as well as the eventual tax procedures including reward eligibility and fee liability, and all reporting requirements. As reasonably assessed by the committee, 5 years is a likely minimum scenario from introduction to implementation.

After implementation, various metrics can be used to evaluate effectiveness of the policy. Evaluation could benefit from measures of virgin resin sold at various fee-and-reward rates, compared with recycled material sold. After the policy has been in place for a few years, actual effects of specific fees and rewards on prices and quantities of virgin materials and on prices and quantities of recycled materials need to be compared with projected effects, along with actual and projected effects on environmental damages from extraction, landfills, and plastic litter.

As referenced in its statement of task, the committee did not have sufficient data available to recommend specific monetary values of such fees and rewards, so this policy requires further study. Nevertheless, sufficient information is available to outline this policy structure and approach (see Figure 5-15). The committee provides this summarizing conclusion and policy option:

Conclusion 5-1: A revenue-neutral policy that applies a fee for using virgin plastic resins and a reward for using recycled plastic resins would increase the cost-competitiveness of recycled materials relative to virgin inputs and would enhance end markets for recyclable materials.

Key Policy Option 5-1: The U.S. Congress could enact a new revenue-neutral fee-and-reward policy to increase the competitiveness of recycled materials relative to virgin inputs. It would encourage the use of recycled plastic resins in the manufacturing of plastic packaging and single-use products. This policy could comprise two levers:
- First, the Department of the Treasury, in partnership with the U.S. Environmental Protection Agency (EPA), could implement a new fee on the use of virgin plastic resins used in product packaging and in the manufacturing of consumer products, and a corresponding reward for the use of postconsumer recycled plastic resins in those same manufacturing processes. If implemented, this new fee and reward should be paid and received by domestic manufacturers that use plastic resins in their manufacturing processes, should be weight-based, and should be of sufficient value to encourage the use of recycled plastic resins. Market parity can facilitate economic competition between recycled plastic resin and virgin resin.

---

[8] Superfund excise taxes on virgin plastic feedstocks may also help raise these prices (see Box 5-9), but Superfund does not direct those revenues toward rewarding recycled plastic use.

- Second, the Department of the Treasury, in partnership with EPA and U.S. Customs and Border Protection could impose a new border adjustment fee on fully manufactured imported plastic packaging and single-use products, to be paid by the importer of those products.

If pursued, this policy should be revenue neutral for the federal government, such that the total annual sum of fees collected equals the total annual sum of rewards distributed. Furthermore, the Department of the Treasury, in partnership with EPA and other relevant parties, would need to study and identify the appropriate levels of fees and rewards to fully encourage the use of recycled plastic resins while minimizing motivations for changing manufacturing locations.

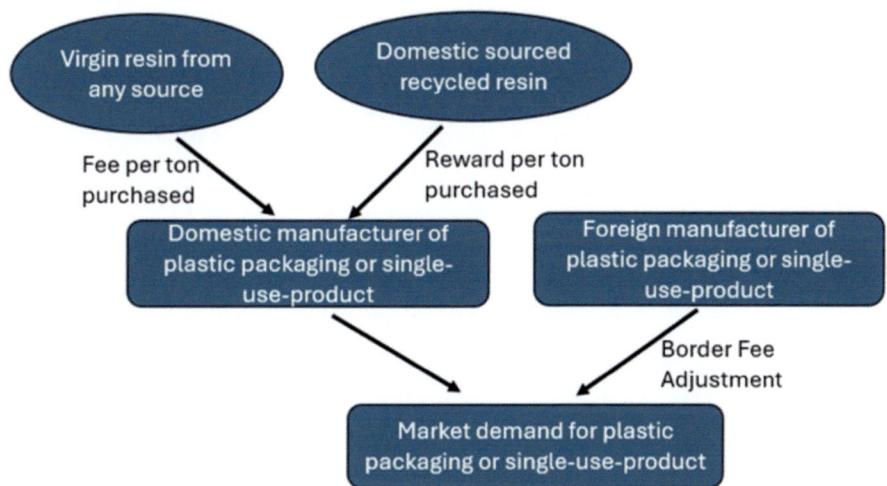

FIGURE 5-15 Conceptual map of the proposed fee-and-reward program.

5.5.2 Federal Funding for MSW Recycling and Research

Research could help improve product recyclability and develop new uses for recycled materials, contributing to higher and more stable prices. It could also make existing markets more robust and environmentally beneficial. Some innovations might open markets for materials that are currently difficult or impossible to recycle such as certain types of plastic or make recycling feasible in areas of the country where end markets are currently too thin. Federal research funding could drive advancements in sorting and processing technologies, improve the environmental impact of recyclables, and potentially develop new uses for currently underutilized materials, such as certain plastics. Ideally, such innovation would make the recycling system more sustainable without modifying its current incentives. A model for this goal is in the electricity sector, where innovative activity has made renewable energy production desirable based on market prices alone, without a need to account for environmental benefits.

Research funding would address two fundamental problems. First, only some recovered materials can be sold for high values in end markets. If lower recovery values make the net cost of recyclables processing higher for local governments and businesses, they may face difficulty paying for recycling. Local governments may scale back the availability of curbside recyclables collection or limit access to recycling in rural areas, where it may be too costly to transport materials. Limited end-market options also contribute to volatility in recyclable prices, which makes local governments unwilling to bear the financial risk. In some worst-case conditions, households, firms, and governments spend time and money to separate recyclables, only to have the material landfilled or incinerated anyway; these episodes are not only wasteful but also undercut faith in recycling and willingness to contribute. Second, private incentives for research and development (R&D) in recycling are insufficient and not well targeted toward improving environmental outcomes. All facets of the economy suffer from insufficient investments in R&D because the knowledge it creates is a public good (see Chapter 3). In addition to this fundamental market failure, private R&D does not tend to focus on finding end uses for materials that reduce environmental costs. This difficulty is illustrated by the recent tensions between desirable market properties of "advanced" or "chemical" recycling of plastics and concerns about their environmental implications (Gracida-Alvarez et al., 2023). Publicly funded research could help steer innovative activity toward end uses with more desirable environmental properties.

The recommended research program could easily fit within existing federal research infrastructure. Private–public partnerships (e.g., Department of Energy's REMADE) might be possible for some research areas, such as improved design for recycling. Given the national or even global goals that such a program would address, the federal government would be the best level for establishing the program. This policy could be implemented soon through existing funding mechanisms, although the timeline for its effects on the recycling system will be longer. Intermediate metrics for success of this effort might be research articles in journals and new patent activity. A long-term metric would be sustained prices that generate profit for recyclable materials in applications that also generate environmental improvements relative to landfilling or incinerating of these materials. The federal government could also partner with nonprofits, such as the Environmental Research and Education Foundation, that fund research in MSW management.

Conclusion 5-2: Advancing research and development in technology areas relevant to recycling and adopting new technologies in the MSW recycling system can help achieve multiple policy objectives for recycling:
- *enhancing end markets for recyclable materials,*
- *increasing the cost-competitiveness of recycled materials relative to virgin inputs,*
- *improving the cost-effectiveness of recycling collection and processing,*
- *decreasing contamination of post-consumer recycling streams, and*
- *enhancing social and environmental benefits associated with recycling.*

Conclusion 5-3: Increased recycling collection may have little benefit without end uses for the collected materials that are environmentally sound and economically valuable. Thus, increased collection needs to be combined with support for end markets, with attention to the environmental implications of end uses.

Key Policy Option 5-2: Federal agencies that fund research related to recycling, including the U.S. Environmental Protection Agency, the Department of Energy, and the National Science Foundation, could enhance investments in research related to recycling systems and recyclable materials. When pursued, this research should prioritize environmentally sound and economically valuable end uses for recycled commodities and other approaches to increase end-use values nationally and internationally. Examples include recyclable design for consumer products, and technologies to reduce contamination of the recycling material stream. Funding from Congress to support this endeavor could include public–private partnerships in manufacturing innovation to increase opportunities for recyclable materials end uses.

5.5.3 The Basel Convention

The Basel Convention on the Control of Transboundary Movements of Hazardous Wastes and Their Disposal is an international agreement aimed at controlling hazardous wastes and their effects on human health and the environment. The United States was an original signatory to the Basel Convention in 1990 and the U.S. Senate provided bipartisan advice and consent approval for the agreement in 1992. However, for a range of legal and political reasons outlined by Yang and Fulton (2017), Congress has never taken the final step to ratify the agreement. Yang and Fulton (2017) make a compelling legal case for why the United States should ratify the Basel Convention. Beyond the legal argument, ratification has benefits for recyclable materials.

First, the scope of the Basel Convention covers a wide range of trade in wastes beyond just those deemed as hazardous, including those that are covered by this committee's statement of task. Annex II of the Convention covers household garbage and "other wastes" that are not deemed to be hazardous but are often recyclable or must be disposed of in environmentally safe ways. International trade in these materials is not specifically covered under EPA's Resource Conservation and Recovery Act or other U.S. domestic law. Ratification of the Basel Convention could help fill loopholes with respect to coverage of trade in nonhazardous recyclable materials.

Second, plastic waste and scrap amendments to the Basel Convention were adopted in 2019 and began to be enforced by parties as of January 1, 2021. Currently, 190 countries and the European Commission are ratified members. Under the new amendment, many plastics now fall under the Convention's prior notice and consent rules. Not being a party to the convention may inhibit the United States's ability to trade waste and scrap plastics with most countries. Furthermore, the Convention's rule that member countries may not engage in trade of waste and scrap with nonmember countries (such as the United States) could limit markets for importing and exporting waste and scrap of recyclable materials beyond plastics, if member countries increasingly enforce the rule. Ratifying of the Basel Convention would ensure that the United States is at the table and can advocate for its interests in future amendments to the Convention; ratifying would also ensure access to trade in recyclable materials markets with 190 member nations.

An important note and potential disadvantage regarding ratification is that in 2019, the Convention adopted the "Basel Ban Amendment," which prohibits OECD countries from exporting hazardous wastes as defined by the Convention to non-OECD countries for final disposal or for reuse, recycling, or recovery operations. However, recent international policy changes, including China's National Sword policy and other countries' subsequent bans on imports of plastics, have limited the impact this amendment would have on U.S. trade should the United States ratify the Basel Convention.

Conclusion 5-4: Full participation in international agreements can help facilitate open markets for trade of recyclable material and grow end markets for recyclable materials collected in recycling programs, as well as enhance social and environmental benefits associated with recycling. In particular, full U.S. ratification of the Basel Convention on the Control of Transboundary Movements of Hazardous Wastes and Their Disposal would better align the United States with international law regarding the transport of wastes covered under the Convention, and ratification would help to facilitate open markets for trade of these recoverable and recyclable materials.

Key Policy Option 5-3: The United States could ratify the Basel Convention because of its assistance in closing loopholes regarding trade in nonhazardous recyclable materials. If pursued, the U.S. Congress should provide necessary domestic authorities and relevant agencies to regulate the full scope of the Basel Convention's provisions.

REFERENCES

AF&PA (American Forest & Paper Association). 2025. Paper & Cardboard Recycling. https://www.afandpa.org/priorities/recycling.

ASTM. 2022. Standard Practice for Coding Plastic Manufactured Articles for Resin Identification. Standard D7611/D7611M-21. https://www.astm.org/d7611_d7611m-21.html.

Allaway, D. 2024. Presentation to the Committee on Costs and Approaches for Municipal Solid Waste Recycling Programs on April 16, 2024. National Academies of Sciences, Engineering, and Medicine.

Amini, H.R., and D.R. Reinhart. 2011. Regional prediction of long-term landfill gas to energy potential. *Waste Management* 31(9–10):2020–2026. https://doi.org/10.1016/j.wasman.2011.05.010.

Andini, E., P. Bhalode, E. Gantert, S. Sadula, and D.G. Vlachos. 2024. Chemical recycling of mixed textile waste. *Science Advances* 10(27):eado6827. https://doi.org/10.1126/sciadv.ado6827.

Arbex, M., and Z. Mahone. 2024. Materials, Technology and Growth: Quantifying the Costs of Circularity. Department of Economics Working Papers. McMaster University. https://sites.google.com/site/zacharymahone/research.

Avery, E., E. Nduagu, E. Vozzola, T.W. Roux and R. Auras. 2025. Polyethylene packaging and alternative materials in the United States: A life cycle assessment. *Science of The Total Environment* 961:178359.

Badgett, A., and A. Milbrandt. 2021. Food waste disposal and utilization in the United States: A spatial cost benefit analysis. *Journal of Cleaner Production* 314:128057.

Baldé, C., R. Kuehr, T. Yamamoto, R. McDonald, E. D'Angelo, S. Althaf, G. Bel, O. Deubzer, E. Fernandez-Cubillo, V. Forti, V. Gray, S. Herat, S. Honda, G. Iattoni, D.S. Khetriwal, V.L di Cortemiglia, Y. Lobuntsova, I. Nnorom, N. Pralat, and M. Wagner. 2024. International Telecommunication Union (ITU) and United Nations Institute for Training and Research (UNITAR).

Balkevicius A., M. Sanctuary, and S. Zvirblyte, S. 2020. Fending off waste from the west: The impact of China's Operation Green Fence on the international waste trade. *World Economy* 43(10):2742–2761. https://doi.org/10.1111/twec.12949.

Basushi, R., K. Bhuwalka, E. Moore, and R. Kirchain. 2023. Overview of Recycled Plastic Supply and Demand: Identifying the Critical Market Bottlenecks for Closing the Loop. Appendix E in Recycled Plastics in Infrastructure National Academies.

Bening, C.R., J.T. Pruess, and N.U. Blum. 2021. Towards a circular plastics economy: Interacting barriers and contested solutions for flexible packaging recycling. *Journal of Cleaner Production* 302:126966.

Bishop, G., D. Styles, and P.N.L. Lens. 2020. Recycling of European plastic is a pathway for plastic debris in the ocean. *Environment International* 142:105893. https://doi.org/10.1016/j.envint.2020.105893.

Bradshaw, S.L., H.A. Aguirre-Villegas, S.E. Boxman, and C.H. Benson. 2025. Material recovery facilities (MRFs) in the United States: Operations, revenue, and the impact of scale. *Waste Management* 193:317–327.

Brewer, B. 2025. *Recycling envy: Ten American cities with excellent practices*. https://www.bustedcubicle.com/outside/top-american-cities-recycling.

Butler, J., and P. Hooper. 2005. Dilemmas in optimising the environmental benefit from recycling: A case study of glass container waste management in the UK. *Resources, Conservation and Recycling* 45(4):331–355.

Chowdhury, H., R. Naim, N. Islam, and F. Alam. 2022. Challenges and prospects of plastic recycling: A review. *Journal of Polymers and the Environment* 30(3):816–831. https://doi.org/10.1007/s10924-021-02154-7.

Dai, L., R. Ruan, S. You, and H. Lei. 2022. Paths to sustainable plastic waste recycling, *Science* 377(6609).

Department of State. 2025. *Basel Convention on Hazardous Wastes*. Office of Environmental Quality. https://www.state.gov/key-topics-office-of-environmental-quality-and-transboundary-issues/basel-convention-on-hazardous-wastes.

EIA (Energy Information Agency). 2024. *How much oil is used to make plastic?* https://www.eia.gov/tools/faqs/faq.php?id=34&t=6.

Ellen MacArthur Foundation. 2020. *The Global Commitment 2020 Progress Report*. https://www.ellenmacarthurfoundation.org/global-commitment/overview.

EPA (U.S. Environmental Protection Agency). 2018. *Frequent questions about the Hazardous Waste Export-Import Revisions Final Rule*. https://www.epa.gov/hwgenerators/frequent-questions-about-hazardous-waste-export-import-revisions-final-rule.

EPA. 2020. *Historical Recycled Commodity Values*. https://www.epa.gov/sites/default/files/2020-07/documents/historical_commodity_values_07-07-20_fnl_508.pdf.

EPA. 2024a. *Facts and Figures About Materials, Waste and Recycling*. https://www.epa.gov/facts-and-figures-about-materials-waste-and-recycling Current National Picture, accessed March 3, 2024).

EPA. 2024b. *Comprehensive Procurement Guideline (CPG) Program*. https://www.epa.gov/smm/comprehensive-procurement-guideline-cpg-program.

European Commission. 2020. *A New Circular Economy Action Plan for a Cleaner and More Competitive Europe*. https://ec.europa.eu/environment/strategy/circular-economy-action-plan_en.

Eurostat. 2018. *Economy-wide material flow accounts – Handbook*. Luxembourg: Publications Office of the European Union.

GAO (Government Accountability Office). 2023. *Superfund: Status of the Trust Fund and Site Remediation Efforts*. https://www.gao.gov/assets/a239185.html.

Geyer, R., B. Kuczenski, T. Zink, and A. Henderson. 2016. Common misconceptions about recycling. *Journal of Industrial Ecology* 20(5):1010–1017. https://doi.org/10.1111/jiec.12355.

Glass Packaging Institute. 2025. https://www.gpi.org/facts-about-glass.

Glass Recycling Coalition. 2024. *Survey Findings: Consumer Expectations for Glass Recycling Remains High*. https://www.glassrecycles.org/news/2023-glass-recycling-survey-findings-released.

Goodwin, L. 2023. *The Global Benefits of Reducing Food Loss and Waste, and How to Do It*. https://www.wri.org/insights/reducing-food-loss-and-food-waste.

Gracida-Alvarez, U.R., P.T. Benavides, L. Uisung, and M. Wang. 2023. *Life-cycle analysis of recycling of post-use plastic to plastic via pyrolysis*. https://doi.org/10.1016/j.jclepro.2023.138867.

Igini, M. 2023. *Thailand Announces Ban on Plastic Waste Imports by 2025*. https://earth.org/thailand-ban-plastic-imports.

IRS (Internal Revenue Service). 2023. *Superfund chemical excise taxes*. https://www.irs.gov/businesses/small-businesses-self-employed/superfund-chemical-excise-taxes.

Ishimura, Y., D. Ichinose, and K. Nomura. 2024. *Plastic waste trade and international agreements*. https://ssrn.com/abstract=4845055.

Katz, C. 2019. Piling up: How China's ban on importing waste has stalled global recycling. *Yale Environment 360*. https://e360.yale.edu/features/piling-up-how-chinas-ban-on-importing-waste-has-stalled-global-recycling.

Kellenberg, D. 2012. Trading wastes. *Journal of Environmental Economics and Management* 64(1):68–87. https://doi.org/10.1016/j.jeem.2012.02.003.

Kim, Y., J. Kang, and H. Chun. 2022. Is online shopping packaging waste a threat to the environment? *Economics Letters* 214:110398.

King County Solid Waste Division. 2023. *NextCycle Washington: Circular Accelerator Program Overview*. https://kingcounty.gov.

Kiran, E.U., A.P. Trzcinski, W.J. Ng, and Y. Liu. 2014. Bioconversion of food waste to energy: A review. *Fuel* 134:389–399. https://doi.org/10.1016/j.fuel.2014.05.074.

Kumar, V. 2017. Recycling of waste and used papers: A useful contribution in conservation of environment: A case study. *Asian Journal of Water, Environment and Pollution* 14(4):31–36.

Le Pera, A., M. Sellaro, G. Grande, E. Bencivenni, and M. Migliori. 2023. Effect of quality of separately collected glass, paper plus cardboard and lightweight packaging waste on environmental, energetic and economic sustainability of the material recovery facility operations. *Journal of Cleaner Production* 425:1238973.

Lorang, E., A. Lobianco, and P. Delacote. 2023. Increasing paper and cardboard recycling: Impacts on the forest sector and carbon emissions. *Environmental Modeling & Assessment* 28:189–200. https://doi.org/10.1007/s10666-022-09850-5.

Mann, R. 2024. Environmental justice, plastic, and tax. *Human Rights Magazine* 50(1/2). https://www.americanbar.org/groups/crsj/publications/human_rights_magazine_home/environmental-energy-climate-justice/environmental-justice-plastic-and-tax.

McCarthy, J. 1994. The trade implications of recycled content in newsprint: The US view. In *Life-Cycle Management and Trade*. Paris: Organisation for Economic Co-operation and Development.

Metcalf, G.E. 2018. The impact of removing tax preferences for US oil and natural gas production: Measuring tax subsidies by an equivalent price impact approach. *Journal of the Association of Environmental and Resource Economists* 5(1):1–37.

Milbrandt, A., K. Coney, A. Badgett, and G.T. Beckham. 2022. Quantification and evaluation of plastic waste in the United States. *Resources, Conservation and Recycling* 83. https://doi.org/10.1016/j.resconrec.2022.106363.

Milbrandt, A., J. Zuboy, K. Coney, and A. Badgett. 2024. Paper and cardboard waste in the United States: Geographic, market, and energy assessment. *Waste Management Bulletin* 2.1:21–28.

Miller, C. 2024. Presentation to the Committee on Costs and Approaches for Municipal Solid Waste Recycling Programs. National Academies of Sciences, Engineering, and Medicine.

Moore, J., D. Folkinshteyn, and J.P. Howell. 2022. The potential for exchange-traded futures on recycled materials to improve recycling efficiency. *Investment Management and Financial Innovations* 19(3):93–104. https://doi.org/10.21511/imfi.19(3).2022.09.

NAPCOR (National Association for PET Container Resources). 2023. 2023 US PET Bottle Recycling Rate Reaches Highest Level in Decades; Recycled Pet Content in US Bottles Reaches Highest Level Ever. https://napcor.com/news/2023-pet-bottle-recycling-reach-new-heights.

NASEM (National Academies of Sciences, Engineering, and Medicine). 2023. *Recycled plastics in infrastructure: Current practices, understanding, and opportunities*. Washington, DC: The National Academies Press. https://doi.org/10.17226/27172.

National Solid Wastes Management Association. 1991. *Recycling in the States: 1990 Review*. https://p2infohouse.org/ref/30/29711.pdf.

NERC (Northeast Recycling Council). 2021. *Domestic Recycled Paper Capacity Increases*. https://nerc.org/documents/Summary%20of%20Announced%20Increased%20Capacity%20to%20use%20Recycled%20Paper%20May%202021.pdf.

NERC. 2023. *Recycled glass used as alternative daily cover in the Northeast US and Quebec*. https://nerc.org/documents/Glass/ADC%20Report%20Final.pdf.

NERC. 2024. *Northeast Materials Recovery Facilities (MRFs) Commodity Values Report*. https://www.nerc.org/quarterly-mrf-reports.

ODEQ (Oregon Department of Environmental Quality). 2024. *2022 Oregon Material Recovery and Waste Generation Rates Report*. https://www.oregon.gov/deq/recycling/Documents/2022MRWGRatesReport.pdf.

OECD (Organisation for Economic Co-operation and Development). 2022. *Plastic pollution is growing relentlessly as waste management and recycling fall short, says OECD*. https://www.oecd.org/en/about/news/press-releases/2022/02/plastic-pollution-is-growing-relentlessly-as-waste-management-and-recycling-fall short.html.

OECD. 2024. *Data Explorer: Environment and Climate Change Technology and Innovation*. https://data-explorer.oecd.org.

Olawade, D.B., O.Z. Wada, O.T. Ore, A. Clement David-Olawade, D.T. Esan, B.I. Egbewole, and J. Ling. 2024. Trends of solid waste generation during COVID-19 Pandemic: A review. *Waste Management Bulletin* 1(4):93–103. https://doi.org/10.1016/j.wmb.2023.10.002.

Palmer, K., and M. Walls. 1997. Optimal policies for solid waste disposal: Taxes, subsidies, and standards. *Journal of Public Economics* 65(2):193–205. https://doi.org/10.1016/S0047-2727(97)00028-5.

Plastics Europe. 2021. Plastics – the Facts 2021: An Analysis of European Plastics Production, Demand and Waste Data. Pub. L. 117-169.

RCC (Retail Compliance Center). 2023. *Mandatory Recycled Content Laws for Packaging*. https://www.rila.org/getmedia/f4b51c8f-80c7-46c0-932b-6a40f8fe079c/Mandatory-Recycled-Content-Laws-for-Packaging-11-23.pdf?ext=.pdf.

Reinhart, D., J. Chen, and M. Golgoz. 2023. Life-Cycle Assessments of Residential Curbside Collection Programs for Recycling Quality Improvement. Final report to the Hinkley Center for Solid and Hazardous Waste Management.

Resource Recycling. 2022. *From Green Fence to Red Alert: A China Timeline*. https://resource-recycling.com/recycling/2018/02/13/green-fence-red-alert-china-timeline.

Runsewe, T., O. Bafail, and N. Celik. 2020. Performance Analysis of Waste Collection Programs in Material Recovery Facilities. IIE Annual Conference Proceedings 1–6. https://www.proquest.com/scholarly-journals/performance-analysis-waste-collection-programs/docview/2522430103/se-2.

Sigman, H. 1995. A comparison of public policies for lead recycling. *Rand Journal of Economics* 26:452–478.

Sigman, H., and R. Strow. 2024. China's waste import ban and US solid waste management: Effects of the loss of a waste haven. *Journal of the Association of Environmental and Resource Economists* 11(6):1527–1557. https://doi.org/10.1086/729899.

SP Global Insights. 2024. Mexico joins plastic sustainability battleground in US West Coast. https://www.spglobal.com/commodity-insights/en/news-research/blog/chemicals/072624-mexico-joins-plastic-sustainability-battleground-in-us-west-coast.

Staub, C. 2024. RPET imports driving 'perfect storm' for reclaimers. Resource Recycling, March 26, https://resource-recycling.com/plastics/2024/03/26/rpet-imports-driving-perfect-storm-for-reclaimers.

Taylor, R., H. Williams, and S. Zhang. 2025. *Plastic Waste Trade & Coastal Litter: Evidence from Citizen Science Data*. https://ssrn.com/abstract=4918045.

The Aluminum Association. 2024. *Circular Economy Solution*. https://www.aluminum.org/Recycling.

Timpane. 2024. Presentation to the Committee on Costs and Approaches for Municipal Solid Waste Recycling Programs on April 16, 2024. National Academies of Sciences, Engineering, and Medicine.

UNEP (United Nations Environment Programme). 2021. *Turning Off the Tap: How the World Can End Plastic Pollution and Create a Circular Economy*. United Nations Environment Programme.

Wayman, E. 2023. *Recycling rare-earth elements is hard—but worth it*. https://www.snexplores.org/article/recycling-rare-earth-elements-hard-reuse-greener-technology.

Yang, T., and C.S. Fulton. 2016. The Case for U.S. Ratification of the Basel Convention on Hazardous Waste (February 28, 2016). Santa Clara Univ. Legal Studies Research Paper No. 1-15. http://dx.doi.org/10.2139/ssrn.2688173.

Washington State Department of Ecology. 2023. *NextCycle Washington – Supporting a Circular Economy*. https://ecology.wa.gov.

Washington State Recycling Association. 2022. *Statewide Initiatives in Recycling Market Development: Washington's Approach*. https://wsra.net.

6
Behavioral Considerations and Social Impacts of Recycling Programs

---

Summary of Key Messages

- **Barriers to recycling:** While consumer surveys consistently find high support among respondents for recycling and its programs, they also highlight barriers—mainly a lack of convenience and confusion over what materials can be recycled.
- **Successful recycling programs:** To be successful, residential recycling services must be convenient, easy to use, and low in cost to residents and business owners. These factors alone, however, are not always sufficient to increase recycling.
- **Labeling:** Inconsistent and misleading packaging labels, including the use of the chasing arrows symbol and resin identification codes, are significant causes of consumer uncertainty and misunderstanding.
- **Predictors of recycling behavior:** Studies indicate that the greatest predictors of recycling participation are residence type and income.
- **Targeted interventions:** Behavioral interventions for promoting recycling and decreasing contamination are most effective when they target a specific barrier to recycling for a given population of consumers.
- **Policy support:** Evidence suggests that consumers and voters would support policies that place greater responsibility for recycling onto product manufacturers.
- **Research needs:** More regular collection and reporting of direct observations of household and commercial behavior related to recycling are needed to support recycling policy decision-making. As one example, new and more rigorously collected data on household time costs are needed to perform recycling cost-benefit analyses more accurately.

This chapter explores the science and policies relevant to household attitudes and behaviors around recycling. It also explores the social impacts of recycling. Understanding how and why individuals engage in recycling practices is crucial for designing policies that effectively increase participation rates and improve recycling outcomes. Household recycling behavior is shaped by various factors, including the availability and accessibility of recycling programs, convenience, public awareness, and economic incentives. Additionally, the presence of social norms and community engagement can further influence participation in recycling efforts.

6.1 RESIDENTIAL PROGRAM AVAILABILITY

Household recycling behavior is influenced by a combination of structural and psychological factors, including program availability, ease of access, education, and personal motivation. While many residents express strong support for recycling, participation rates often lag behind access due to barriers such as unclear recycling guidelines, a lack of convenient options, and the perceived effort required. This section explores the availability of residential recycling programs, a contributing factor in household recycling decisions.

6.1.1 Availability of Residential Recycling Programs

As discussed in Chapter 2, U.S. residents can recycle their household materials in two ways (Sustainable Packaging Coalition, 2021):

1. Curbside recycling collection at their home by public or private service providers, either automatically or on an opt-in basis or subscription basis.
2. A publicly or privately operated drop-off recycling location within their municipality.

Alternatively, households can dispose of recyclable materials in general trash bins intended for the landfill or incinerator. Intermediate diverters (sometimes referred to as scavengers) may collect recyclable material from curbside recycling or trash containers and drop it off at recycling centers to receive payment for those materials. Scavengers' role in diverting material from the waste stream to the recycling stream is important, because otherwise those materials would have ended up in landfills or incinerators. Finally, materials recovery facilities (MRFs) obtain recyclable materials from curbside haulers or drop-off centers (see Figure 6-1).

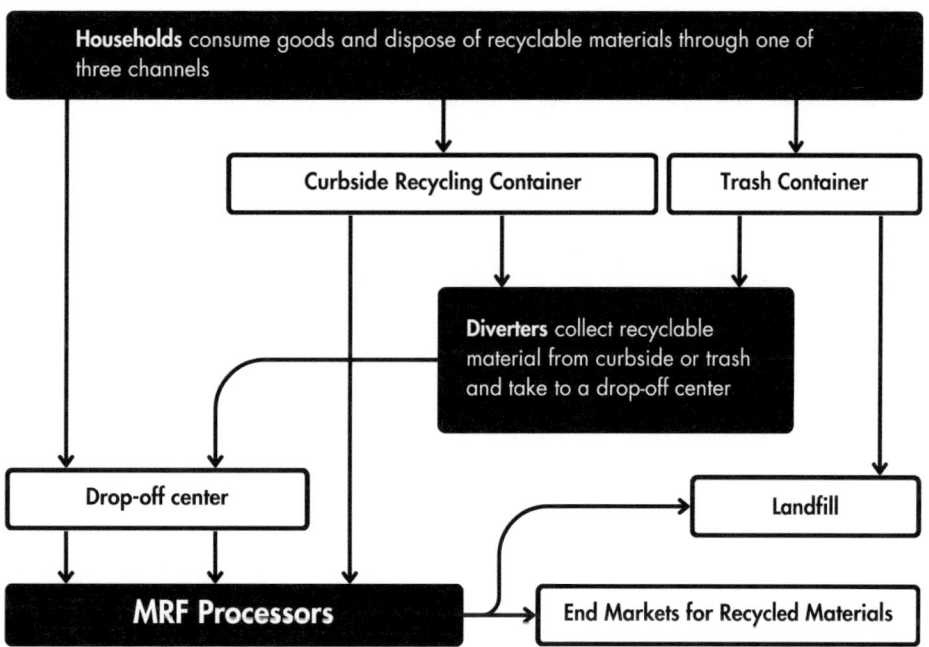

FIGURE 6-1 Recyclable materials disposal channels.
NOTE: MRF = materials recovery facility.

The availability of these residential recycling programs varies substantially across the United States, which becomes important to consider in the attitudes of individuals toward recycling, as discussed in the next section. A study by the Sustainable Packaging Coalition (2021) on the availability of recycling programs in 1,950 U.S. communities found that approximately 34 percent of the sampled population had only curbside collection available, 25 percent had both curbside programs and drop-off programs available, 32 percent had only drop-off programs available, and 9 percent had no recycling program available.[1] Recycling program availability was found to be strongly correlated with housing type: only about 3 percent

---

[1] The Sustainable Packaging Coalition (2021) defines the availability of recycling as a resident having access to either one or more of consumer recycling services at their place of residence.

of respondents living in single-family homes reported no access to recycling programs, and nearly one in four respondents who live in multifamily housing reported no access to recycling programs (Sustainable Packaging Coalition, 2021).

Other organizations estimate lower rates of recycling access. For instance, The Recycling Partnership (2024) estimates recycling access to be only 73 percent across all households in the United States. However, this lower rate partly reflects the organization's definition of *access*, which does not count access to deposit-return systems. Yet, similar to other studies, The Recycling Partnership (2024) also found differences by housing type, where access for single-family households is 85 percent and for multifamily households it is 37 percent. Additionally, while 73 percent of households have access, only 59 percent of those who have access participate in recycling (43 percent of all households participate) (The Recycling Partnership, 2024).

Availability of recycling also varies by type of product. While metal, paper, and plastics are commonly assumed to be recyclable, what can actually be recycled depends on the characteristics of products and not just their material. The Sustainable Packaging Coalition (2021) assessed the availability of residential recycling programs by product type. Products with the highest recycling availability included aluminum beverage cans (89 percent), corrugated boxes (88 percent), steel food cans (87 percent), polyethylene terephthalate (PET) bottles (87 percent), high-density polyethylene (HDPE) bottles (87 percent), and paperboard boxes (84 percent). Materials with the lowest recycling availability included aluminum foil and foil packaging (37 percent); rigid polystyrene packaging (45 percent); PET cups (52 percent); and PET clamshells, tubs, and trays (54 percent).

Figure 6-2 shows substantial heterogeneity in recycling participation by state, with participation rates lagging access rates in all states. The *recycling participation rate* is the percentage of all households that participate in recycling; analogous rates can be defined for composting, incineration, and landfill (The Recycling Partnership, 2024). The lowest recycling participation rates are 20 percent (Mississippi) and 21 percent (Louisiana); the highest are 62 percent (California) and 59 percent (Oregon), and the national average recycling participation rate is 43 percent (The Recycling Partnership, 2024). The fraction of municipal solid waste (MSW) that gets landfilled follows similar patterns to the recycling rate. For instance, the South-Central region has the highest landfilling rate, 83 percent, whereas the Northeast has the lowest landfilling rate, 47 percent (The Recycling Partnership, 2024).

Figure 6-3 depicts the fate of recyclable materials in the residential recycling process in number of tons. The materials are differentiated by the amount recycled, lost at the MRF, and lost to trash. On average, 21 percent (10 million tons) of residential recyclable materials are recycled, 3 percent are lost to processing at the MRF, and the remaining 76 percent of recyclable materials are thrown out by households as trash (The Recycling Partnership, 2024). However, as indicated by Figure 6-3, heterogeneity of recycling rates by material is considerable: the highest rates nationally are for cardboard (32 percent) and aluminum cans (30 percent), while the lowest rates are for film and flexible materials, bulky rigid plastics, and plastic types 3–7 (at or below 1 percent).

6.1.2 Availability of Residential Food Waste Programs

In the United States, the availability of residential food waste programs is much lower than the availability of recycling programs. A 2023 survey of cities, counties, and solid waste authorities estimated that 12 percent of U.S. households had access to residential food waste collection programs (BioCycle, 2023). However, the number of residential food waste programs has been growing over the past decade. A prior version of this survey found that less than 3 million U.S. households had access to residential food waste programs in 2013–2014; over the next 10 years, this number grew five times to roughly 15 million households (BioCycle, 2023; see Figure 6-4).

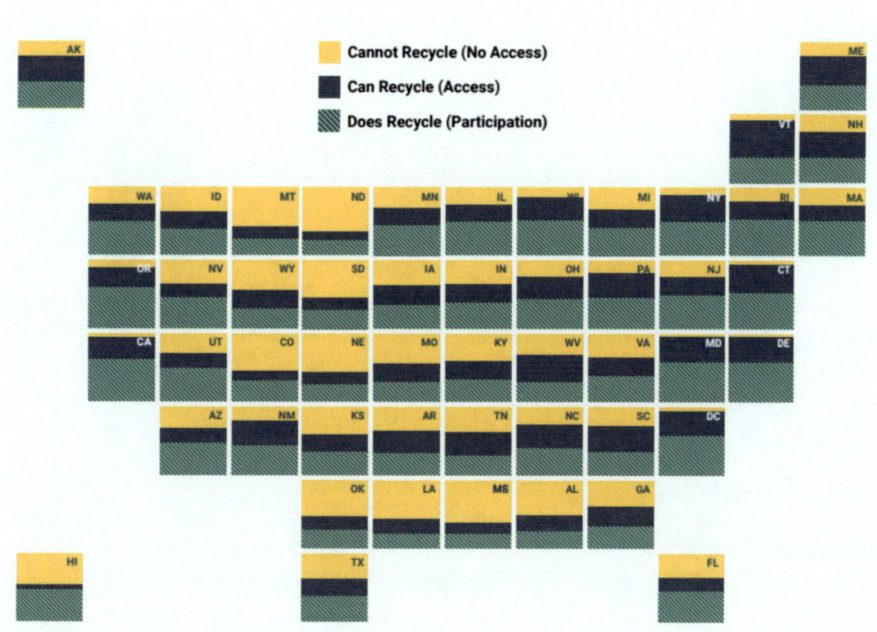

FIGURE 6-2 State-by-state levels of recycling access and participation.
NOTES: Access data do not include access to a deposit-return system. Curbside recycling participation data were derived from a community survey. Participation data for drop-off systems and on-property multifamily recycling were unavailable and presumed rates of 30 percent and 50 percent were used, respectively. A subscription uptake rate of 30 percent was assumed, based on engagement with communities and data from previously published studies.
SOURCE: The Recycling Partnership, 2024.

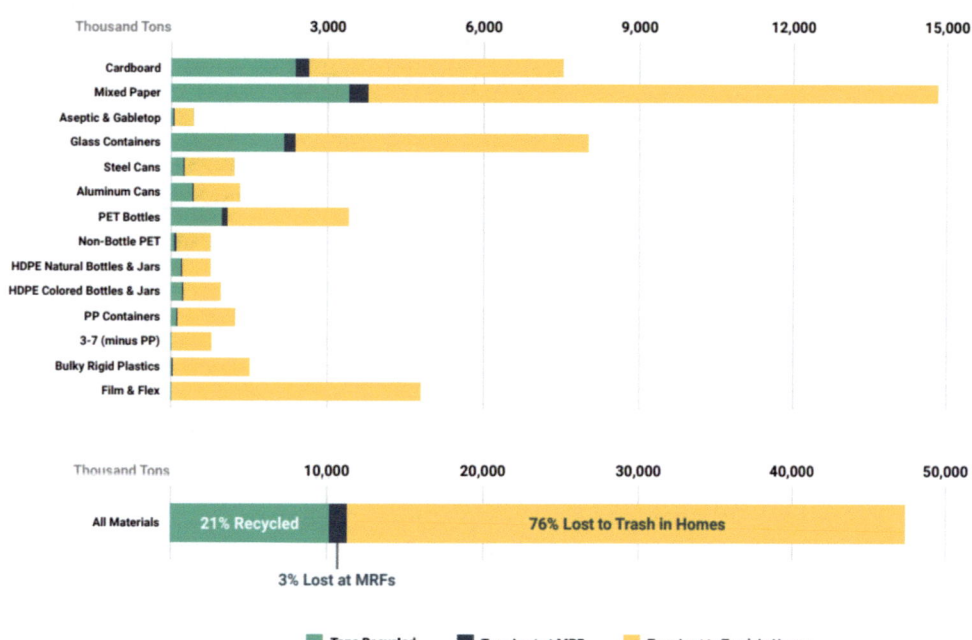

FIGURE 6-3 Fate of materials in residential recycling, tons per year.
NOTE: HDPE = high-density polyethylene; MRF = materials recovery facility; PET = polyethylene terephthalate; PP = polypropylene.
SOURCE: The Recycling Partnership, 2024.

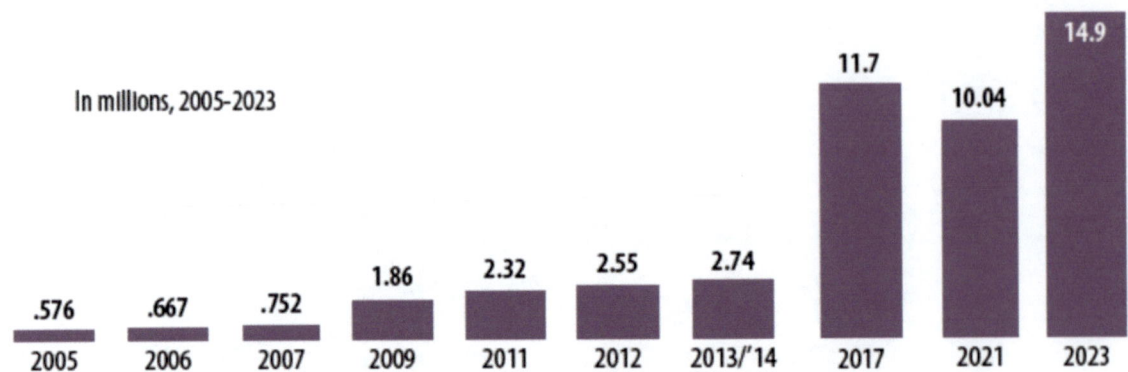

FIGURE 6-4 Number of U.S. households with food waste collection, 2005–2023.
SOURCE: BioCycle, 2023.

Most residential food waste programs are provided via curbside collection. Of the 400 programs tracked in BioCycle (2023), 230 offered curbside only, 139 offered drop-off only, and 31 offered both curbside and drop-off (see Figure 6-5).

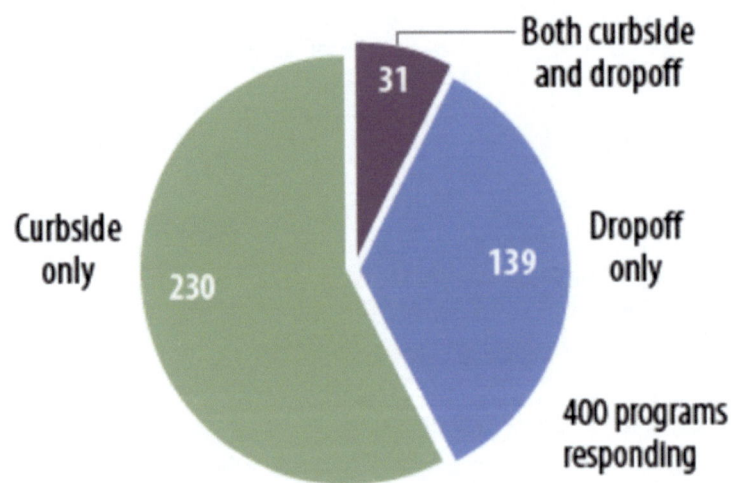

FIGURE 6-5 Food waste collection methods.
SOURCE: BioCycle, 2023.

BioCycle (2023) also describes the uneven spread of residential food waste programs across states. California has over 100 programs, and Illinois, New York, and Minnesota have over 40 programs each. However, 26 states have no residential food waste programs (see Figure 6-6).

6.2 ATTITUDES TOWARD AND BARRIERS TO RECYCLING

Much survey data and academic evidence have focused on recycling, and this section reviews those findings. However, less data and evidence are available on attitudes toward composting and anaerobic digestion options for residents to recycle organic waste (such as biomass, manure, leaf or yard waste, and food waste).

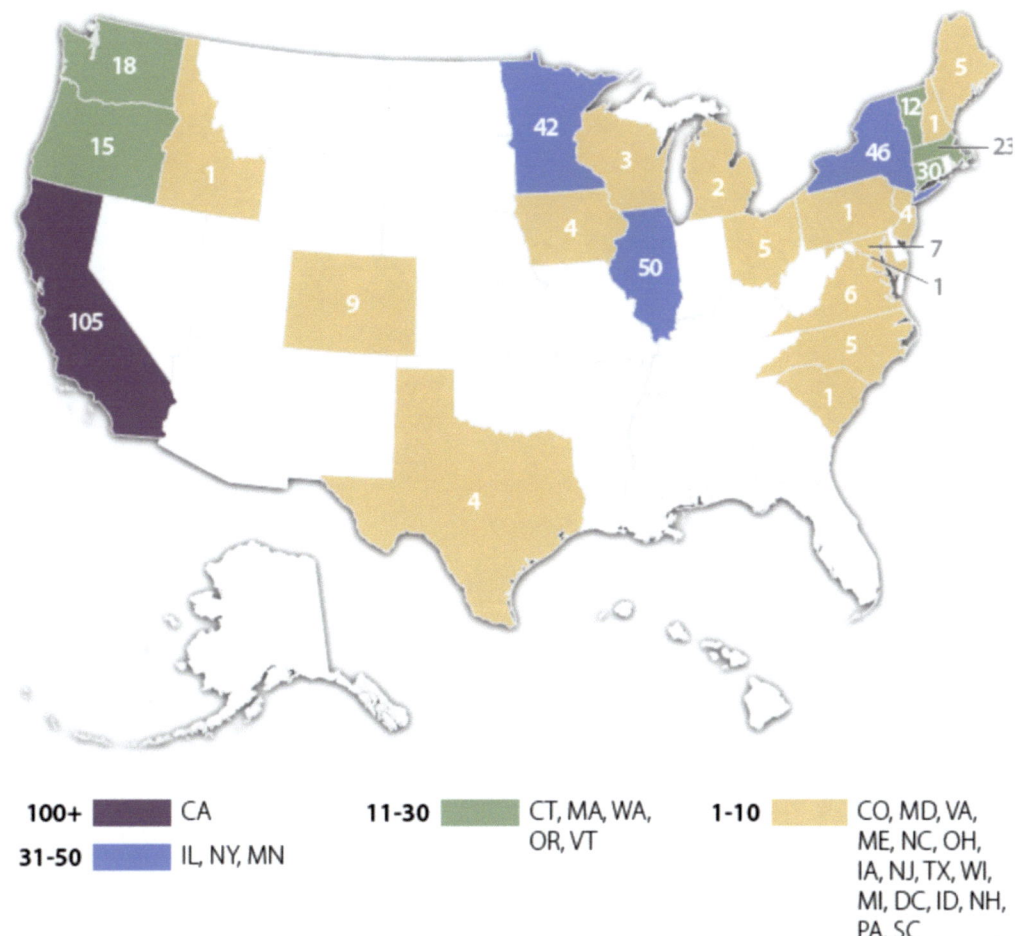

FIGURE 6-6 Residential food waste collection programs by state.
SOURCE: BioCycle, 2023.

6.2.1 Overview of Current Consumer Attitudes Toward Household Recycling

Polling companies, consulting firms, and nongovernmental organizations have conducted numerous representative consumer surveys to gauge individuals' attitudes and beliefs toward recycling. These surveys consistently find high support among respondents for recycling and recycling programs. However, they also highlight barriers to recycling, such inconvenience and confusion over what materials can be recycled.

The World Economic Forum, SAP, and Qualtrics conducted a joint survey in 2021 that questioned people from different parts of the globe about their attitudes toward recycling, among other sustainability topics (World Economic Forum, 2021). This survey collected 11,686 responses across 28 countries. Globally, 84 percent of respondents reported that it is "extremely or very important for them to personally recycle when they can." Across the eight regions surveyed, responses were relatively consistent, varying between 74 percent in East Asia and the Pacific and 93 percent in Latin America and Caribbean (World Economic Forum, 2021; see Figure 6-7). North America falls in the middle, at 80 percent. In North America, the top barriers respondents reported that kept them from recycling more were "lack of programs/services to enable recycling" (with 30 percent reporting this barrier) and "inconvenience of recycling" (with 28 percent reporting this barrier). Approximately 50 percent of respondents in North America also reported that they would be willing to avoid products that are hard to recycle (World Economic Forum, 2021).

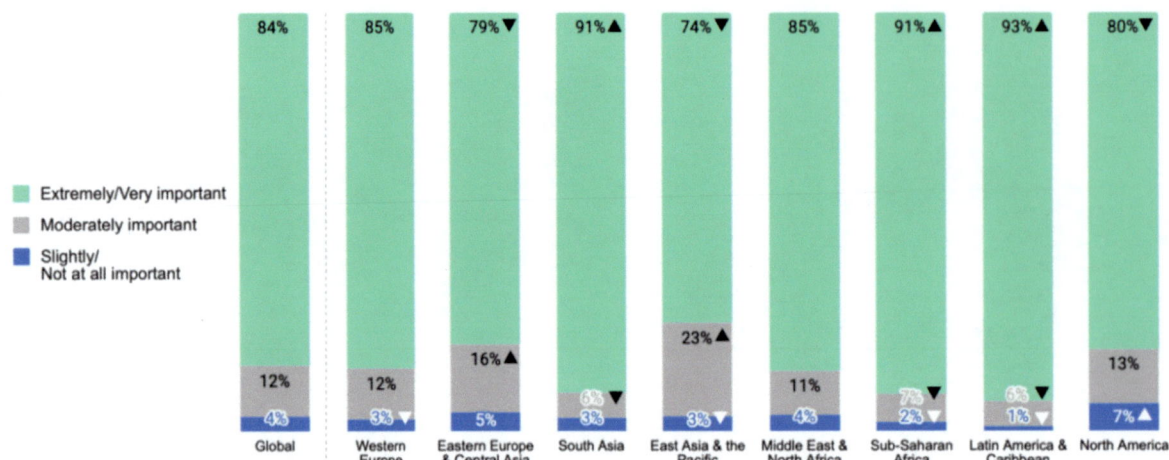

FIGURE 6-7 Importance of recycling.
SOURCE: World Economic Forum, 2021. CC BY-NC-ND 4.0.

A 2014 survey conducted by Harris Poll on behalf of Recycled Materials Association surveyed 2,013 U.S. adults about their recycling attitudes and opinions; 68 percent of respondents reported that "recycling is the right thing to do" (Harris Poll, 2014). Reasons chosen for recycling included "it's the socially responsible thing to do" (55 percent), it is critical to reduce landfill space" (54 percent), "it is critical to conserve natural resources" (49 percent), and "it's critical to reduce energy consumption" (40 percent). This survey also found that the vast majority of respondents say they have recycled, with 43 percent saying they always recycle, 26 percent saying they often recycle, 24 percent saying they sometimes recycle, and only 6 percent saying they never recycle (Harris Poll, 2014).

6.2.2 Barriers to Household Recycling

While surveys reveal broad support for recycling in principle, they also highlight the difficulties people have with recycling in practice. A survey of 1,000 U.S. adults, conducted by Reputation Leaders (2023), found that 43 percent of respondents throw away most items instead of recycling them. Approximately one-third of all respondents reported that the biggest challenge is that they are not always sure which packaging can and cannot be recycled. For those reporting that they do not recycle, other barriers reported were inconvenience (22 percent), having to use a separate bin for recycling (20 percent), and not believing that recycling will make a difference or that the benefits of recycling do not outweigh the costs (17 percent) (Reputation Leaders, 2023). Reschovsky and Stone (1994) found that household storage space matters, with households much more likely to report that they recycle if they have adequate storage space, though this may be influenced by the availability of curbside recycling as opposed to use of drop-off centers.

In terms of confusion over which products are recyclable, according to a survey by McKinsey & Company (2020), two-thirds of respondents are not confident that they know which products are recyclable, and 17 percent find it difficult to know if something is recyclable. A separate survey conducted by Corona Insights (2020) on behalf of the World Wildlife Fund surveyed 1,098 adults using the nationally representative AmeriSpeak panel. This survey found that availability and uncertainty of what is accepted were the top reasons for not recycling more household plastic waste. About 14 percent of respondents reported no access to recycling, with nonmetro communities being more likely not to have access (Corona Insights, 2020; see Figure 6-8).

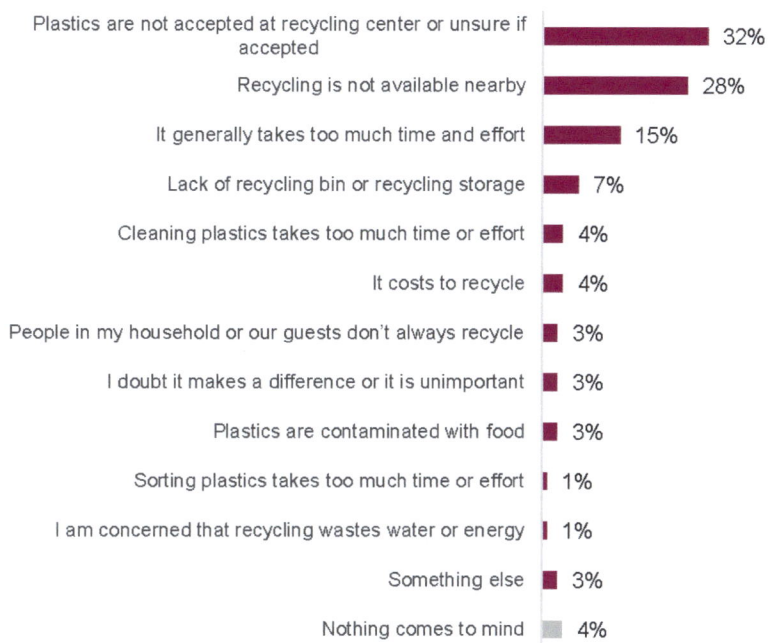

FIGURE 6-8 Barriers to recycling more household plastic waste.
NOTES: Results are coded from open-ended responses.
SOURCE: Corona Insights, 2020.

Since one of the most common barriers to recycling reported by U.S. households is time costs, studies have estimated the time it takes for households to sort and dispose of their recycling. Jakus and colleagues (1996) used survey data (of self-reported times to recycle one unit of a material) to estimate an average of 36.38 (range: 3.67–102.4) seconds per unit to recycle newspaper and 53.75 (range: 6.64–154.1) seconds per unit to recycle glass at drop-off centers. For the weekly time requirement per household for the activities associated with preparing recyclables for curbside collection, the U.S. Environmental Protection Agency (EPA, 1974) calculated 15.9 minutes per week, while Schaumberg and Doyle (1994) assumed households take 5–15 minutes per week. For translating time costs for nonwork hours into dollar values, a common practice is to use half the U.S. average hourly wage (Small, 2013), which was $35.07 × 0.5 = $17.54 per hour in July 2024 (U.S. Bureau of Labor Statistics, 2024). Alternatively, the federal minimum wage of $7.25 per hour could be used to value time. Table 6-1 shows how the time costs of recycling would vary under different scenarios, based on household data from the U.S. Census Bureau (2024).

TABLE 6-1 Time Costs of Recycling Under Different Scenarios

Household Time Spent Recycling	Household Time Cost, Using Federal Minimum Wage	Household Time Cost, Using Half Average Hourly Wage	Time Cost for All U.S. Households, Using Federal Minimum Wage	Time Cost for All U.S. Households, Using Half Average Hourly Wage
5 minutes/week	$2.62/month	$6.33/month	$4.12 billion/year	$9.97 billion/year
15 minutes/week	$7.85/month	$18.98/month	$12.37 billion/year	$29.91 billion/year

NOTE: Total U.S. households was 131,332,360 according to the latest American Community Survey 1-year estimates by the U.S. Census Bureau (https://data.census.gov/profile/United_States?g=010XX00US).
SOURCE: U.S. Census Bureau, 2024.

However, the assumption that households spend 5–15 minutes per week on recycling may be an underestimate. An informal poll conducted online by Earth911 (2018) found that 27 percent of respondents spend 10–30 minutes on recycling, 25 percent spend 30 minutes to 1 hour, and 25 percent spend more than 1 hour, suggesting the values reported in Table 6-1 are lower than current time costs. New and more rigorously collected data on household time and costs are needed, both on average and by household characteristics.

Some surveys on recycling have focused on product labels. A June 2021 survey of 1,300 U.S. consumers conducted by The Recycling Partnership (2023) asked consumers about their use of and attitudes toward recycling information on products and product labels. Consumers were found to rely heavily on product recycling labels, with 78 percent of consumers reporting that they look at recycling information on products to try to sort them correctly. Approximately 82 percent of consumers trust the recycling information found on products to be accurate, and 71 percent feel deceived and discouraged when a product claims to be recyclable when it is not. Among consumers that look at product recycling labels, 63 percent report still being confused about whether an item is recyclable (The Recycling Partnership, 2023).

Surveys have also asked respondents what actions they would be willing to take to improve recycling rates and what government policies they would support. The 2014 Harris Poll survey on behalf of Recycled Materials Association found that respondents, on average, would be willing to spend 13 percent more for a product if they knew it was easy and convenient to recycle (either via curbside collection or drop-off center), and 55 percent said they would be willing to spend more for a product if they knew it was made of recycled materials (Harris Poll, 2014). Among the respondents, 90 percent agreed that "recycling collection sites need to be more readily accessible to consumers." And 68 percent of respondents agreed that "manufacturers and/or retailers should pay for recycling programs when they are not already available to consumers" (Harris Poll, 2014).

Other studies have found that, in the absence of information on environmental impact, consumers are willing to pay less for products made of recycled and remanufactured material compared with products made of new material, because they believe that recycled materials are lower in quality (Michaud and Llerena, 2011; Pretner et al., 2021). Once informed of the environmental benefits, consumers no longer have a lower willingness to pay for products made of recycled materials.

Similarly, a Corona Insights (2020) survey found that 88 percent of respondents agreed that "the recycling system for plastic in the United States needs improvement." Most respondents (67 percent) believed that individuals that use plastic, businesses that produce/sell plastic, and government share responsibility for reducing plastic waste. However, when asked to rank these groups' responsibility, businesses were ranked as the most responsible (Corona Insights, 2020). The survey also revealed that respondents slightly preferred subsidies for reusing and recycling items over banning nonrecyclable items. Taxes and fees for using nonrecyclable plastics were preferred the least (Corona Insights, 2020; see Figure 6-9).

6.2.3 Heterogeneity in Consumer Attitudes and Barriers

Consumer surveys reveal diverse beliefs, barriers, and behaviors around recycling at the household and individual levels. Researchers have examined how these beliefs vary across sociodemographic, geographic, and psychographic characteristics.

With respect to sociodemographic characteristics, Huber and colleagues (2023) analyzed annual survey data from 2005 to 2014 for 145,000 participants in the Knowledge Networks-GfK Knowledge Panel. The authors found that the individual characteristic that is most predictive of recycling support is educational attainment—individuals with college degrees are more likely to recycle than those with no college degree. Huber and colleagues (2023) also find that recycling rates grow with age and are higher among individuals who are White, who vote Democrat, and who are women. In terms of household characteristics, the authors found that the most important household characteristic is home type, with households living in a single-family house having higher recycling rates than those living in other types of dwellings. Recycling also increases with household income (Huber et al., 2023). To a smaller degree, higher recycling rates are associated with owning a home, being married, and not having children under the age of 6 years. These findings have implications for targeted recycling programs and support (e.g., in counties with lower household incomes). Huber and colleagues' (2023) findings are supported by other studies, as compiled by Shaw and colleagues (2014):

Individual characteristics that increase likelihood of recycling:
- Women (Ando and Gosselin, 2005; Barr, 2007; Harris Poll, 2014; Oates and McDonald, 2006)
- Older adults (Barr, 2007; Harris Poll, 2014; Jenkins et al., 2003; Nixon and Saphores, 2009; Sidique et al., 2010)
- Larger households (e.g., Ando and Gosselin, 2005; Nixon and Saphores, 2009)
- Higher education (Owens et al., 2000; Sidique et al., 2010)
- White (Johnson et al., 2004)

Individual characteristics that decrease likelihood of recycling:
- African Americans (Johnson et al., 2004; Nixon and Saphores, 2009)
- Foreign-born Latinos (Johnson et al., 2004)
- Renters (Nixon and Saphores, 2009; Owens et al., 2000)

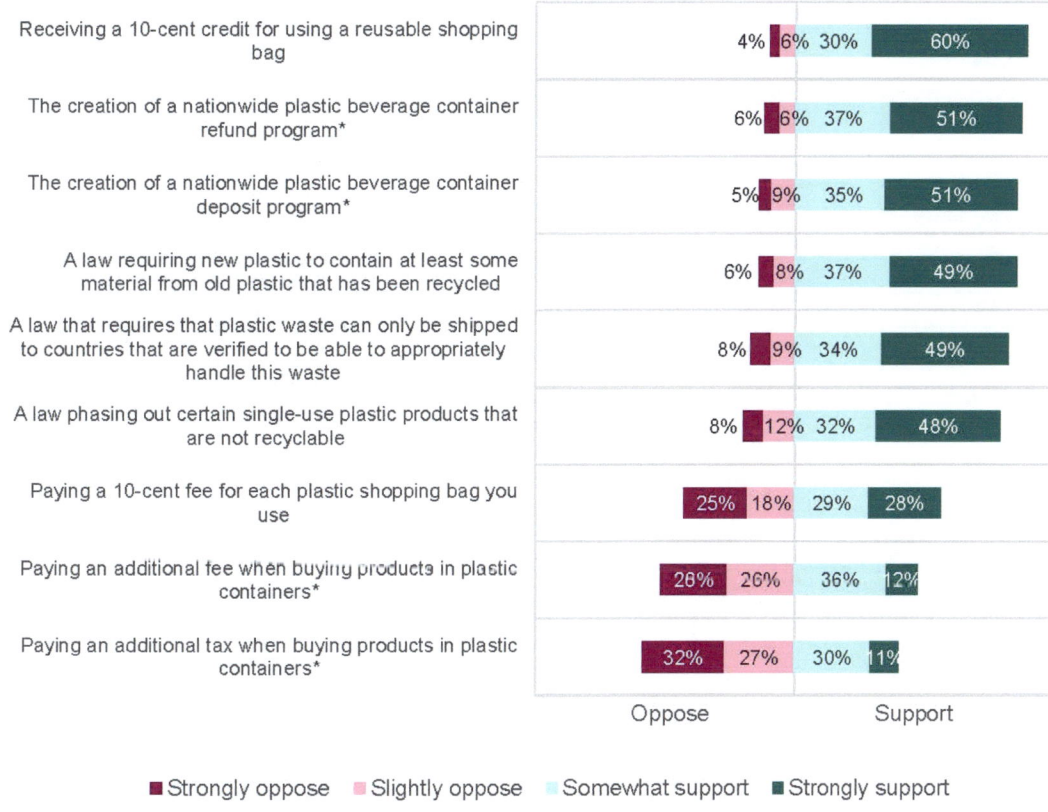

*Paired items were each shown to one-half of respondents in order to compare different wordings.

FIGURE 6-9 Support for regulations.
SOURCE: Corona Insights, 2020.

With respect to geographic characteristics, recycling participation is higher in Pacific Coast and Northeast states than in the South and Midwest states (see Figure 6-2). Recycling rates also vary by county- and state-level characteristics. Recycling rates increase with a county's median income, population, percent White, and population density (Huber et al., 2023). At the state level, recycling rates increase with tipping fees per ton, the presence of container deposit-return and mandatory recycling laws, and state spending per person (Huber et al., 2023).

With respect to psychographics, The Recycling Partnership (2022) conducted a qualitative research study—with in-depth interviews of 24 participants and an online survey of 2,506 adult participants—to identify distinct types of recyclers or audience segments. They identified five types of recyclers based on their attitudes and beliefs and the barriers they face with recycling (see Table 6-2).

TABLE 6-2 Five Types of Recyclers

Type of Recycler	Percentage	Description
Eco activators	25%	"They care about the environment, research and share how to recycle with others, and have overcome inconveniences to do so. They are hopeful and take pride in their local program but ideally, they want greater systemwide investment, and clear information. They are social and civic minded and encourage others to be as well."
Committed followers	24%	"This straight-forward group feels good about doing their duty for their community, especially since it helps reduce waste and is easy for them to do. They are confident they know exactly what to do and feel they need little additional support but would benefit from reassurance. They are civic minded but private."
Discouraged self-doubters	18%	"They're not confident in their knowledge of recycling and feel negatively toward it, possibly due to key obstacles like confusion about what and how to recycle, which may be helped with frequent reminders. They lack confidence, clarity, and an understanding of why recycling matters."
Detached abiders	16%	"They recycle because it's convenient and, in many cases, because it is required or feels like it is. They're not confident in what to do and their obstacles lead to negativity. It is as if they are keeping score. They want to know how recycling benefits them and their community and don't want to make an extra effort."
Conflicted and overwhelmed	16%	"This diverse segment agrees that recycling has some benefits, but they experience many obstacles. In the context of their busy life, recycling feels important conceptually, but the practical steps required to do it fall short of being worthwhile. This conflict leaves them frustrated and worried. They feel judged whether they do or don't recycle. They need more information and support on specific 'to-dos' from their municipality. It needs to be easier, and they need to see others doing it as well."

SOURCE: The Recycling Partnership, 2022.

Geiger and colleagues (2019) performed a meta-analysis of 91 studies on individual and household recycling and classified the most robust predictors of recycling across studies. The authors found that behavior-specific factors (e.g., past recycling behaviors and personal norms toward recycling) were better predictors of recycling than general factors (such as general knowledge about environmental concerns and general environmental attitudes) (Geiger et al., 2019):

Factors predictive of recycling behavior:
- past recycling behaviors—whether a person has recycled in the past
- personal norms toward recycling—feelings of moral obligation to engage in recycling
- perceived behavioral control over recycling—degree to which an individual perceives themself as being able to engage in a certain behavior
- recycling self-identity—degree to which a person sees themself as a person who recycles their waste
- descriptive norms to engage in recycling—extent to which people think other people recycle their waste
- attitudes toward recycling—extent to which people evaluate recycling favorably

Factors less predictive of recycling behavior:
- general knowledge about environmental problems—extent to which people know about the causes and consequences of environmental problems, or know which behaviors cause such problems
- general environmental attitudes—extent to which an individual is concerned about the environment in general
- general personal norm—feelings of moral obligation to engage in pro-environmental behavior generally

Geiger and colleagues (2019) also considered contextual factors such as home ownership, housing type, number of recycling facilities in the neighborhood, possession of a recycling bin at home, distance to a drop-off recycling location, and the size of the neighborhood. The authors found that possessing a recycling bin at home was a strong predictor of recycling.

Similarly, in an earlier meta-analysis of 63 studies, Miafodzyeva and Brandt (2013) found that personal norms toward recycling (i.e., moral concern), convenience (i.e., how easy it is to understand and use a recycling program), and information (i.e., an individual's specific recycling knowledge) are strong predictors of recycling.

Jacobsen and colleagues (2022) conducted a systematic review of empirical research published between 2015 and 2020 on plastic packaging waste recycling. They classified drivers and barriers to recycling as (1) consumer motivation, (2) consumer ability, and (3) consumer opportunity. Jacobsen and colleagues (2022) summarized their systematic review as follows:

> Overall, research suggests that consumers' PPW [plastic packaging waste] recycling is driven by their environmental concern and that environmentally related messages can increase their motivation to recycle. Research on consumers' ability to recycle mainly uncovered a negative impact of insufficient consumer knowledge on how to source-separate correctly and also found that recycling can be increased through better communication. The research on consumers' opportunity to recycle find that the design of the waste sorting system and its built-in convenience (in terms of time and effort to recycle) are strong determinants of consumers' PPW recycling. (p. 73)

Multiple studies in this systematic review found that differences in the waste collection system account for a substantial portion of differences in recycling rates across municipalities: recycling rates are higher with curbside collection than drop-off centers (Hage et al., 2018; Hahladakis et al., 2018; van Velzen et al., 2019), and recycling rates are higher with greater density of drop-off centers (Hage et al., 2018; Oliveira et al., 2018).

6.2.4 Attitudes of Local Government About Recycling

Fewer academic studies and opinion surveys have been conducted on the attitudes and beliefs of local government leaders about recycling. However, a handful of examples do exist, such as the Michigan

Public Policy Survey conducted by the Center for Local, State, and Urban Policy (Horner et al., 2022a,b, 2023). In Michigan, 86 percent of local leaders report that recycling is somewhat or very important to their community. Additionally, 67 percent of local leaders from jurisdictions with at least some recycling services are satisfied with the current recycling opportunities available, while only 24 percent of local leaders from communities with no access to recycling services are satisfied with their lack of services. Local leaders were more likely to say their community would choose lower taxes and fewer recycling services instead of higher taxes and more services, but many such leaders expressed uncertainty (Horner et al., 2022a,b, 2023; see Figure 6-10).

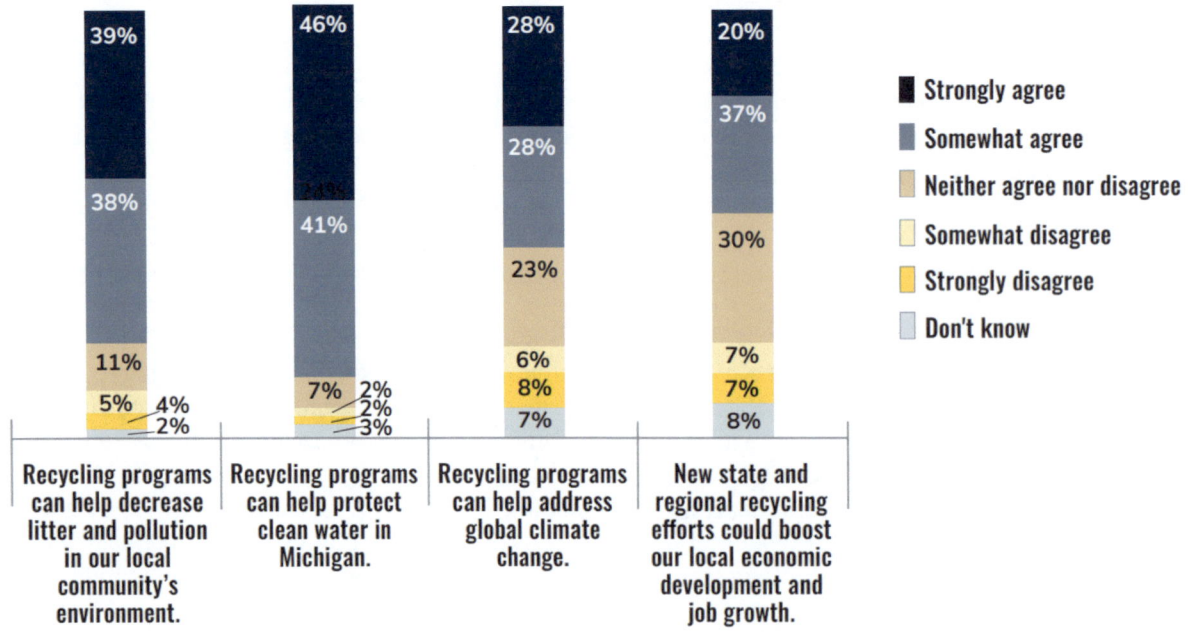

FIGURE 6-10 Local officials' agreement with statements regarding recycling benefits.
SOURCE: Horner et al., 2023.

6.2.5 Willingness to Pay for Recycling Programs

People may have different views regarding the benefits of a proposed program and its anticipated outcomes. In order to assess consumers' valuation of recycling programs, several studies estimate *willingness to pay*, or the maximum price a consumer is willing to pay for a product or service. Willingness to pay estimates, and their variations across groups, can inform decisions about what policies or projects to implement. Estimates of average household willingness to pay for recycling services vary from $1.47 to $28.35 per month (in 2017 USD; see Table 6-3).

The methodologies applied to generate the results in Table 6-3 have been used for many years by scientists, economists, and planners to inform decisions about public perspectives on the benefits and costs of the decision. These methodologies are often used to place a value on public goods, services, or programs for which market transactions are not available, and for which social preferences are expressed indirectly through political decisions. Examples include a proposed program's associated taxes, budgets, regulations, approvals, and mandates.

The methodologies chosen can result in nuanced results. For example, the Jakus et al. (1996) and Tiller et al. (1997) willingness to pay studies examined drop-off centers in Williamson County, Tennessee. This rural area has households dispersed across wide distances, making curbside recycling pick-up services prohibitively costly and unavailable. Similarly, garbage collection services are limited for some residents.

As a result, many households rely on "convenience centers" for both garbage disposal and recycling. Jakus and colleagues (1996) employed a two-stage model to estimate recycling demand while accounting for selection bias. They found an average willingness to pay of approximately $10.10 for drop-off recycling services. In contrast, Tiller et al. (1997) used a contingent valuation method (CVM) to assess consumers' hypothetical willingness to pay for additional drop-off centers. Their findings revealed an average willingness to pay ranging from $7.08 for nonrecyclers without curbside trash collection to $20.51 for recyclers with access to curbside trash pick-up.

TABLE 6-3 Literature on Willingness-to Pay Estimates

Authors	Year	Program Type	2017 USD/Month	N	Setting	Method
Jakus et al.	1996	drop-off	$10.00	284	Williamson County, TN	revealed preference, stated/observed behavior
Lake et al.	1996	curbside	$7.61	285	Hethersett, South Norfolk, UK	CVM-dichotomous choice
Tiller et al.	1997	drop-off	$7.08–$20.51	481	Williamson County, TN	CVM-dichotomous choice with follow-up
Aadland and Caplan	1999	curbside	$3.13	401	Ogden, UT	CVM-ordered interval choice
Kinnaman	2000	curbside	$12.04	100	Lewisburg, PA	CVM- dichotomous choice
Caplan et al.	2002	curbside and green waste pick-up	$9.17–$13.75	350	Ogden City, UT	CVM-contingent ranking
Aadland and Caplan	2003	curbside	$8.30–$9.69	1,000	Utah	CVM-double-bounded dic. choice and revealed pref. stated behavior
Blaine et al.	2005	curbside	$1.47–$3.20	2,000	Lake County, OH	CVM-single-bounded referendum and CVM-payment card
Aadland and Caplan	2006	curbside	$4.05–$7.64	4,000	40 Western U.S. cities	CVM-double-bounded dic. choice and revealed pref. stated behavior
Jamelske and Kipperberg	2006	upgrade to automatic/ single stream	$4.11–$4.13	301	Madison, WI	CVM-double-bounded dic. choice
Bohara et al.	2007	curbside	$7.47	400	Logan, UT	CVM-single-bounded dic. choice
Karousakis and Birol	2008	curbside	$5.82/material	188	London, UK	Stated preference choice experiment
Troske et al.	2009	curbside	$2.71	600	Lexington, KY	CVM-dichotomous choice
Gillespie and Bennett	2013	curbside (fortnight collection)	$11.33	600	Brisbane, AU	Stated preference choice modeling
Berck et al.	2017	drop-off	$28.35	1,005	California	Stated preference choice modeling

NOTE: CVM = contingent valuation method.

The use of CVM to calculate survey estimates of willingness to pay has generated controversy regarding theoretical issues of (a) how to represent the options and how project attributes can be substitutes for other commodities, (b) methodological issues of replication validity, (c) evidence that willingness-to-pay estimates are highly influenced by the options and scales used in the valuation questions, and (d) ethical issues regarding informed consent for subjects as to how their responses may or may not be used to support decisions. Examples of these concerns, along with counterarguments favoring the use of well-designed and well-implemented willingness-to-pay studies, are presented in Box 6-1. Many argue that CVM and willingness-to-pay surveys—when conducted with clear and simple options, replication, and pretesting of option descriptions—are the most effective way to gather estimates of economic benefits applicable across a broad sample of a population. As a result, government agencies often use these methods to develop and evaluate their management options.

BOX 6-1
Opposition and Support for Contingent Valuation and Willingness to Pay

The following arguments are findings and conclusions from various studies with arguments for or against contingency valuation methods and willingness to pay.

ARGUMENTS IN OPPOSITION

Willingness to pay fails to capture the complex interaction among economic and ethical attributes:

Among [contingent valuation studies] is evidence that modified lexicographic preferences, where the substitutability of environmental quality with other commodities is rejected, can be common. Human value formation with respect to the environment combines ethical and economic aspects in a more complex way than most economists have assumed. (Spash, 2000, p. 1433)

Lack of quality control, reporting standards, and validity tests:

The currently dominant survey protocols and practices are inadequate. This is most evident from robust findings that the valuations are heavily influenced by the response options or response scales used in the valuation questions. The widely accepted survey guidelines do not require the validity tests and reporting standards that would be needed to make the uncertainty of the results transparent. The increasing use of inadequate survey results by policymakers threatens to undermine trust in environmental valuation, environmental policies, and political institutions. (Schläpfer, 2021, p. 1)

Lack of informed consent by subjects regarding use of study results:

Agreeing to participate in a study eliciting environmental values means agreeing to abide by the commitment implied by any proposal that one accepts or rejects in it. That might mean anything from addressing the gist of an issue to expressing an explicit willingness to pay for an environmental change. By soliciting such participation, investigators promise to provide the information that participants need in order to evaluate the proposals being presented. This paper proposes a standard for providing such information that must be met in order to conduct valid and ethical value-elicitation studies. Namely, investigators must secure the informed consent of participants. (Fischhoff, 2000, p. 1439)

ARGUMENTS IN SUPPORT

Contingent valuation estimates are methodologically sound and fit for purpose:

Such surveys are a practical alternative approach for eliciting the value of public goods, including those with passive use considerations. . . . Although discussions of contingent valuation often focus on whether the method is sufficiently reliable for use in assessing natural resource damages in lawsuits, it is important to remember that most estimates from contingent valuation studies are used in benefit–cost assessments. I put forward an affirmative case for contingent valuation and address a number of the concerns that have arisen. (Carson, 2012, pp. 28–29)

continued

> **BOX 6-1** *continued*
>
> The method has undergone high-level review and approval, and is often used:
>
> The CVM [contingent valuation method] was debated in a symposium sponsored by the Exxon Corporation, which led to the creation of a government panel– established by the National Oceanic and Atmospheric Administration (NOAA) and chaired by two Nobel laureates in economics – to assess the scientific validity of the CVM. The NOAA panel offered its approval of CVM subject to a set of best-practice guidelines that influenced the development of the methodology. Nowadays, the CVM is a method widely used in academic research and in environmental valuation studies. (Ferreira and Marques, 2015, p. 112)
>
> The contingent valuation method is acceptable, but research is needed on ethics, protest behavior, reliability, and construct validity:
>
> The contingent valuation method (CV) has become a recognised tool for estimating monetary non-market values. Despite the pragmatic acceptance of CV in policy evaluation, the application of CV-based estimates in decision-making remains controversial, as critics argue that CV suffers of hypothetical bias and question its accuracy to reflect non-market values via willingness to pay (WTP) estimates. . . . Our findings suggest that WTP estimates are suitable to infer the economic value of environmental assets . . . capable of eliciting rational behaviour from individuals that is free of hypothetical bias. However, hypothetical bias might be lurking in a less researched area, that of market participation, which calls for a rigorous treatment of protest behaviour. (Perni et al., 2021, p. 1)

6.2.6 Approaches for Comparing Willingness to Pay and Recycling Costs

The analysis in this report is limited to a direct comparison of the range of costs and willingness-to-pay values reported in the literature for curbside programs.[2] As shown in Table 6-3, for the 11 curbside studies reviewed (spanning locations in the United States, the United Kingdom, and Australia), the estimated values of willingness to pay ranged from $1.47 to $28.35 per month per household (in 2017 USD). In comparison, our review of curbside cost estimates in Chapter 4 found costs in North Carolina municipalities ranging from $2.79 per month to $3.75 per month per household (in 2021 USD). These cost estimates overlap the range of reported willingness to pay in Table 6-3. Similarly, the average U.S. household time costs estimated in Table 6-1, ranging from $2.62/month to $18.90/month (in 2024 USD), are within the range of willingness-to-pay values. This comparison provides moderate (though not definitive) evidence that the elicited benefits in these cases generally exceed reported costs, though perhaps not by much. This finding for curbside recycling is similar to that of Aadland and Caplan (2006) who studied 40 western U.S. cities and found that the average unit cost and the average willingness to pay were nearly identical, though with a high degree of variability around each mean.

6.3 STRATEGIES FOR PROMOTING BEHAVIOR CHANGES

Municipalities and researchers have applied various psychological intervention strategies for promoting household recycling. In a meta-analysis of randomized field studies, Varotto and Spagnolli (2017) grouped 70 interventions into six types, based on previous work by Osbaldiston and Schott (2012, p. 272):

1. **Prompts and information:** providing information on recycling (factual, persuasive, or merely reminders) to targeted individuals to encourage recycling behavior. Information can be delivered face-to-face, though written information (in print or online) is more common as it can reach a considerable number of people with low effort and costs.

---

[2] The willingness-to-pay estimates capture the maximum price an individual is willing to pay for a certain recycling feature, the perceived benefits, and willingness to pay. It does not include external benefits.

2. **Feedback:** providing either individuals or groups with information regarding their recycling behavior along with a comparison to a predefined standard, to show the difference between the standard and their current recycling behavior.
3. **Commitment:** asking individuals to commit to produce a certain behavior or reach a certain goal.
4. **Incentives:** any kind of benefit received by consumers as a result of their participation in a recycling program (e.g., monetary rewards, refund and unit pricing programs, gifts, prizes, lottery tickets, discount coupons).
5. **Environmental alterations:** making recycling more convenient and easier to perform by modifying the physical environment, for instance by increasing bin proximity or number, changing bin appearance, or providing home equipment for sorting waste.
6. **Social modeling:** any kind of passing of information via demonstration or discussion in which the initiators indicate that they personally engage in the behavior.

Varotto and Spagnolli (2017) found that all six types of interventions increased recycling behavior on average, but the magnitude of the behavioral change varied substantially across interventions. Social modeling was by far the most effective intervention technique, followed by environmental alterations; the remaining groups were about equally effective.

Social Modeling

The assumption behind social modeling is that people learn through observation of the behavior of their peers. The field studies that examined social modeling recruited community members who already participated in a recycling program to act as block leaders. These block leaders were then tasked with modeling proper recycling behaviors and informing and convincing their nonrecycling neighbors to also participate in recycling. Varotto and Spagnolli (2017) pull from Burn (1991) to posit that social modeling techniques are effective because they (1) communicate to individuals that their neighbors are recycling, engendering a social recycling norm, and (2) may lead individuals to perform the behavior out of a desire for social approval. While social modeling has low costs compared with other techniques, its effectiveness depends on the extent to which block leaders are present and willing to participate, and the extent to which residents see themselves as part of the community.

Environmental Alterations

The second most effective strategy was environmental alterations. Varotto and Spagnolli (2017) suggest that the effectiveness of environmental alterations may be due to the reduced amount of effort required to recycle and thus the perceived costs of recycling. However, no single bin size and collection frequency suits all households, and households report wanting to choose their bin sizes depending on their waste habits and home storage space (Willman, 2015).

Prompts and Information

Prompts and information are the most commonly tried interventions because of the relative ease and low-cost nature of disseminating written information via fliers, brochures, and websites that advocate recycling and explain how, why, and when to carry it out. When lack of information is the main barrier to household recycling (i.e., households are motivated to recycle but do not know how to recycle), then simple dissemination of information can lead to changes in recycling behavior (McKenzie-Mohr, 2011). This situation often occurs at the start of a new recycling program or when the existing program changes or is particularly complex (NRC, 2002). However, lack of information may not be the main barrier to recycling (e.g., the program may be difficult to use). In that case, households can choose not to read the information provided, and information interventions may not lead to behavioral change. The type of information provided also matters. Beyond informing households how to recycle and what can be recycled, some initiatives provide information on the broader benefits of recycling. While some studies found no additional effects of highlighting the benefits of recycling (Klaiman et al., 2017), other studies found positive effects of this information (Winterich et al., 2019). Together, these studies suggest that providing information may not always increase recycling rates, but at least it does not seem to reduce recycling rates, nor is it particularly costly to implement.

In the domain of food waste, Schäufele-Elbers and colleagues (2024) found that guests at a European hotel who were made aware of the food waste problem were responsive to information nudges on the topic. Messages were posted at the buffet providing arguments for reducing personal food waste, such as "Use instead of waste—Reduce food waste for a sustainable future," and "1/3 of all food never reaches the human stomach. You can help reduce this by wasting less." Guests reduced their average rate of food waste generation significantly during the study period (Schäufele-Elbers et al., 2024).

Incentives

Several monetary incentives aimed at encouraging recycling have been used in the United States, as surveyed in Kinnaman and Fullerton (2000) and Kinnaman (2000). Chapter 3 describes the use of corrective ("Pigovian") taxes as a market-based instrument for promoting recycling behavior, essentially imposing a fine when recycling behavior is not correct.

Another monetary incentive is a deposit-return system, which requires consumers to pay a small deposit for each eligible beverage container at the time of purchase; consumers can get the deposit back as a refund when the container is properly returned to a drop-off recycling center. The idea behind this model is that the potential for a financial return will encourage recycling. See Box 3-3 in Chapter 3 for more information on incentives via deposit-return systems and Chapter 4 for deposit-return system policies and examples of their implementation.

While many of the policies mentioned above and discussed in previous chapters are market based, a few additional policies are available that center on household-level behaviors. These include pay-as-you-throw policies and fines for illicit dumping and burning. The idea with these methods is to make garbage collection relatively more expensive than recycling appropriately.

Pay-as-You-Throw Policies

Most U.S. households pay for garbage collection through annual property taxes or a fixed monthly fee. Thus, they do not pay for each additional bag of garbage they add to their weekly collection. However, some municipalities have pay as-you-throw systems, where households pay for the amount of waste they produce by volume, weight, number of bags (using specially labeled bags or stickers), or frequency of collection (Gradus et al., 2019). Empirical studies of pay-as-you-throw programs have found that they are associated with decreases in household disposal of MSW and increases in recycling, such as in Virginia (Fullerton and Kinnaman, 1996), Minnesota (Sidique et al., 2010), Massachusetts (Starr and Nicolson, 2015), Italy (Bucciol et al., 2015), the Netherlands (Dijkgraaf and Gradus, 2017), and South Korea (Lee, 2023).

Fines for Illicit Burning and Dumping

However, charging a price per bag of garbage may unintentionally provide incentives for illicit burning and dumping (Fullerton and Kinnaman, 1995). It may also increase the amount of "wish-cycling," when households use their recycling bin for waste that is not recyclable, in order to avoid paying the price for landfill waste. This behavior in turn increases contamination in the recycling stream, making it more expensive for MRFs to sort. Thus, another policy option is using fines to punish illicit dumping and burning and improper recycling sorting. In a randomized field experiment, Vollaard and van Soest (2024) found that informing households about fines for not separating their recycling led to more than a 10 percent reduction in residual waste.

However, any such fine would require monitoring, enforcement, and administration. If these behaviors cannot be easily monitored or controlled, then a household subsidy for recycling might provide more effective incentives for recycling. While subsidies are popular with consumers, as discussed above, governments must find a way to raise funds to pay the subsidy, which often entails using other taxes, which could have their own social costs. Additionally, subsidies may raise the amount of wish-cycling (to gain the subsidy even for nonrecyclable materials). For these reasons, deposit-return programs, discussed at length in Chapter 4, have been shown to be preferable to pricing garbage collection or subsidizing recycling collection in most cases (Palmer and Walls, 1997).

Commitment

Varotto and Spagnolli (2017) identified commitment (e.g., someone following through on their intended behavior or goal to do something) as a highly individual motivator. Some of the eight studies they examined (e.g., Werner et al., 1995) indicated that garnering commitment is a more effective strategy than providing information or incentives. However, this strategy would be difficult to implement on a large scale.

Feedback

More recent studies demonstrate and compare the effectiveness of different types of feedback campaigns (see Box 6-2; Hewitt et al., 2023; McKie et al., 2024; Schäufele-Elbers et al., 2024).

Another type of feedback is bin tagging, where haulers put a tag on curbside recycling bins that are contaminated with nonrecyclable material. Box 6-3 discusses an example of how bin tagging is used in Seattle. McKie and colleagues (2024) compared recycling quality (as measured by lower contamination rates) for households that received recycling information only against those that were also penalized for excessive contamination or missorting. Households that were also penalized for these errors were subject to temporary loss of their recyclable collection service. The study found that those who were subject to penalties reduced contamination more than those who were only provided with information (McKie et al., 2024).

6.3.1 State-Level Policies and Household Behavior

State-level recycling policies vary substantially across the United States (see Figure 6-12). Cecot and Viscusi (2022) categorized state laws into five types: goal laws, plan laws, opportunity laws, mandatory recycling laws, and deposit laws. *Goal laws* are aspirational laws that advocate that local governments set a recycling goal but have no concrete policy mechanism that will assist in meeting that goal. *Plan laws* require municipalities to develop a plan for meeting their recycling goals and to evaluate their current recycling programs. Plan laws are the most common form of recycling law, implemented in 15 states. *Opportunity laws*, the third level of stringency, include laws that require local governments to implement policies that provide recycling opportunities for households to engage in recycling but do not mandate that all residents recycle. Finally, *mandatory laws* require all residents to separate their recyclable products from other household waste and appropriately recycle those products. Additionally, some states have *deposit laws* (i.e., deposit-return systems, discussed in Chapter 4).

Viscusi and colleagues (2022) examined how moving between states with more or less stringent recycling laws impacts a household's recycling rate. Moving to a state with a deposit-return system increased the number of material types recycled by households by 41 percent, while moving to a state with a strict recycling law (either mandatory or opportunity law) increased the number of materials recycled by 9 percent. Moving out of a state with a deposit-return system decreased the number of materials recycled by 13 percent, while moving out of a state with a strict recycling law impacted only recycling of plastic, which decreased by 12 percent (Viscusi et al., 2022).

BOX 6-2
Social Feedback Case Study

In New York City, Hewitt and colleagues (2023) studied recycling rates among residents of multifamily buildings. They compared families that received feedback on recycling only in their building (noncomparative feedback) to families who received feedback regarding recycling both in their own building and in another nearby building (comparative feedback; see Figure 6-11). Both buildings exhibited increased recycling, providing evidence for the effectiveness of feedback that appeals to social norms. However, the comparative feedback information in the left panel of Figure 6-11 showed a greater apparent impact than the noncomparative feedback in the right panel (Hewitt et al., 2023).

FIGURE 6-11 Sample intervention graphic: Comparative social feedback (left panel) versus noncomparative feedback (right panel).
SOURCE: Hewitt et al., 2023. CC BY NC ND.

BOX 6-3
Case Study: Bin Tagging as Feedback in Seattle

Given the constantly changing population it serves, Seattle Public Utilities engages in education and incentive efforts for Seattle's diverse and growing population; it uses the motto "Educate, educate, educate." If a household's curbside material in a recycling bin is contaminated, educational material is left and the bin is tagged as contaminated. Then, if the household fixes the contamination in the bin, the items will be collected later and a fee is charged. Seattle Public Utilities has a group of inspectors that will visit the household to educate and explain why the item was tagged. On the third time a household's items get tagged, the utility issues a fine, announced to the household in a mailer. However, fines are rarely applied, as education typically results in less contamination and correct recycling behavior.

SOURCE: McKie et al., 2024.

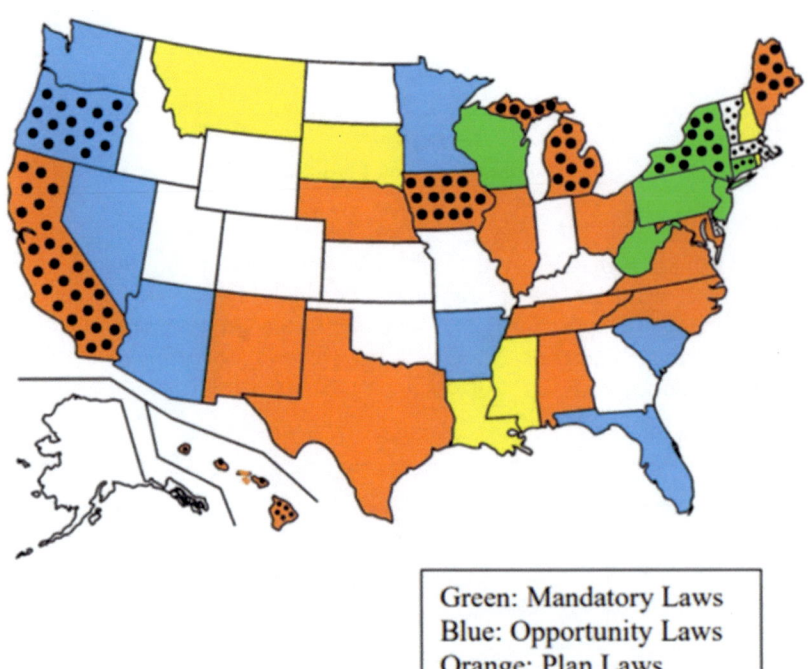

FIGURE 6-12 U.S. recycling laws.
SOURCE: Cecot and Viscusi, 2022.

In a related study, Viscusi and colleagues (2023) examined factors beyond state policy that lead to changes in household recycling behavior. They found that increased market prices for cans and glass were associated with increases in recycling, but the same was not true for increased market prices for plastic and paper. Viscusi and colleagues (2023) hypothesized that the financial return that municipalities can reap from cans and glass, relative to plastic and paper, may "provide an inducement to promote recycling as a revenue source, and the resources obtained by selling the recycled materials may enable the municipality to promote household recycling efforts" (p. 4). Additionally, they found that declines in recycling are associated with large positive or negative household income shocks greater than 20 percent, moving into an apartment, getting married, and having a baby (Viscusi et al., 2023).

6.3.2 Increasing Recycling Access Through Business-Led Programs

As noted in previous chapters, access to municipal recycling collection and drop-off locations varies substantially across regions in the United States, across residential housing types, and across material types. Private-sector businesses sometimes fill in accessibility gaps by providing alternative collection programs. For instance, Ridwell[3] provides a subscription service to collect items from subscribers' doorsteps that are traditionally not accepted in municipal recycling programs, such as multilayered plastics, plastic films, batteries, and lightbulbs. And Terracycle[4] collects hard-to-recycle materials via mail; customers purchase from the company a Zero Waste Box, which they can fill and ship back to the company. NexTrex[5]

---

[3] See https://www.ridwell.com.
[4] See https://www.terracycle.com/en-US.
[5] See https://nextrex.com.

provides drop-off locations for recycling plastic bags and film in major retailers such as Kroger, Albertsons, Meijer, and Kohl's. More research and data are needed on the extent to which these business-led programs increase recycling access and recycling rates.

Nespresso is an example of a company trying to foster recycling partnerships between households and businesses. In New York City (and Jersey City in the near future), households may dispose of their aluminum Nespresso capsules into the blue recycling bins. In addition, these aluminum capsules can be recycled across the entire country if consumers order a bag that they fill with used capsules and the filled bags are mailed via the post service for recycling.[6]

6.3.3 Available, Convenient, and Accessible Infrastructure

To assess whether recycling is convenient for consumers requires an understanding of how consumers define convenience for a given recycling option. Consumer recycling convenience for a curbside program includes the time and space needed to sort and store materials. Consumer recycling convenience for drop-off programs involves transportation costs, as most of the existing drop-off centers are located outside city limits. However, some drop-off recycling centers have the benefit that consumers get paid when they redeem their recyclables (e.g., states with container deposit-return laws).

For drop-off options, identifying attributes of a visit that consumers value is key. Berck and colleagues (2021) found that consumers in California generally find drop-off recycling centers convenient when they are close to home, open at convenient times, and have short lines. Drop-off recycling centers in California seem to meet this definition for typical users, who tend to have lower levels of income and education. Those who do not choose to recycle at drop-off centers tend to be more affluent and hence may not find the money gained at drop-off centers to be worthwhile. Instead, they choose to recycle through curbside collections and at businesses (Berck et al., 2021).

Beatty and colleagues (2007) consider what would happen to overall recycling rates in California if access to curbside pick-up services were extended to more consumers. Using a panel regression framework, where material recycled is regressed on share of the population with access to curbside services, they found that marginal gains from extending these services would be small, as they would mostly induce consumers switching from drop-off to curbside recycling (Beatty et al., 2007). Best and Kneip (2019) found that a curbside scheme in Germany had no effect on paper recycling but increased recycling participation by 10–25 percentage points for plastic and packaging. In the United Kingdom, Abbott and colleagues (2017) found that the findings on the trade-off between recycling via curbside or noncurbside methods are ambiguous.

Berck and colleagues (2024) simulated the elimination of government-subsidized recycling drop-off centers, finding that closing them would not significantly alter consumer well-being for any major demographic group and would have little impact on whether households chose to recycle, given households could switch to nonsubsidized recycling drop-off centers and/or curbside recycling.

6.3.4 Community and Household Campaigns to Address Inconsistencies Across Programs

As discussed earlier in this chapter, misleading product labels, such as the chasing arrows symbol, are being used on products that are not accepted by most MRFs in the United States (EPA, 2023). For instance, many consumers are confused by on-product resin identification codes, which use the chasing arrows symbol around a number 1 through 7, as an indicator of a product's recyclability. However, resin identification codes indicate only the type of resin (e.g., PET, high-density polyethylene [HDPE], polypropylene) and not whether a product is widely recyclable. Section 6.7.1 discusses federal guidelines to restrict the use of the chasing arrow symbol on products and packaging.

---

[6] See https://www.nespresso.com/us/en/circularity.

> **BOX 6-4**
> **Case Study: Seattle's Online Recycling Search Tools**
>
> Seattle offers recycling information for households with the "Where Does It Go" online search tool; this tool provides, in many languages, information about how to dispose of household waste items properly (see Figure 6-13).
>
>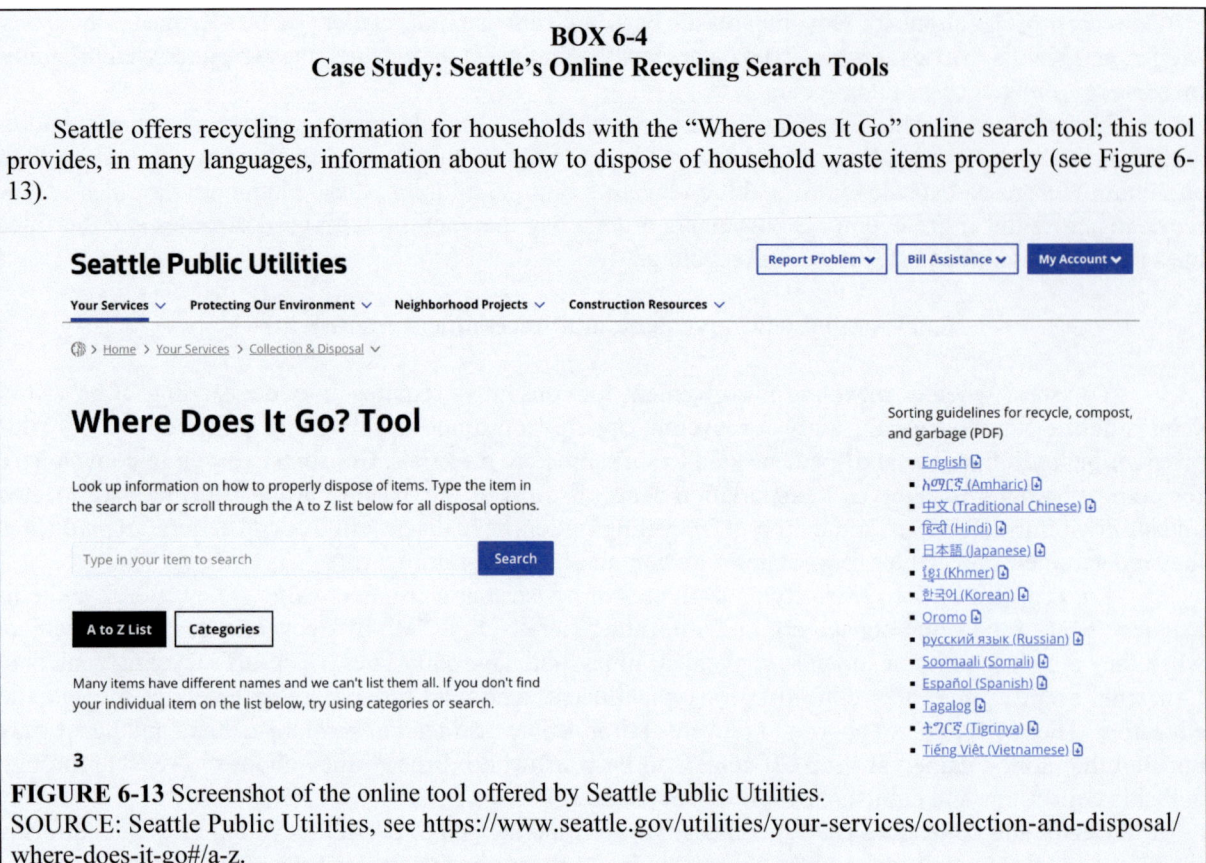
>
> **FIGURE 6-13** Screenshot of the online tool offered by Seattle Public Utilities.
> SOURCE: Seattle Public Utilities, see https://www.seattle.gov/utilities/your-services/collection-and-disposal/where-does-it-go#/a-z.

Confusion also arises because the items that can and cannot be collected for recycling may differ from state to state, city to city, and even household to household. For instance, in Champaign, Illinois, households contract individually with one of four curbside collection haulers, with some haulers taking a larger variety of recyclable materials than others (City of Champaign, n.d.). In the bordering town of Urbana, Illinois, all curbside recycling collection is handled through one service, which has a list of recyclable materials that differs from those in Champaign (City of Urbana, n.d.). Complicating matters further, the largest employer in Champaign-Urbana, the University of Illinois, operates a Waste Transfer Station and has its own list of recyclable materials. For instance, plastic shopping bags can be put out for recycling collection in Urbana, but not for some of the haulers in Champaign or at the University of Illinois. Plastics with resin numbers 3–7 are collected for recycling in Urbana and Champaign, but not at the University of Illinois. Section 6.7.1 later suggests a policy option for supporting and evaluating a national recycling label standard that would provide accurate information on which products are widely recyclable across the United States and which products require checking locally.

One avenue for addressing this confusion is community-based information campaigns. This approach could mean offering funding to local governments to develop materials about how residents can recycle in their specific neighborhood and residence type. For instance, the State of Delaware (n.d.) developed on online tool called Recyclopedia, where households can enter their address and find out what can be recycled in their neighborhood and how to recycle it. The State of Florida (n.d.) has a website called Rethink, Reset, Recycle; it discusses various types of recyclable materials and points to the contact information of county recycling coordinators for questions and concerns about what residents can recycle. Local governments, especially in municipalities where residents all have the same access to recycling, also distribute information on what can and cannot be recycled; for example, see the "What Can I Recycle?" website provided by Hamilton County, Ohio (Hamilton County Environmental Services, n.d.). Box 6-4 provides a case study on the online tools provided in Seattle, Washington.

Private companies are also beginning to provide more information to households. For instance, Recycle Coach[7] is a mobile app that provides users with fast information about their local recycling program, including personalized recycling schedules, pick-up reminders, and information on what goes where. Recycle Coach has the goal of making recycling education fun, simple, and engaging. However, as with many crowdsourced mobile apps, Recycle Coach works best when more users verify that the local information is correct. Thus, these types of mobile apps may not be effective in places where recycling rates are currently low.

6.3.5 Targeted Programs for Different Demographics

As discussed earlier in this chapter, surveys and academic studies consistently find differences in recycling rates and recycling attitudes across demographic groups. For instance, groups found to be less likely to recycle include renters, those with lower levels of formal education, smaller households, younger adults, men, foreign-born Latinos, and African Americans (Shaw et al., 2014). Efforts to calibrate campaigns to the specific characteristics of the groups they target (e.g., providing informational material in Spanish in neighborhoods with high Spanish-speaking populations; using forms of media preferred by different demographics) can strengthen campaigns and make them more effective (Varotto and Spagnolli, 2017). While some groups (especially older adults and homeowners) with medium to high recycling rates are motivated to recycle for environmental reasons and for concern about neighborhood cleanliness, those with low recycling rates are more motivated by convenience (Shaw et al., 2014) and understanding how to recycle (Varotto and Spagnolli, 2017). In neighborhoods with high turnover of residents, social modeling and norm initiatives may be less effective, since social modeling works best when residents see themselves as part of the community (Schultz et al., 1995). Thus, ascertaining convenience and information needs for low-recycling populations, as well as social norms, is an important first step for campaigns targeting these groups.

Rural recycling is characterized by long hauling distances, sometimes four times those of urban and suburban areas. Low population density in rural areas results in lower waste generation and high usage of burn barrels. While recycling efforts in rural areas predominantly uses collection boxes scattered throughout large areas of rural counties, maintenance problems associated with those boxes have driven most programs to switch to staffed convenience centers, where both recyclables and mixed waste are collected for transport to more distant landfills. However, rural programs face challenges in finding markets for the materials they have collected, because of low market value and long distances to markets. Thus, efforts are necessary to find and create new markets close to home (Link and Stoke, 2021).

Funding can be tailored to enhance incentives for different demographic groups. For example, in the National Strategy for Reducing Food Loss and Waste and Recycling Organics (White House, 2024), substantial funding is available for tribal communities and communities with environmental justice concerns. More research specifically examining how groups with low recycling rates respond to recycling campaigns is needed. For instance, Lakhan (2016) examined how first-generation ethnic minorities respond to different types of recycling promotion and education campaigns used by municipalities in Ontario, Canada. Lakhan (2016) found that none of the recycling campaigns tested were able to increase recycling awareness or change recycling behavior among first-generation ethnic minorities. Participants reported that the campaigns were excessively complex and confusing, and they were skeptical of what municipalities did with the waste after it was collected. This study shows that (1) municipalities may need to rethink and redesign recycling initiatives to better engage minority communities and (2) more research is needed on what types of campaigns are effective in minority communities (Lakhan, 2016).

---

[7] See https://www.recyclecoach.com/solutions/home.

6.4 OVERVIEW OF SOCIAL IMPACTS

Social impact generally refers to the range of consequences, positive or negative, that affects communities (Jena and Ahmed, 2024). The social impacts of waste management span public health, community well-being, and economic opportunities, encompassing both intended and unintended effects on society (Aktar, 2023). Current waste management literature often focuses on environmental and economic aspects, leaving gaps in understanding of some broader social dimensions, such as equity and inclusion (Douglas, 2012; Martuzzi et al., 2010).

Frameworks such as the United Nations (n.d.) Sustainable Development Goals (SDGs) highlight the need for standardized social impact assessments across contexts, as discussed below, but no single framework is specific to waste management. Insights from environmental justice and social cost perspectives shed light on issues of equity and public health, but these perspectives rarely address the full scope of social impacts in waste systems (Murphy, 2010; Sharma et al., 2021). The World Bank's framework—emphasizing inclusion, cohesion, resilience, and process legitimacy—suggests a holistic approach that can guide future assessments of waste management's social effects (Haregu et al., 2016; WHO, 2023).

The following sections explore these frameworks, underscoring the need for a more integrated approach to evaluating the social dimensions of waste management comprehensively.

6.4.1 Defining Social Impact

The term *social impact* refers to the consequences of any action that affect society, which can be positive or negative, intended or unintended (Freudenburg, 1986). The term has multiple definitions, with no consensus reached on a universal definition (Marc and Ponikvar, 2022). Scholars note that the social impacts of waste management are predominantly linked to the environment and economy and that other dimensions, such as equity and well-being, are not considered separately. The lack of a clear definition for *social impacts* trickles down to social waste management scholarship (Hird, 2022). Some of the well-studied impacts of waste management include those on health and livelihoods (Ma and Hipel, 2016). For instance, waste management practices can impact the health of individuals living near landfills because of the release of hazardous substances into the environment. Waste management systems can also impact livelihoods by creating or diverting jobs from recycling plants or their automation.

Social impacts can be complex and diverse. As such, they need to be studied using common language and assessment criteria. International frameworks, such as the SDGs and the European Pillar of Social Rights, provide a set of guidelines and targets that help to standardize evaluation of social impacts across different contexts (European Commission, n.d.; Marc and Ponikvar, 2022). As discussed below, these frameworks emphasize various social dimensions, including poverty alleviation, education, health, and equality, as well as the environment or, more specifically, waste management. In addition, they aim to guide countries and organizations in achieving sustainable and equitable development. However, no single framework can measure all social impacts of waste.

Academic literature on environmental justice explores the relations between environment and society (Chowkwanyun, 2023). It categorizes social impact into two broad areas: *equity*, which concerns the distribution of resources, and *procedural justice*, which relates to the fairness of processes. Environmental justice waste studies tend to focus on the geospatial patterns of landfills or hazardous waste sites and the socioeconomics of nearby communities (Heiman, 1996).

The social cost literature, originating from environmental economics, also offers insights about social impacts. It assigns a monetary value to externalities, or indirect social impacts such as emissions. Typically, this is limited to the social cost of public health or livelihoods. Impacts related to labor in recycling and the well-being of informal recyclers are understudied (Dauvergne and LeBaron, 2013). Like practitioner literature, academic scholarship on the social impact of waste has been limited to an environmental lens instead of a societal lens.

6.4.2 Social Impact Frameworks

The World Bank posits that no single measure of social impact is universally accepted (Ballon Fernandez and Cuesta Leiva, 2024). It suggests that any holistic measure should account for four main dimensions: inclusion, cohesion, resilience, and process legitimacy. These dimensions provide a holistic framework for assessing social impacts, including those related to waste management. An *inclusive* society ensures access for all to markets, services, and spaces, allowing everyone to thrive. A *resilient* society can withstand shocks and stresses. A *cohesive* society has a shared purpose and trust, enabling collective action toward common goals. *Process legitimacy* refers to the fairness and credibility of the policies and programs implemented, which is crucial for public acceptance and participation (Ballon Fernandez and Cuesta Leiva, 2024).

Despite its comprehensive nature, this framework has not yet been applied to waste management. Scholars have an opportunity to define social dimensions for waste and to create a framework that can be used to explore the social aspects of waste and recycling. Table 6-4 summarizes some ways that the key social dimensions identified by Ballon Fernandez and Cuesta Leiva (2024) can be applied to waste management for a more comprehensive assessment. The indicators in the table were adapted from their original study for waste management and recycling. These indicators provide examples of how the social impact of waste management has been studied in the practitioner literature or academic literature (Ballon Fernandez and Cuesta Leiva, 2024). For example, The Recycling Partnership (2024) summarized how access to recycling differs across communities in the United States.

TABLE 6-4 Social Dimensions in Waste Management

Dimension	Indicator	Examples of Studies
Inclusion	Access to recycling programs and recycling infrastructure across different socioeconomic groups	The Recycling Partnership (2024)
	Proportion of the population with regular waste collection services	The Recycling Partnership (2024)
	Involvement of marginalized groups in decision-making related to waste management	Petts (2002)
	ADA-compliant and multilingual recycling information on product labels	Anton et al. (2020)
Cohesion	Level of information shared among households	
	Ease of coordination among stakeholders in the waste life cycle	
	Community participation rates in waste reduction and recycling initiatives	Folz and Hazlett (1990); National Recycling Survey (1990)
	Degree of collaboration between various stakeholders in the waste management system	Lintz (2015)
Resilience	The flexibility of waste management systems to adapt to changes in waste volume or composition	EPA (2024)
	The ability of recycling markets to continue operating during extreme disruptions	
	Risks from the waste management system	
Process Legitimacy	Public participation in waste management systems	Bernstein (2004); EPA (2023) (Note: Limited to hazardous waste); Wiedemann and Femers (1993)
	The level of transparency in the waste management policy formulation process	
	Public access to information regarding waste management practices and policies	The Recycling Partnership (2024)
	Existence of formal grievance mechanisms for stakeholders to raise concerns about waste management.	

SOURCE: Generated by the committee, adapted from Ballon Fernandez and Cuesta Leiva, 2024.

Additional frameworks have been developed for assessing and measuring the social aspects of development and well-being, which could be applied to the study of the social impacts of recycling and other waste management:

1. Global Indicator Framework for the SDGs: Adopted in 2017, the SDGs framework consists of 17 goals and 169 targets, measured through 232 individual indicators. This framework is a comprehensive approach to sustainable development, covering a wide range of social issues such as poverty, education, gender equality, and health. It includes Goal 12, which focuses on ensuring sustainable consumption and production patterns. Indicator 12.4 specifically targets the environmentally sound management of chemicals and all wastes throughout their life cycle. Indicator 12.5 aims to reduce waste generation substantially through prevention, reduction, recycling, and reuse.
2. OECD Framework to Measure Well-Being: Through its Better Life Initiative, the Organisation for Economic Co-operation and Development (OECD) has developed a methodology to measure each OECD country's distance from the SDG targets. The framework aims to support countries in understanding their progress and identifying critical focus areas. It includes environmental quality as a key dimension of well-being, with waste management as a critical component. Indicators related to waste include recycling rates and waste generation.
3. Equal Measures 2030 SDG Gender Index: Launched in 2018, this index measures progress on the gender equality aspects of the SDGs. It aggregates gender-related goals into a single measure, providing insights into the status of women and girls in various countries. While this index focuses on gender equality, it includes SDG targets that relate to waste management, such as SDG 6 (clean water and sanitation) and SDG 11 (sustainable cities and communities), which encompass waste treatment and pollution reduction.
4. European Pillar of Social Rights (EPSR): The EPSR is a set of 20 principles and rights essential for fair and well-functioning labor markets and welfare systems in the European Union. Introduced in 2017, the EPSR is monitored through the Social Scoreboard, which assesses member states' progress in relation to the Pillar. While the EPSR primarily addresses social and employment issues, waste management can be related to its principles indirectly, through the promotion of a circular economy and sustainable work environments, which are part of the broader EU sustainability strategy.
5. Social Progress Index: Developed by the Social Progress Imperative, this index measures the capacity of a society to meet the basic human needs of its citizens, enhance the quality of their lives, and create conditions for individuals to reach their full potential. The index includes 12 components that focus on actual life outcomes in areas ranging from shelter and nutrition to rights and education. This index includes components such as access to basic knowledge and environmental quality. Waste indicators may be reflected in the measurement of pollution levels and the effectiveness of waste management in contributing to a healthy environment.
6. Happy Planet Index: Introduced by the New Economics Foundation in 2006, this index describes human well-being and environmental impacts. It combines subjective well-being, life expectancy, inequality of outcomes, and ecological footprint to show how efficiently residents of different countries use environmental resources to lead long, happy lives. This index's ecological footprint component directly relates to waste indicators, as it measures the resources used by a country, including waste production and management. It reflects how waste impacts the overall sustainability of a country's lifestyle.
7. Sustainable Society Index: Developed by the Sustainable Society Foundation, this index broadly covers sustainability based on the three-pillar model of human, environmental, and economic well-being. The human well-being dimension, representing the social pillar, includes basic needs, personal development, health, and a well-balanced society. The index comprises 24 indicators across seven categories, providing a comprehensive assessment of a society's sustainability. The human well-being dimension of this index can be associated with waste indicators through the

basic needs category, which would include sanitation and waste disposal facilities, as well as through the health category, which is affected by waste management practices.

Each of these frameworks offers a unique perspective on assessing waste management from a social perspective and emphasizes the importance of considering a wide range of social dimensions. Each serves as a valuable tool for policymakers, researchers, and practitioners for measuring and improving social outcomes, ensuring that development efforts lead to a more equitable and sustainable future for all. In their review of several additional frameworks, Gionfriddo and Piccaluga (2024) emphasized that these models lack scientific rigor and flexibility.

While frameworks can be helpful individually, it is essential to develop a standardized framework to be used across multiple settings and scales, including firms and local governments. Such standardization would enable scholars, practitioners, and policymakers to share a consistent understanding of the social dimensions critical to these systems.

6.4.3 Social Impacts Considerations for Value Chain Actors

Social impact studies on waste management have traditionally concentrated on the direct impacts at the point of waste treatment, such as health outcomes from toxic leakage and livelihood effects from recycling centers. While important, these do not account for the entire waste life cycle.

A product's waste life cycle includes choices and outcomes with significant social implications, including product design choices, consumer purchase and disposal choices, material recovery, material treatment, and material reuse (Hafsa et al., 2022). Hafsa and colleagues (2022) describe how decisions at each of these value chain stages impact end of life waste.

Building on the waste life cycle concept, each value chain stage has social impacts. For instance, the way products are designed can either facilitate or hinder recycling and reuse, impacting the volume and toxicity of waste and, ultimately, public health (Fullerton and Wu, 1998). Retailers influence consumer choices through packaging options and marketing, which can either promote sustainability or contribute to a throw-away culture. Consumer purchasing decisions—often based on price, convenience, or brand loyalty—directly affect the amount and type of waste generated. Additionally, the way consumers dispose of products—whether through recycling, composting, or simply discarding—can be influenced by social norms, education, and the accessibility of waste management infrastructure (as described earlier in this chapter). Finally, social impacts considerations are relevant to waste treatment methods and the functioning of recycling markets (see Chapter 5). These markets are shaped by a complex interplay of social, economic, and policy factors that determine who participates and who benefits. A comprehensive approach to social waste studies that encompasses the entire waste value chain would provide a more holistic understanding of the social impacts at each stage and offer insights into more effective interventions for sustainable waste management.

It would be important, however, to consider the key social dimensions for each value chain actor. To identify some research opportunities for holistic social impact research on waste management and recycling, the committee combined the key social dimensions identified by Ballon Fernandez and Cuesta Leiva (2024) with the waste value chain decision-making framework shown in Table 6-5. Potential research questions are identified for each value chain stage and social dimension. For example, at the product design stage, what sort of considerations can influence inclusion, resilience, social cohesion, and process legitimacy, which will trickle down into waste management? These questions indicate that while the problem-framing remains similar, the nature of the value chain stage impacts the sort of action that can be taken to promote a healthy social impact.

6.5 COMMUNITY IMPACTS

Recycling programs play a pivotal role in enhancing community cohesion and resilience, with impacts that reach beyond environmental benefits to foster social and economic value. Effective recycling initiatives encourage local participation and shared responsibility, helping to establish recycling as a community norm. Through outreach and education tailored to diverse populations, these programs promote widespread engagement and support long-term behavioral shifts toward sustainability. In smaller communities, such targeted efforts have proven particularly effective, where interpersonal connections can amplify community involvement.

TABLE 6-5 Potential Research Questions for Assessing the Recycling Value Chain and Its Social Impact

	Inclusion	**Resilience**	**Social Cohesion**	**Process Legitimacy**
Product design: The design must be recyclable, compostable, or reusable.	How does inclusive product design impact the use and acceptance of waste-based products by various demographic groups?	What design principles are most effective in creating products whose waste is adaptable to changing environmental conditions?	How can product design for waste strengthen social ties and collective action within communities?	What are the ethical considerations in product design for waste, and how do they impact process legitimacy?
Consumer behavior: The consumer must place it in the appropriate material recovery channel.	What barriers do marginalized communities face in accessing recycling facilities, and how can these be overcome?	How do consumers adapt their waste disposal and recycling behaviors in response to changes in local recycling policies or infrastructure?	How does community involvement in recycling initiatives influence individual consumer behavior toward waste disposal?	What factors influence consumers' trust in the recycling process and their willingness to participate in proper waste sorting?
Material recovery: Convenient and efficient channels are necessary for successful material recovery.	How can material recovery channels be designed to accommodate the needs of diverse populations, including people with disabilities, low-income households, and non-English speakers?	What roles do consumer education and awareness play in maintaining resilient material recovery practices?	How do social networks and community organizations contribute to the promotion of effective recycling practices?	What is the role of clear communication and information dissemination in establishing the legitimacy of material recovery channels?
Material treatment: It must be profitable to recycle or compost recovered material.	What barriers do small businesses and informal waste pickers face in participating in profitable recycling ventures, and how can these barriers be reduced?	What strategies can be implemented to make recycling businesses resilient to market fluctuations and changes in demand for recycled materials?	What impacts do local recycling cooperatives have on fostering social cohesion and improving economic outcomes?	What roles do certifications and standards play in legitimizing recycling operations and enhancing profitability?

SOURCES: Generated by the committee, using dimensions from Ballon Fernandez and Cuesta Leiva, 2024, and Hafsa et al., 2022.

6.5.1 Awareness and Education Initiatives

In a Pew Research Center poll (2016), 28 percent of Americans reported that their community's social norms strongly encourage recycling and reuse, while 22 percent reported that most people in their community do not encourage recycling. The remaining 48 percent reported being somewhere in the middle (Pew Research Center, 2016).

Communication strategies related to implementing waste management systems need to focus on raising awareness while allowing for the consistent and effective flow of information between local authorities and the local community. In addition, education efforts are needed at recycling facilities to ensure that the quality of recycling materials meets end-user buyers' quality requirements (see Box 6-5).

BOX 6-5
Case Study: Quality Disconnect Between MRFs and Wisconsin End Users

Some materials recovery facilities (MRFs) are not paying adequate attention to the quality of the collected materials they are trying to market, as seen in examples of steel cans leaving recycling facilities (the MRFs) and going to Wisconsin foundries (the end users). The recycled materials arrived contaminated with shredded paper, engine parts, plastics, and the materials that were picked up when the cans were processed (Burgert, 1993). Local governments could engage in education to improve MRF management, especially in rural areas, to make sure they can meet the quality requirements of end users' recycled can markets, which is the most profitable material for rural areas. The Wisconsin Department of Natural Resources[a] has implemented initiatives aimed at improving recycling programs that can help address the quality and marketability of recyclables such as the Wisconsin Recycling Markets Directory.[b]

[a] See https://dnr.wisconsin.gov/topic/Recycling.
[b] See https://www.uwgb.edu/recycling.

6.5.2 Reducing Public Health Risks

Research has documented reduced exposure to litter, leakage, or waste in general due to recycling behavior. For instance, Deeney and colleagues (2023) conducted a meta-analysis of studies focusing on consumer plastics in high-income countries. Their evidence suggests that the health risks of plastic use and improper disposal could be reduced by increasing recycling. And other studies have shown that leakage from landfills can be toxic to groundwater and nearby populations, particularly where landfill design and operating requirements are insufficient (Alslaibi et al., 2011; EPA, 2001; Regadío et al., 2012); thus, diverting recyclables from landfills can improve groundwater quality and human health of nearby residents.

6.5.3 Creating Value from Recycled Materials

Local communities have developed efforts to reduce the need to dispose of recyclable materials and create alternative uses for recycled or remanufactured materials (see Boxes 6-6 and 6-7). Examples include using old newspapers for soil preparation and animal bedding (Bond, 2017) and reusing glass as aggregate in paving materials (Harrison et al., 2020).

BOX 6-6
Case Study: The Minnesota-Based Wood From the Hood

Minnesota uses its recycling development program to support local businesses that use recycled materials from local waste in their business model. An example is Wood From the Hood (2024), which "was born with a simple idea: Reclaim discarded trees from local neighborhoods to create beautiful, high-quality hardwoods."

6.5.4 Creating Jobs

Since 2014, MRF operators have increasingly adopted automation, robotics, and artificial intelligence in mechanical recycling processes, most commonly in sorting. While several of these technologies replaced human labor, in some cases automation was implemented because a shortage of sorters (Pyzyk, 2019).

BOX 6-7
Case Study: Community-Centered Recycling in Phoenix

The City of Phoenix launched the Circular Plastics Microfactory in partnership with Arizona State University (ASU), Goodwill of Central and Northern Arizona, and Hustle PHX (Kass, 2024). This initiative aims to recycle plastic waste into valuable products such as skateboards, furniture, and durable plastic components.

The facility consolidates multiple processes (plastic collection, processing, and remanufacturing) in one location. Plastic waste sourced from Goodwill serves as raw material for future projects, ensuring a steady supply chain. ASU plays a pivotal role in the design process through its Circular Living Lab, which tests and develops products made from recycled plastics. Hustle PHX contributes by supporting minority entrepreneurs through financial capital and training programs. This partnership underscores the potential of shared goals and collective action in creating a sustainable future.

The microfactory addresses pressing environmental challenges while fostering economic growth and job creation for the community. The facility currently employs 10 workers, equipping them with valuable skills and generating economic opportunities within the community.

This approach exemplifies the power of collaboration between institutions committed to sustainability. By integrating sustainability practices with economic resilience, the Circular Plastics Microfactory serves as a model for other cities, proving that local initiatives can drive global change while benefiting communities directly.

While waste disposal has traditionally involved labor in the garbage collection and landfilling industries, Morris and Morawski (2011) found that the number of jobs created by disposing of material pales in comparison with the job creation potential in the circular economy through reuse, recycling, and remanufacturing.

6.6 PURSUING FAIRNESS IN ACCESS TO BENEFITS OF RECYCLING

Environmental justice considerations in recycling involve ensuring that all communities have fair access to waste management services, including convenient recycling facilities and infrastructure for waste sorting and recovery. Environmental justice includes addressing disparities in service availability, such as differences between public and private facilities, and ensuring that decision-making processes in waste management are inclusive and transparent. Expanding the extensive research on environmental justice to include the impact of recycling and waste management policies on vulnerable communities can help create more effective recycling systems.

6.6.1 Environmental Justice in Recycling Services

Environmental justice in the context of recycling and waste management focuses on two primary approaches: equity and justice. These approaches aim to address the fair distribution of waste management resources and just processes for their allocation. Despite the importance of these issues, scholarship has been limited on how environmental justice intersects with waste and recycling, compared with areas such as emissions and infrastructure. This section explores the current trends in environmental justice related to waste management, the existing gaps in the literature, and directions for future research.

Equity in waste management includes the fair allocation of segregated recycling collection centers, the establishment of recycling markets, the placement of MRFs, and the availability of services for collecting hard-to-recycle waste. It ensures that all communities, regardless of income or race, have access to facilities and services that allow them to manage waste responsibly. Without equity, low-income and marginalized communities often face the brunt of poor waste management practices, leading to increased exposure to pollution and health hazards.

Procedural justice in waste management focuses on fairness of the processes used to allocate resources and manage waste, including transparent decision-making, inclusive participation, and accountability in setting up waste management systems. Procedural justice ensures that community members have a voice in decisions that affect their environment, particularly in marginalized communities that have historically been excluded from such processes. By incorporating procedural justice, waste management practices can be made more democratic and reflective of the needs and rights of all community members.

According to Chowkwanyun (2023), the environmental justice literature has focused predominantly on issues of income and race, environmental emissions, and infrastructure (e.g., schools, highways). The existing literature has effectively identified reasons for disproportionate exposure to environmental harms and issues of mitigation and maldistribution. However, relatively little literature has been devoted to environmental justice issues specific to MSW and recycling. Several critical areas are in need of further research:

1. Policy decisions and regulations: Given a gap in understanding of how policy decisions and regulatory frameworks shape waste management practices, the literature often fails to address how different policies may lead to unequal outcomes in waste distribution and exposure.
2. Demographic transformations: Changes in population demographics, such as urbanization and aging, have significant impacts on waste generation and management. The environmental justice literature has not adequately explored how these transformations affect waste management practices and the resulting effects on justice.
3. Migration patterns: The movement of populations, whether due to economic opportunities or environmental displacement, affects waste management. But little research has tried to understand the intersection between migration and waste management from an environmental justice perspective.
4. Political arrangements: The role of local, state, and national political structures in shaping waste management policies and practices is another area not thoroughly examined. Political arrangements can influence how resources are allocated and which communities are prioritized.
5. Corporate behavior: The actions of private corporations, which often manage waste facilities, play a crucial role in waste management practices. The environmental justice literature has not fully addressed how corporate behavior influences environmental justice outcomes, particularly in terms of waste management.

To address these gaps, future research needs to focus on broader social impacts and historical accounts of policies that explain modern-day problems. Understanding the historical evolution of recycling infrastructure and policies can shed light on current disparities in waste management. By examining how past policies and practices have led to present-day issues, researchers can better identify pathways for more equitable waste management systems. The following research questions could guide such studies:

- How have historical waste management policies impacted different communities over time?
- How have race and socioeconomic status influenced the placement of landfills, incinerators, and recycling centers? What historical policies or practices led to these patterns?
- What are the long-term health impacts on communities that were historically exposed to poor waste management practices?

Environmental justice research has often been closely related to litigation, which has been a crucial tool for addressing disparities. Ongoing litigations or government accounts of waste management practices can provide valuable insights into the systemic issues that lead to unequal outcomes. For instance, the United Church of Christ (1987) highlighted that communities of color disproportionately bore the burden of exposure to toxic waste. And a study by the U.S. Government Accounting Office (1995) identified spatial patterns of landfill and hazardous waste locations along race and income. These studies underscore the importance of using litigation and governmental research to uncover and address environmental injustices in waste management. Some questions to consider:

- What waste-related litigation has led to significant policy or regulatory changes?
- How do federal and state policies differ in their approach to environmental justice litigation?
- Are disparities in litigation outcomes based on the demographics of the plaintiffs? How do these disparities reflect broader societal inequities?

Addressing these gaps requires a focus on equitable distribution of waste management resources and just processes for decision-making. By expanding the scope of environmental justice research to include waste management, policymakers and scholars can develop more comprehensive strategies for ensuring that all communities are treated fairly in the distribution and management of waste.

6.6.2 Sources for Environmental Justice–Related Data

Two data sources have been developed for understanding the social impacts of waste management. After describing each one, the committee offers suggested questions for guiding research using the data offered in these sources.

EPA's Interactive Recycling Infrastructure and Market Opportunities Map

EPA's Interactive Recycling Infrastructure and Market Opportunities Map[8] offers a detailed overview of waste management facilities across the United States, supporting the goals of the National Recycling Strategy and the Bipartisan Infrastructure Law. This map provides information on existing recycling infrastructure, per capita waste generation, and recycling rates for postconsumer materials, along with other market-related factors.

By mapping these details, the tool aims to bolster both primary and secondary markets for recyclable materials, promote cleaner communities by reducing landfill waste, and support climate change initiatives by diverting waste from landfills. The map covers all stages of the recycling process from waste generation to end use. It identifies the locations of 15 types of waste management facilities, including MRFs, composting sites, electronics recyclers, recycling facilities for various specific materials, anaerobic digesters, municipal solid waste landfills, transfer stations, and secondary processors for glass and wood. For each facility, the map provides comprehensive data such as the facility's name, address, contact information, type, and feedstock, where available.

Additionally, the map offers insights into the estimated tons of generated and recycled materials, categorized by zip code and material type. It includes data on 16 types of recyclable materials, such as aluminum, cardboard, electronics, various plastics, food waste, glass, steel cans, tires, paper, textiles, yard trimmings, and wood.

The data presented in this tool are based on the best available information from 2021 to 2022. The map serves as a valuable resource for understanding the current landscape of recycling infrastructure and opportunities for market development in the United States.

Complementing EPA's waste management map with demographic information and historical accounts offers promising research opportunities to explore the social impacts of recycling infrastructure and waste management practices. By correlating facility locations with socioeconomic indicators such as income levels, education, and employment, researchers can examine how economic factors influence recycling rates and access to infrastructure, potentially revealing disparities between affluent and lower-income communities. Integrating health statistics can provide insights into whether areas with more robust recycling programs and better waste management infrastructure experience improved health outcomes, especially concerning waste-

---

[8] See https://www.epa.gov/circulareconomy/recycling-infrastructure-and-market-opportunities-map.

related illnesses. Additionally, studying the evolution of recycling practices and infrastructure over time can reveal trends and the historical impact of waste management policies on different communities. Analyzing historical pollution and landfill data can further illustrate past environmental burdens and assess how advancements in recycling have addressed these issues over the years.

Plastic Litigation Tracker

In 2022, the New York University School of Law launched the Plastics Litigation Tracker,[9] which provides a database of the past and pending cases involving plastic products or pollution. The database includes 44 distinct cases dating back to 1971. The tracker enables environmental justice scholars to identify examples of waste-related litigation and observe how they may lead to policy changes.

For example, in 2018, *Smith v. Keurig Green Mountain, Inc.* (No. 4:18-cv-06690),[10] a federal court approved a class-action settlement involving claims that a company misled customers about the recyclability of its single-use coffee pods. The lawsuit argued that the company violated laws in California and Massachusetts by falsely advertising the pods as recyclable, which could contribute to plastic waste and climate change. The settlement includes a $10 million payment for customers, legal fees, and other costs. The company must also add a disclaimer, "Check locally—Not recycled in many communities," on packaging and ads. Any leftover settlement funds will go to Ocean Conservancy and Consumer Reports.

In Minnesota, Attorney General Keith Ellison filed a lawsuit against Reynolds Consumer Products and Walmart for misleading consumers by advertising Hefty recycling bags as recyclable, even though they are made from low-density polyethylene, which cannot be processed at recycling facilities (No. 62-CV-23-3104 [Minn. Dist. Ct. 2024]). As a result, recyclable items placed in these bags end up in landfills. In August 2024, the companies reached a settlement, agreeing to halt the sale of the bags in Minnesota for 2.5 years. Afterward, they will label the bags as "not recyclable." The companies will also pay $216,670, covering profits from the bags, and Reynolds will implement antigreen washing training and revise its marketing review process.

These litigations show that many states are concerned about recycling labels on products and following the "polluter pays" principle. In both cases the firms were asked to retract their labels and use information that correctly portrayed whether the product could be recycled locally (in those states).

6.7. KEY POLICY OPTIONS

6.7.1 National Recycling Labels and Regulations for Products and Packaging

One of the primary objectives of MSW recycling programs is to make them easy to understand and use. Surveys show high levels of confusion on the part of consumers as to what can be recycled and how to dispose of recyclables "correctly." Labels on products and their packaging are sometimes inconsistent with local rules about what consumers can recycle. For an important example, in the 1980s, the American Society of the Plastic Industry developed resin identification codes (1–7) to indicate the type of plastic used in a product or packaging. In addition to the plastic code number, these codes use the chasing arrows symbol, which is an internationally recognized symbol for recycling. The arrows lead consumers to believe the product or packaging can be recycled, even though the codes indicate only the type of plastic and technical feasibility of recycling and not whether it can be recycled in any given location.

While several U.S. cities offer consumer-oriented programs to improve information and reduce confusion in recycling, more could be done at the national level to evaluate existing efforts and develop consistent messaging to reduce confusion at the point of recycling decisions. Specifically, the Federal Trade Commission and EPA could support the adoption of on-product recycling labels that reflect the recyclability of products more accurately, as well as the variability of recycling across the United States. The same labels

---

[9] See https://plasticslitigationtracker.org.
[10] See the Plastics Litigation Tracker at https://plasticslitigationtracker.org/?keywords=18-cv-06690.

should be used for online purchases. An example of a state law is California's SB 343 (Allen, Chapter 507, Statutes of 2021), which "directs CalRecycle to publish data about the types of materials actually recycled in California. Manufacturers and other interested parties must use that information as part of their assessment of whether products can be considered recyclable for labeling purposes. The law outlaws manufacturers and others from selling products or packaging labeled as recyclable unless the items are regularly collected and processed for recycling in the state" (CalRecycle, n.d.).

National labeling standards have been enacted in Australia and New Zealand through the Australasian Recycling Label (ARL).[11] This label was developed by the nonprofit Australian Packaging Covenant Organization. The Australian National Government (2023) supports the ARL with educational materials (see, e.g., Department of Climate Change, Energy, Environment and Water, n.d.) and by investing "$5 million to support 20,000 small to medium-sized enterprises (SMEs) to improve the sustainability of their packaging and adopt the Australasian Recycling Label (ARL) on their packaging through the SME ARL Program" (para. 13). The Australian Packaging Covenant Organisation assumes responsibility for the ARL, and since 2022, has owned and operated the label with the support of other nonprofits, such as Planet Ark and PREPDESIGN.

Similarly, EPA, in partnership with producers, could support national recycling label standards; the How2Recycle labels presented in Table 6-6 were created by the Sustainable Packaging Coalition. If EPA considers a new label instead, it would be important to pretest to ensure that it is accurate and clear to consumers across a full range of educational and demographic backgrounds. Additional costs for this program include:

- Funding regular, nationwide analyses at MRFs to identify which materials are commonly collected, sorted, sold, or transferred for recycling in the United States. This data collection would be similar to what was required in California under SB 343.
- Monitoring the proper use of the new label, as well as any improper use of chasing arrow symbols.
- Creating funds for businesses, especially small and medium enterprises, to adopt the new label.

Regulating the use of the chasing arrow symbol would eventually reduce consumer confusion and restore trust in on-package recycling information. The label could be developed in a similar fashion to the U.S. Department of Agriculture's USDA Organic label to clarify and coordinate market information. Additionally, it can be used in conjunction with other policy options as part of an extended producer responsibility program. A trade-off for this policy is the monitoring it would require on the part of the federal government.

Implementation of this policy could begin with a 2-year initial product labeling study at the national level. It could draw on waste characterization studies already being conducted or funded by states (e.g., California), federal agencies (e.g., Department of Energy, Bioenergy Technology Office), and industry groups (e.g., Sustainable Packaging Coalition) to identify data gaps. The study would then collect data to fill in identified gaps in evidence available. New prohibitions against mislabeling products could be enforced 2 years after completing the initial study. Future product studies would then be needed every 5 years after the first study.

A consumer survey could be developed to assess the effectiveness of the policy implementation, investigating consumer recognition of new labels and their clarity and effectiveness in communicating the intended information (e.g., Boyer et al., 2021; Donato and Adıgüzel, 2022; Fischhoff et al., 1998). Curbside audits could collect measures of contamination before and after the policy is implemented. Regular monitoring of misuse of recycling labels would also be needed.

---

[11] See https://arl.org.au/about.

TABLE 6-6 How2Recycle Labels

Widely Recyclable	Recycle packages with the Widely Recyclable label through curbside or drop-off programs. Packages with this label are accepted by 60% of America's recycling facilities and 50% of Canada's.	 
Check Locally	Recycling programs vary across communities. Always check locally to see what your area accepts.	 
Store Drop-off	In the United States, plastic bags, wraps, and films are not accepted in *most* curbside or drop-off programs. However, many of these packages are eligible for Store Drop-off recycling.	 
Not Yet Recyclable	Dispose of packages with the Not Yet Recyclable label in the trash. Less than 20% of Americans and Canadians can recycle this package or significant challenges exist in sortation, reprocessing, or end markets. Check online for regular updates to the list of recyclable products.	 

NOTE: The labels shown here will be updated in 2025; the version displayed reflects the state at the time this report was written.
SOURCE: How2Recycle.info.

Conclusion 6-1: Reforming product labeling regulations and practices to provide accurate information (i.e., to prevent mislabeling) on what products are or are not recyclable would achieve multiple policy objectives, including clarifying information for consumers, decreasing contamination, and increasing efficiency of recycling systems.

Recommendation 6-1: The Federal Trade Commission (FTC) should revise its *Guides for the Use of Environmental Marketing Claims* so that resin identification codes no longer use the chasing arrows symbol. Additionally, FTC should prohibit use of the chasing arrows symbol

or any other indicator of recyclability on products and packaging unless the items are regularly and widely collected and processed for recycling across the United States. Furthermore, with or without a mandate to do so, producers should adopt and use updated resin identification symbols that do not include the chasing arrows symbol.

Key Policy Option 6-1: The U.S. Environmental Protection Agency (EPA), in partnership with producers could support and evaluate national recycling label standards—through education, outreach, and funding—such as the How2Recycle symbols created by the Sustainable Packaging Coalition. Additionally, the U.S. Congress, through EPA, could provide funding for small- to medium-sized companies that lack capability for transitioning to a new national recycling label standard.

6.7.2 Funding Social Modeling Programs

Social norms are beliefs or behaviors held by individuals in a social network or in a perceived identity group with whom individuals wish to affiliate. This group identification can be local (e.g., family, friends, coworkers) or regional, national, or global, especially as facilitated by the internet. Social norms may be used to design communications that address the concerns and values of a target population.

As discussed earlier in this chapter, Varotto and Spagnolli (2017) found that social modeling was by far the most effective intervention for increasing recycling behaviors. Many of the programs studied recruited community members who were already participating in a recycling program to act as leaders in their community. These "block leaders," as they are often known, are tasked with modeling proper recycling behaviors and informing and convincing their nonrecycling neighbors to participate in recycling. Block leaders may organize meetings or may have more informal face-to-face interactions with those in their community. Other means of education and communication, such as a website or library display, may support block leaders' efforts and provide further opportunities for social modeling.

Varotto and Spagnolli's (2017) findings are consistent with other studies of recycling behavior (Burn, 2006) and studies in other domains, such as energy and water conservation, where social modeling with the block leader approach is used to promote targeted behaviors (Abrahamse and Steg, 2019; Champine, 2023; Fiorillo and Senatore, 2024; Geiger et al., 2019; Niemiec et al., 2021). The block leader approach, and social modeling more generally, are consistently found to be effective at influencing individual behavior. The intuition behind social modeling programs is that the desire to comply with social norms often provides a strong motivation for individual and group behavior.

The objectives of the proposed policy are to advance social and environmental goals (including distributional effects); increase community engagement in recycling; and reduce household confusion, in order to ultimately reduce contamination. Face-to-face programs and sharing advice have been shown to be very effective at influencing beliefs and behaviors. These programs are typically run among neighbors or other residential groups (e.g., homeowners' association, local charitable organization), or in work and office settings.

An advantage of social modeling is its low cost, as compared with other techniques. A disadvantage is that it depends on the extent to which block leaders are present and willing to participate and the extent to which residents see themselves as part of the community. Face-to-face programs require ongoing effort to recruit and retain block leaders and to provide programming and feedback to block leaders. The ability to maintain and spread these programs over a large area and for a long time has not (as far as we know) been demonstrated. Additionally, the broader impact may be limited by the participant makeup. Residential groups tend to attract participants who are already environmentally conscious and already recycle, with less potential for significant upward shifts by a more representative demographic. Workplace groups may include a mix of those who volunteer and those that are "voluntold," and a range of workplace concerns and agendas may distract from the recycling purpose. Thus, rather than relying solely on residential and workplace groups, block leaders can likely increase their impact by going out into the community and reaching people in "third places"—informal public gathering places, such as cafes, coffee shops, community and recreation centers, beauty parlors, general stores, bars, and church groups (Oldenburg, 1989).

Following a call for proposals (from state or municipal agencies administering the grants), the following timeline is suggested: 1 year (Year 0) to gather a team, appoint project leaders, and write and submit a responsive proposal. In Year 1 (following funding): convene first diagnostic social modeling group and block leader to develop and iterate on materials and procedures for group meetings. Have participants report on their commitments, planned behavior, and actual recycling behavior. In Year 2 use the results from Year 1 to develop revised materials and procedures for a second group (or two). Compare the level of success achieved in Year 2 with that in Year 1. Then in Year 3, write and submit community reports and a website describing the study and its achievements. Begin to explore opportunities for spawning multiple groups across the MSW service area.

To assess the effectiveness of the policy implementation, consumer surveys could be developed to investigate consumer recycling knowledge. Curbside audits could collect measures of contamination before and after the policy is implemented. Data on material recovery before and after the policy is implemented could be collected from MRFs in areas that implemented it compared with control MRFs in areas without social modeling programs.

Conclusion 6-2: Social modeling programs are effective interventions for enhancing recycling behavior and establishing positive recycling norms in communities. Policies that promote social modeling programs can achieve various objectives for recycling. They can clarify information for consumers, decrease contamination, increase the cost-effectiveness of recycling collection and processing, and enhance the social and environmental benefits associated with recycling.

Recommendation 6-2: The U.S. Environmental Protection Agency should provide grants for state, municipal, local, and tribal governments for enhancing and expanding local social modeling programs, especially in disadvantaged communities and communities with high numbers of multifamily dwellings. Local governments, in turn, should implement or support social modeling programs, potentially through partnership with local nonprofits or other community-based groups, to engage directly with community members to promote positive social norms and recycling practices.

Key Policy Option 6-2: The U.S. Congress could reauthorize and further appropriate funds to the Consumer Recycling Education and Outreach Grant Program, authorized in the Infrastructure Investment and Jobs Act, to support social modeling programs.

Key Policy Option 6-3: To ensure sufficient data are available to inform policy decisions on recycling, the U.S. Environmental Protection Agency (EPA) could support studies to update or otherwise fill important data gaps and research needs. These options include:
- Regularly collect and report direct observations of household and commercial behavior related to recycling. In addition to filling knowledge gaps, these data would complement top-down modeling in the recycling system and enable empirical study of the impact of public policy. As part of these efforts, EPA could consider a periodic household and commercial survey for waste and recycling akin to the Energy Information Administration's Residential Energy Consumption Survey.
- Track household time spent on recycling to support more complete and accurate estimates of the economic and social costs of recycling and to ensure that lifecycle assessment models are as updated and as accurate as possible.

REFERENCES

Aadland, D., and A.J. Caplan. 1999. Household valuation of curbside recycling. *Journal of Environmental Planning and Management* 42(6):781–799.

Aadland, D., and A.J. Caplan. 2003. Willingness to pay for curbside recycling with detection and mitigation of hypothetical bias. *American Journal of Agricultural Economics* 85(2):492–502.

Aadland, D., and A.J. Caplan. 2006. Curbside recycling: Waste resource or waste of resources? *Journal of Policy Analysis and Management: The Journal of the Association for Public Policy Analysis and Management* 25(4):855–874.

Abbott, A., S. Nandeibam, and L. O'Shea. 2017. The displacement effect of convenience: The case of recycling. *Ecological Economics* 136:159–168.

Abrahamse, W. 2019. *Encouraging pro-environmental behaviour: What works, what doesn't, and why.* Academic Press.

Aktar, N. 2023. Unveiling the impact of solid waste management on health and poverty alleviation in Dhaka City. *Global Journal of Human-Social Science.* https://www.academia.edu/download/105564005/5_Unveiling_the_Impact.pdf.

Alslaibi, T., Y. Mogheir, and S. Afifi. 2011. Assessment of groundwater quality due to municipal solid waste landfills leachate. *Journal of Environmental Science and Technology* 4. https://doi.org/10.3923/jest.2011.419.436.

Ando, A., and A. Gosselin. 2005. Recycling in multifamily dwellings: does convenience matter? *Economic Inquiry* 43:426–438. https://doi.org/10.1093/ei/cbi029.

Antón, E., N.B. Soleto, and J.A. Duñabeitia. 2020. Recycling in Babel: The Impact of Foreign Languages in Rule Learning. *International Journal of Environmental Research and Public Health* 17(11):3784. https://doi.org/10.3390/ijerph17113784.

Ballon, F., P. Marcela, and J.A.C. Leiva. 2024. Measuring social sustainability: A multidimensional approach. *Policy Research Working Paper Series.*

Barr, S. 2007. Factors influencing environmental attitudes and behaviors: A U.K. case study of household waste management. *Environment and Behavior* 39(4):435–473. https://doi.org/10.1177/0013916505283421.

Beatty, T.K.M., P. Berck, and J.P. Shimshack. 2007. Curbside recycling in the presence of alternatives. *Economic Inquiry* 45(4):739–755.

Berck, P., G. Englander, S. Gold, S. He, J. Horsager, S. Kaplan, M. Sears, A. Stevens, C. Trachtman, R. Taylor, and S.B. Villas-Boas. 2021. Recycling policies, behavior and convenience: Survey evidence from the CalRecycle program. *Applied Economic Perspectives and Policy* 43(2):641–658. https://doi.org/10.1002/aepp.13117.

Berck, P., M. Sears, R.L.C. Taylor, C. Trachtman, and S.B. Villas-Boas. 2024. Reduce, reuse, redeem: Deposit-refund recycling programs in the presence of alternatives. *Ecological Economics* 217:108080. https://doi.org/10.1016/j.ecolecon.2023.108080.

Bernstein, J.D. 2004. *Social Assessment and Public Participation in Municipal Solid Waste Management.* Washington, DC: World Bank Group.

Best, H., and T. Kneip. 2019. Assessing the causal effect of curbside collection on recycling behavior in a non-randomized experiment with self-reported outcome. *Environmental and Resource Economics* 72. https://doi.org/10.1007/s10640-018-0244-x.

BioCycle. 2023. 2023 BioCycle Residential Food Waste Collection Access Study. Industry Report. https://www.biocycle.net/residential-food-waste-collection-access-in-u-s.

Blaine, T.W., F.R. Lichtkoppler, K.R. Jones, and R.H. Zondag. 2005. An assessment of household willingness to pay for curbside recycling: A comparison of payment card and referendum approaches. *Journal of Environmental Management* 76(1):15–22.

Bond, C. 2017. "Preparing Newsprint for Bedding." Ohioline, Ohio State University Extension.

Boyer, R.H., A.D. Hunka, M. Linder, K.A. Whalen, and S. Habibi. 2021. Product labels for the circular economy: Are customers willing to pay for circular? *Sustainable Production and Consumption* 27:61–71.

Bucciol, A., N. Montinari, and M. Piovesan. 2015. Do not trash the incentive! Monetary incentives and waste sorting. *Scandinavian Journal of Economics* 117(4):1204–1229.

Burn., S. 2006. Social psychology and the stimulation of recycling behaviors: The Block Leader Approach. *Journal of Applied Social Psychology* 21(8):611–629. http://dx.doi.org/10.1111/j.1559-1816.1991.tb00539.x.

CalRecycle. 2025. *Accurate Recycling Labels.* https://calrecycle.ca.gov/wcs/recyclinglabels.

Caplan, A.J., T.C. Grijalva, and P.M. Jakus. 2002. Waste not or want not? a contingent ranking analysis of curbside waste disposal options. *Ecological Economics* 43(2–3):185–197.

Carson, R.T. 2012. Contingent valuation: A practical alternative when prices aren't available. *Journal of Economic Perspectives* 26(4):27–42. https://doi.org/10.1257/jep.26.4.27.

Cecot, C., and W.K. Viscusi. 2022. The hierarchy and performance of state recycling and deposit laws. *Vermont Journal of Environmental Law* 23(4):319–348. https://irp.cdn-website.com/ee52edf5/files/uploaded/Cecot%26Viscusi_State%20Recycling%20and%20Deposit%20Laws%5B29%5D.pdf.

Champine, V.M., M.S. Jones, and R.M. Niemiec. 2023. Encouraging social diffusion of pro-environmental behavior through online workshop-based interventions. *Conservation Science and Practice* 5(10):e13016.

Chowkwanyun, M. 2023. Environmental justice: Where it has been, and where it might be going. *Annual Review of Public Health* 44(1):93–111.

City of Champaign. 2025. *Champaign environmental sustainability/recycling*. Public Works Department. https://champaignil.gov/public-works/recycling.

Corona Insights. 2020. *Public Opinion Surrounding Plastic Consumption and Waste Management of Consumer Packaging*. https://www.merkley.senate.gov/wp-content/uploads/imo/media/doc/Public%20Opinion%20Research%20to%20WWF%202021.pdf.

Dauvergne, P., and G. LeBaron. 2013. The social cost of environmental solutions. *New Political Economy* 18(3):410–430. https://doi.org/10.1080/13563467.2012.740818.

Deeney, M., R. Green, X. Yan, C. Dooley, J. Yates, H.B. Rolker, and S. Kadiyala. 2023. Human health effects of recycling and reusing food sector consumer plastics: A systematic review and meta-analysis of life cycle assessments. *Journal of Cleaner Production* 397:136567. https://doi.org/10.1016/j.jclepro.2023.136567.

Dijkgraaf, E., and R. Gradus. 2017. An EU recycling target: What does the Dutch evidence tell us? *Environmental and Resource Economics* 68(3):501–526. https://doi.org/10.1007/s10640-016-0027-1.

Donato, C., and F. Adıgüzel. 2022. Visual complexity of eco-labels and product evaluations in online setting: is simple always better? *Journal of Retailing and Consumer Services* 67:102961.

Douglas, I. 2012. Urban ecology and urban ecosystems: Understanding the links to human health and well-being. *Current Opinion in Environmental Sustainability* 4(4):385–392. https://doi.org/10.1016/j.cosust.2012.07.005.

Earth911. 2018. Survey Results: How Much Time Do You Spend Recycling? - Earth911 %. https://earth911.com/survey/survey-results-time-spent-recycling.

EPA (U.S. Environmental Protection Agency). 1974. Analysis of Source Separate Collection of Recyclable Solid Waste-Collection Center Studies.

EPA. 2001. Risk assessment guidance for Superfund. Human Health Evaluation Manual. Part D, Standardized Planning, Reporting, and Review of Superfund Risk Assessment. Final Publication 9285.7-47.

EPA. 2016. Public Participation and Citizen Action. https://archive.epa.gov/epawaste/hazard/web/html/index-46.html.

EPA. 2023. Green Guides Review, Matter No. P954501. U.S. Environmental Protection Agency Comments on the Federal Trade Commission's Proposed Rule entitled "Guides for the Use of Environmental Marketing Claims," April 20, 2023. FTC-2022-0077-1366_attachment_1.pdf.

European Commission. 2025. "European Pillar of Social Rights." Employment, Social Affairs and Inclusion. https://employment-social-affairs.ec.europa.eu/policies-and-activities/european-pillar-social-rights-building-fairer-and-more-inclusive-european-union_en.

Fischhoff, B., D. Riley, D.C. Kovacs, and M. Small. 1998. What information belongs in a warning? *Psychology & Marketing* 15(7):663–686.

Folz, D.H., and J.M. Hazlett. 1990. A national survey of local government recycling. *Resource Recycling* 82–85.

Freudenburg, W.R. 1986. Social impact assessment. *Annual Review of Sociology* 12(1):451–478.

Fullerton, D., and T.C. Kinnaman. 1995. Garbage, recycling, and illicit burning or dumping. *Journal of Environmental Economics and Management* 29(1):78–91. https://doi.org/10.1006/jeem.1995.1032.

Fullerton, D., and W. Wu. 1998. Policies for green design. *Journal of Environmental Economics and Management* 36(2):131–148. https://doi.org/10.1006/jeem.1998.1044.

Geiger, J., L. Steg, E. van de Werff, and A.B. Unal. 2019. A meta-analysis of factors related to recycling. *Journal of Environmental Psychology* 64:78–97. https://doi.org/10.1016/j.jenvp.2019.05.004.

Gionfriddo, G., and A. Piccaluga. 2024. "Startups' contribution to SDGs: A tailored framework for assessing social impact." *Journal of Management and Organization* 30(3):545–573. https://doi.org/10.1017/jmo.2024.3.

Gradus, R., G.C. Homsy, L. Liao, and M.E. Warner. 2019. Which US municipalities adopt pay-as-you-throw and curbside recycling? *Resources, Conservation and Recycling* 143:178–183. https://doi.org/10.1016/j.resconrec.2018.12.012.

Hafsa, F., K.J. Dooley, G. Basile, and R. Buch, R. (2022). A typology and assessment of innovations for circular plastic packaging. *Journal of Cleaner Production* 369:133313.

Hage, O., K. Sandberg, P. Söderholm, and C. Berglund. 2018. The regional heterogeneity of household recycling: A spatial-econometric analysis of Swedish plastic packing waste. *Letters In Spatial and Resource Sciences* 11:245–267.

Hahladakis, J. N., P. Purnell, E. Iacovidou, C.A. Velis, and M. Atseyinku. 2018. Post-consumer plastic packaging waste in England: Assessing the yield of multiple collection-recycling schemes. *Waste Management* 75:149–159.

Hamilton County Environmental Services. n.d. https://hamiltoncountyresource.org/residents/what_can_i_recycle.php.

Haregu, T.N., A.K. Ziraba, and B. Mberu. 2016. A review and framework for understanding the potential impact of poor solid waste management on health in developing countries. *Archives of Public Health* 74(1):54. https://doi.org/10.1186/s13690-016-0166-4.

Harrison, E., A. Berenjian, and M. Seifan. 2020. Recycling of waste glass as aggregate in cement-based materials. *Environmental Science & Ecotechnology* 4:100064.

Heiman, M.K. 1996. Race, waste, and class: New perspectives on environmental justice. *Antipode* 28(2):111–121.

Hewitt, E.L., Y. Wang, A.S. Eck, and D.J. Tonjes. 2023. Keeping up with my neighbors: The influence of social norm feedback interventions on recycling behavior in urban multifamily buildings. *Resources, Conservation and Recycling Advances* 18:200156.

Hird, M.J. 2022. *A Public Sociology of Waste*. 1st ed. Bristol: University Press.

Horner, D., N. Fitzpatrick, T. Ivacko, and J. Berger. 2022a. Recycling Issues, Policies, and Practices among Michigan Local Governments. The Center for Local, State, and Urban Policy. https://closup.umich.edu/michigan-public-policy-survey/98/recycling-issues-policies-and-practices-among-michigan-local-governments.

Horner, D., N. Fitzpatrick, and T. Ivacko. 2022b. Michigan local leaders' views on recycling: Current challenges and opportunities for improvement. The Center for Local, State, and Urban Policy. https://closup.umich.edu/michigan-public-policy-survey/99/michigan-local-leaders-views-recycling-current-challenges-and-opportunities#embed.

Horner, D., N. Fitzpatrick, and T. Ivacko. 2023. Michigan local leaders report widespread support for community recycling programs. The Center for Local, State, and Urban Policy. https://closup.umich.edu/michigan-public-policy-survey/111/michigan-local-leaders-report-widespread-support-community-recycling-programs.

Huber, J., W.K. Viscusi, and J. Bell. 2023. Using objective characteristics to target household recycling policies. *Environmental Law Reporter* 53:10804.

Jacobsen, L.F., S. Pedersen, and J. Thogersen. 2022. Drivers of and barriers to consumers' plastic packaging waste avoidance and recycling—A systematic literature review. *Waste Management* 141:763–789. https://doi.org/10.1016/j.wasman.2022.01.021.

Jakus, P.M., K.H. Tiller, and W.M. Park. 1996. Generation of recyclables by rural households. *Journal of Agricultural and Resource Economics* 96–108.

Jena, L., and M.M. Ahmed. 2024. Integrating Innovative Sustainability Practices for Public Health, Economic Benefits, and Community Well-Being. http://junikhyatjournal.in/no_1_Online_24/37_online_july.pdf.

Johnson, C.Y., J.M. Bowker, and H.K. Cordell. 2004. Ethnic variation in environmental belief and behavior: An examination of the new ecological paradigm in a social psychological context. *Environment and Behavior* 36(2):157–186. https://doi.org/10.1177/0013916503251478.

Kass, M. 2024. "First-of-its-kind plastics recycling microfactory to transform waste, create new jobs." *ASU News*, 2024. https://news.asu.edu/20240207-environment-and-sustainability-firstofitskind-plastics-recycling-microfactory-transform.

Kinnaman, T.C. 2000. Explaining the growth in municipal recycling programs: The role of market and nonmarket factors. *Public Works Management and Policy* 5(1):37–51.

Kinnaman, T.C., and D. Fullerton, D. 2000. The economics of residential solid waste management. In H. Folmer and T. Tietenberg (eds.), *The International Yearbook of Environmental and Resource Economics 2000/2001*. Cheltenham, UK: Edward Elgar.

Klaiman, K., D.L. Ortega, and C. Garnache. 2017. Perceived barriers to food packaging recycling: Evidence from a choice experiment of US consumers. *Food Control* 73:291–299.

Lakhan, C. 2016. Effectiveness of recycling promotion and education initiatives among first-generation ethnic minorities in Ontario, Canada. *Social Sciences* 5(2):23.

Lee, S. 2023. "The Benefits and Costs of a Small Food Waste Tax and Implications for Climate Change Mitigation." https://harris.uchicago.edu/files/seunghoon_lee_jmp_foodwaste_sl.pdf.

Link, T., and B. Stoke. 2011. "Intervening in the Rural and Small Community Waste Reduction System," Circular Economies. https://ced.msu.edu/upload/%5Bada%5DFinal%20CERI%20Report%201-19-22.pdf.

Lintz, G. 2016. A conceptual framework for analysing inter-municipal cooperation on the environment. *Regional Studies* 50(6):956–970. https://doi.org/10.1080/00343404.2015.1020776.

Ma, J. and K.W. Hipel. 2016. Exploring social dimensions of municipal solid waste management around the globe—A systematic literature review. *Waste Management* 56:3–12. https://doi.org/10.1016/j.wasman.2016.06.041.

Marc, M., and N. Ponikvar. 2022. How to measure our impact on society: An illustration of social impact analysis. *Dynamic Relationships Management Journal* 11(2):79–96.

Martuzzi, M., F. Mitis, and F. Forastiere. 2010. Inequalities, inequities, environmental justice in waste management and health. *European Journal of Public Health* 20(1):21–25.

McKenzie-Mohr, D. 2011. Fostering sustainable behavior: An introduction to community-based social marketing. New Society Publishers.

McKie, E.C., A. Chandrasekaran, and S. Venkataraman. 2024. How do curbside feedback tactics impact households' recycling performance? Evidence from community programs. *Production and Operations Management* 33(5):1064–1082.

McKinsey & Company. 2020. Sustainability in packaging: US survey insights https://www.mckinsey.com/industries/packaging-and-paper/our-insights/sustainability-in-packaging-us-survey-insights.

Miafodzyeva, S., and N. Brandt. 2013. Recycling behaviour among householders: Synthesizing determinants via a meta-analysis. *Waste Biomass Valorization* 4:221–235. https://doi.org/10.1007/s12649-012-9144-4.

Michaud, C., and D. Llerena. 2011. Green consumer behaviour: An experimental analysis of willingness to pay for remanufactured products. *Business Strategy and the Environment* 20(6):408–420. https://doi.org/10.1002/bse.703.

Morris, J., and C. Morawski. 2011. *Returning to Work: Understanding the Domestic Jobs Impacts from Different Methods of Recycling Beverage Containers*. https://cooplesvaloristes.ca/v2/wp-content/uploads/2015/04/returning-to-work.pdf.

Murphy, B. 2010. *Community Well-Being: An Overview of the Concept*. https://www.nwmo.ca/-/media/Reports---Reports/1681_researchsupportprogram_communitywellbeingoverview.ashx?rev=4e4609ae66c4488ebb6fbb2601c143a7&sc_lang=en.

NRC (National Research Council). 2002. *New Tools for Environmental Protection, Education, Information, and Voluntary Measures*. Washington, DC: The National Academies Press.

Niemiec, R., M.S. Jones, S. Lischka, and V. Champine. 2021. Efficacy-based and normative interventions for facilitating the diffusion of conservation behavior through social networks. *Conservation Biology* 35(4):1073–1085.

Nixon, H., and J.-D. Saphores. 2009. Information and the decision to recycle: Results from a survey of US households. *Journal of Environmental Planning and Management* 52:257–277. https://doi.org/10.1080/09640560802666610.

Oates, C.J., and S. McDonald. 2006. Recycling and the domestic division of labour: Is green pink or blue? *Sociology* 40(3):417–433. https://doi.org/10.1177/0038038506063667.

Oldenburg, R. 1989. *The Great Good Place: Cafés, Coffee Shops, Community Centers, Beauty Parlors, General Stores, Bars, Hangouts, and How They Get You Through the Day.* Paragon House.

Oliveira, V., V. Sousa, J.M. Vaz, and C. Dias-Ferreira. 2018. Model for the separate collection of packaging waste in Portuguese low-performing recycling regions. *Journal of Environmental Management* 216:13–24.

Owens, J., S. Dickerson, and D.L. Macintosh. 2000. Demographic covariates of residential recycling efficiency. *Environment and Behavior* 32(5):637–650. https://doi.org/10.1177/00139160021972711.

Palmer, K., and M. Walls. 1997. Optimal policies for solid waste disposal: Taxes, subsidies, and standards. *Journal of Public Economics* 65(2):193–205. https://doi.org/10.1016/S0047-2727(97)00028-5.

Perni, Á., J. Barreiro-Hurlé, and J.M. Martínez-Paz. Contingent valuation estimates for environmental goods: Validity and reliability. *Ecological Economics* 189(2021):107144.

Petts, J. 2000. Municipal waste management: Inequities and the role of deliberation. *Risk Analysis* 20(6):821–832. https://doi.org/10.1111/0272-4332.206075.

Pew Research Center. 2016. *Perceptions and realities of recycling vary widely from place to place.* https://www.pewresearch.org/short-reads/2016/10/07/perceptions-and-realities-of-recycling-vary-widely-from-place-to-place.

Pretner, G., N. Darnall, F. Iraldo, and F. Testa. 2021. Are consumers willing to pay for circular products? The role of recycled and second-hand attributes, messaging, and third-party certification. *Resources, Conservation and Recycling* 175. https://doi.org/10.1016/j.resconrec.2021.105888.

Pyzyk, K. 2019. "Robots Move In." *Recycling Labor Series.* https://www.wastedive.com/news/recycling-labor-mrf-robots-move-in/568554.

Recycled Materials Association. 2014. *Younger Americans Not as "Green" as Older Generations. Younger Americans Not as "Green" as Older Generations.* ReMA.

Recycling Partnership. 2022. *2022 Foundational Audience Segmentation Research.* https://recyclingpartnership.org/wp-content/uploads/dlm_uploads/2023/06/Audience_Segmentation_Report_Final.pdf.

Recycling Partnership. 2023. *Consumer Research on Recycling Behavior and Attitudes Regarding On-Pack Labeling.* https://recyclingpartnership.org/consumer-research-on-recycling-behavior-and-attitudes-regarding-on-pack-labeling.

Recycling Partnership. 2024. *State of Recycling: The present and future of Residential Recycling in the U.S.* https://recyclingpartnership.org/wp-content/uploads/dlm_uploads/2024/01/Recycling-Partnership-State-of-Recycling-Report-1.12.24.pdf.

Regadío, M., A.I. Ruiz, I.S. de Soto, M. Rodriguez Rastrero, N. Sánchez, M.J. Gismera, M.T. Sevilla, P. da Silva, J.R. Procopio, and J. Cuevas. 2012. Pollution profiles and physicochemical parameters in old uncontrolled landfills. *Waste Management* 32(3):482–497. https://doi.org/10.1016/j.wasman.2011.11.008.

Reputation Leaders. 2023. *Recycling is too much effort for many Americans.* https://www.reputationleaders.com/media/zerowaste2023.

Reschovsky, J.D., and S.E. Stone. 1994. Market incentives to encourage household waste recycling: Paying for what you throw away. *Journal of Policy Analysis and Management* 13(1):120–139.

Schaumberg, G.W., Jr., and K.T. Doyle. 1994. Wasting resources to reduce waste: Recycling in New Jersey (Policy Analysis No. 202). Washington, DC: Cato Institute.

Schäufele-Elbers, I., G. Schamel, and M. Perathoner. 2024. Nudging food waste off the plate? An explorative study investigating the generation of plate waste over time and the effectiveness of information nudges to bridge the attitude-behavior gap. *Journal of Foodservice Business Research* 1–19.

Schultz, P. W., S. Oskamp, and T. Mainieri, T. 1995. Who recycles and when? A review of personal and situational factors. *Journal of Environmental Psychology* 15(2):105–121.

Sharma, H.B., K.R. Vanapalli, B. Samal, et al. 2021. Circular economy approach in solid waste management system to achieve UN-SDGs: Solutions for Post-COVID recovery. *Science of the Total Environment.* https://www.ncbi.nlm.nih.gov/pmc/articles/PMC9622352.

Shaw, B., J. Resick, and J. Van Rossum. 2014. *Research Review of Best Practices in Campaigns to Promote Recycling, Report for the Wisconsin Recycling and Waste Management Program Leaders.* https://dnr.wisconsin.gov/sites/default/files/topic/Recycling/RecyclingBestPracticesWISummary.pdf.

Sidique, S.F., S.V. Joshi, and F. Lupi. 2010. Factors influencing the rate of recycling: An analysis of Minnesota counties. *Resources, Conservation and Recycling* 54(4):242–249. https://doi.org/10.1016/j.resconrec.2009.08.006.

Small, K. 2013. *Urban Transportation Economics.* Routledge.

Starr, J., and C. Nicolson. 2015. Patterns in trash: Factors driving municipal recycling in Massachusetts. *Resources, Conservation and Recycling* 99:7–18. https://doi.org/10.1016/j.resconrec.2015.03.009.

State of Minnesota. 2025. By its Attorney General, Keith Ellison, v. Reynolds Consumer Products, Inc., Reynolds Consumer Products, LLC, & Walmart Inc., No. 62-CV-23-3104 (Minn. Dist. Ct. 2024). Plastics Litigation Tracker. https://plasticslitigationtracker.org/?keywords=No.+62-CV-23-310.

Sustainable Packaging Coalition. 2021. *2020–21 centralized study on availability of recycling.* Industry report. https://sustainablepackaging.org/wp-content/uploads/2022/03/UPDATED-2020-21-Centralized-Study-on-Availability-of-Recycling-SPC-3-2022.pdf.

Tiller, K.H., P.M. Jakus, and W.M. Park. 1997. Household willingness to pay for drop off recycling. *Journal of Agricultural and Resource Economics* 22(2):310–320.

UN (United Nations). 2025. *Sustainable Development Goals.* Department of Economic and Social Affairs. https://sdgs.un.org/goals.

U.S. Bureau of Labor Statistics. 2024. *Average hourly and weekly earnings of all employees on private nonfarm payrolls by industry sector, seasonally adjusted (Table B-3).* https://www.bls.gov/news.release/empsit.t19.htm.

van Velzen, E.T., M.T. Brouwer, and A. Feil. 2019. Collection behavior of lightweight packaging waste by individual households and implications for the analysis of collection schemes. *Waste Management* 89:284–293.

Varotto, A., and A. Spagnolli. 2017. Psychological strategies to promote household recycling. A systematic review with meta-analysis of validated field interventions. *Journal of Environmental Psychology* 51:168–188. https://doi.org/10.1016/j.jenvp.2017.03.011.

Vhowkwanyun, M. 2023. Environmental justice: Where it has been, and where it might be going. *Annual Review of Public Health* 44:93–111.

Viscusi, W.K., J. Huber, and J. Bell. 2022. Quasi-experimental evidence on the impact of state recycling and deposit laws: Household recycling following interstate moves. *American Law and Economics Review* 24(2):614–658. https://doi.org/10.1093/aler/ahac006.

Viscusi, W.K., J. Huber, and J. Bell. 2023. Changes in household recycling behavior: Evidence from panel data. *Ecological Economics* 208:107819.

Vollaard, B., and D. van Soest. 2024. Punishment to promote prosocial behavior: A field experiment. *Journal of Environmental Economics and Management* 124:102899. https://doi.org/10.1016/j.jeem.2023.102899.

White House. 2024. *National Strategy for Reducing Food Loss and Waste and Organic Recycling.* https://www.whitehouse.gov/wp-content/uploads/2024/06/national-strategy-for-reducing-food-loss-and-waste-and-recycling-organics_6.11.24.pdf.

WHO (World Health Organization). 2023. *Assessing the health impacts of waste management in the context of the circular economy.* https://apps.who.int/iris/handle/10665/366667.

Wiedemann, P.M., and S. Femers. 1993. Public participation in waste management decision making: Analysis and management of conflicts. *Journal of Hazardous Materials* 33(3):355–368. https://doi.org/10.1016/0304-3894(93)85085-S.

Willman, K.W. 2015. Information sharing and curbside recycling: A pilot study to evaluate the value of door-to-door distribution of informational literature. *Resources, Conservation and Recycling* 104:162–171.

Winterich, K.P., G.Y. Nenkov, and G.E. Gonzales. 2019. Knowing what it makes: How product transformation salience increases recycling. *Journal of Marketing* 83(4):21–37.

Wood From the Hood. 2024. *Reclaimed Wood Minneapolis.* Wood From the Hood. https://woodfromthehood.com.

World Economic Forum. 2021. *The Climate Progress Survey: Business & consumer worries & hopes.* https://www3.weforum.org/docs/SAP_WEF_Sustainability_Report.pdf.

7
Benefits and Measuring Environmental Impacts and Externalities of Recycling Programs

---

Summary of Key Messages

- **Reducing use of virgin materials, especially those that are non-renewable:** Extraction of non-renewable virgin materials depletes finite natural resources, emits greenhouse gasses, and destroys ecosystems (through deforestation, eutrophication, and acidification). Use of recycled materials in manufacturing reduces environmental damages by avoiding reliance on virgin materials.
- **Reducing use and extending the service life of landfills:** While risk is mitigated through their regulated design, permitting, and monitoring, landfills have the potential for unintended environmental contamination from the release of heavy metals, ammonia and other nitrogen compounds, acids, and salts, as well as organic compounds, through the release of gas or leachate. Recycling various materials can reduce environmental damages, including leaks and other unintended emissions, by reducing the use of landfills. Furthermore, a major benefit of recycling is extending the service life of landfills.
- **Reducing greenhouse gas emissions:** Greenhouse gases are emitted across the recycling process (e.g., collecting, sorting, remanufacturing). However, the largest source of those emissions is fossil-based energy sources in the remanufacturing process. Emissions reductions may be possible with changes to the local recycling infrastructure, recycling processes, and use of renewable energy in the remanufacturing process.

Recycling is one approach to managing materials that can help reduce waste, conserve resources, and limit environmental impacts. In addition to diverting materials from disposal, recycling contributes to broader resource efficiency efforts and supports industries that rely on recovered materials. While costs and logistical challenges are associated with recycling, its benefits are often considered in discussions of sustainability and waste management policies.

7.1 KEY RECYCLING BENEFITS

Recycling offers various potential benefits, including resource conservation, energy savings, waste reduction, and pollution mitigation. These benefits can contribute to more efficient material use and environmental management while also influencing economic and policy decisions. This chapter provides a more detailed discussion about the environmental benefits. The extent of these advantages depends on factors such as material type, recycling infrastructure, and participation rates.

7.1.1 Reducing Resource Depletion and Pollution from Virgin Material Extraction

Resource depletion and pollution are environmental concerns associated with the extraction of virgin materials. Extracting raw materials such as metals, minerals, and fossil fuels to create the products used every day by households and businesses requires extensive mining and harvesting, leading to the depletion of finite natural resources. The concern for natural resource use is especially high for nonrenewable resources, such as fossil fuels, whereas using resources such as paper or food has less impact from a resource management standpoint because these resources can be renewed or replenished. An additional concern is that extracting resources, whether renewable or nonrenewable, often involves clearing vast areas of land that damages ecosystems (Christensen et al., 2020; Psyrri et al., 2024; Ruan and Zou, 2024). What is more, extracting and processing virgin materials produces emissions that can harm human health and degrade natural habitats if not regulated appropriately.

Recovering materials from the existing waste stream reduces the need for virgin material extraction (Anshassi and Townsend, 2024; Erkisi-Arici et al., 2021). This practice not only conserves finite resources but also minimizes the environmental damage caused by mining and drilling activities. Reusing materials already in circulation maintains the physical integrity of Earth's natural resources and the health of ecosystems and global economies. Indeed, some materials—such as metals and glass—are near infinitely recyclable. Others—such as paper and plastics—can be recycled a limited number of times (e.g., paper can typically be recycled only 5–7 times, because the paper fibers are shortened during the recycling process).

7.1.2 Conserving Energy

Energy consumption is a critical aspect of the cradle-to-grave life cycle of a product, beginning from extraction of raw materials and continuing through product disposal (Bian et al., 2023; Yang et al., 2024; Zhao et al., 2024). The equipment used to harvest and mine materials for product creation is typically powered by diesel, gasoline, and other fossil fuels. Once virgin or raw materials are extracted, they must be transported to manufacturing or processing sites, a process that relies heavily on fossil fuels to power trucks, boats (including barges), and trains. At the manufacturing or processing site, fossil fuels continue to play a role, as the machinery and equipment used to process materials into finished products are energy intensive. This is because many processed materials, such as aluminum, steel and glass, require high temperatures to be created from virgin materials.[1] The life cycle energy demand does not end with the product's use; even after it is discarded, energy is required to transport it to end-of-life treatment facilities. Whether the product is sent to a landfill, incinerated, or processed at a materials recovery facility (MRF), energy is needed to power the equipment that manages waste.

Recycling materials conserves the energy that would have been required to harvest and mine virgin raw materials. Recycling further conserves energy by eliminating the need to transport raw materials from extraction sites to manufacturing or processing facilities. Significant energy savings are generally achieved through "closed-loop recycling," in which recycled material is used to produce its original product—examples include using recycled aluminum or steel to manufacture new cans and using recycled glass to manufacture new glass bottles.

On the other hand, open-loop recycling (e.g., using waste-paper as animal bedding or plastic bottles for construction material) often results in less energy savings compared with closed-loop recycling because the secondary products are typically of lower value or functionality than the original item. From a life cycle perspective, which considers all stages from raw material extraction to end-of-life disposal, open-loop recycling introduces additional complexity. The equipment and processes required to convert waste into a new, often unrelated product can be energy- and resource-intensive. In cases where the secondary product replaces a material that is already low-impact or readily available, the offsets in energy use can be minimal—or even a net positive impact if transportation and processing impacts are high. Thus, while open-loop recycling may reduce landfill volume, it may not always produce a net environmental benefit, especially if the new life cycle requires more input than would have been used to produce the virgin equivalent. Evaluating these trade-offs is essential in sustainability assessments such as life cycle assessment, where recycling offsets must account for not only material diversion but also the quality, efficiency, and environmental load of the new product system.

7.1.3 Reducing Waste and Diverting from Landfills

Landfills are the primary method of waste disposal globally, but materials discarded in a landfill are intermixed with soils, dirt, and other waste. This mixing makes the postdisposal recovery of valuable

---

[1] Energy savings are particularly significant for aluminum and steel production from recycled material rather than processing virgin material from their respective ores (Waste Trade, n.d.).

recyclable commodities—such as metals, plastics, and paper—challenging and often unfeasible economically (Jain et al., 2023; Suknark et al., 2023; Zhi et al., 2023). Likewise, in waste-to-energy (WTE) facilities, where materials are incinerated, the opportunity for closed-loop recycling is effectively lost. Incineration reduces most materials to ash, making them unsuitable for direct reuse in manufacture. However, WTE facilities that burn waste also generate electricity, displacing fossil fuel–generated electricity. They also recycle significant amounts of metals from the WTE ash, eliminate methane emissions from landfilled waste, and reduce the mass of landfilled waste by 70–80 percent.

In contrast, recycling is a form of material recovery that supports both waste reduction and landfill diversion (Galavote et al., 2024b; Huang, 2024; Klemeš et al., 2010; Mueller, 2013). By collecting and processing recyclable materials before they are discarded, recycling preserves the inherent qualities of these materials and enables them to be reused in manufacturing new products. In addition, recycling minimizes waste by diverting materials from disposal facilities, and it helps conserve landfill space and capacity of WTE furnaces. This conservation is particularly important because landfills are often challenging to permit and site. From a financial perspective, recycling offers those operating WTE facilities a means to recover valuable commodities such as aluminum and steel cans. These materials have a higher market value when recycled directly rather than being recovered after damage from incineration.

7.1.4 Preventing Pollution

Materials disposal through landfilling or incineration involve potential emissions to the air, soil, and water as the waste decomposes or combusts. Landfills pose possible environmental concerns because of the release of metals, ammonia, and organic compounds in the form of gas or leachate[2] (de Oliveira et al., 2022; Lott et al., 2024; Ma et al., 2023; Reinhart et al., 2020). Leachate often contains high concentrations of heavy metals and, organic chemicals (measured as total suspended solids and total organic carbon). Other concerning emergent chemicals are perfluoroalkyl and polyfluoroalkyl substances (PFAS) (Lott et al., 2024; Reinhart et al., 2010; Robey et al., 2024). Additionally, these chemicals can be present in landfill gas (de Oliveira et al., 2022; Galavote et al., 2024a; Ma et al., 2023). Diverting materials from landfills through recycling or other means can reduce the generation of these pollutant chemicals. Stringent regulations govern landfills and similar waste treatment facilities—such as those outlined in 40 Code of Federal Regulations (CFR)—but possibilities remain for leaks or unintended emissions into the environment. These emissions can affect local ecosystems adversely and pose risks to human health.

Materials themselves can also become pollutants, as evidenced in the widespread concern about marine plastic debris (Jambeck et al., 2015), much of which stems from river and coastal communities in the Global South (Geyer et al., 2017; Lebreton and Andrady, 2019; Meijer et al., 2021; Schmidt et al., 2017). The exact origin of these plastics is often unclear; they may either be imported plastic waste from the Global North, sold to facilities in the Global South for reprocessing and production, or they could have originated from within the local communities (Plastic Pollution Coalition, 2019), many of which lack adequate waste collection and infrastructure systems. Better waste and recycling management practices are critically needed, as well as improved infrastructure to prevent materials from becoming pollutants.

7.2 RECYCLING RATES: CURRENT SUSTAINABILITY METRICS

The recycling rate is one of the main metrics used to assess progress toward achieving better impacts on the environment and sustainability (i.e., economic, environmental, and social goals). Intended to measure the success of a waste diversion program, the traditional recycling rate is calculated as the recycled weight divided by total municipal solid waste (MSW) generated (including the weight of materials both for recycling and disposal). The point at which data are used to estimate the recycling rate can vary. For example, the recycled amount may be measured as recyclables are collected at the curb, or after the materials

---

[2] When waste is exposed to precipitation, contaminants leach from the wastes, which creates a liquid mixture called leachate.

pass through a MRF where contaminates are removed. In that case, the data report the recycled commodities sold to end markets.

However, inherent issues with this metric remain (Anshassi et al., 2018). First, it does not distinguish between the environmental, economic, or social impacts of recycling different materials (e.g., 1 ton of recycled glass is considered equivalent to 1 ton of recycled aluminum cans, despite their vastly different impacts). Second, this measure neglects any impacts to social welfare (e.g., jobs produced, recycling participation). Third, the metric fails to account for the benefits of manufacturing or production improvements, such as lightweighting, or to consider the benefits of source reduction and waste minimization (because the recycled mass or volume is included in both the numerator and denominator). Fourth, the recycling rate is used for any waste component regardless of the availability of collection programs, sorting equipment, or end-markets. Fifth, this recycling rate can rise or fall over time for reasons unrelated to recycling. For example, the positive effects of economic growth or new technologies can appear to have negative effects on the recycling rate, because greater total household consumption raises the denominator, even without changing their recycling in the numerator.

These limitations highlight the need for a different assessment method for measuring recycling benefits and progress, especially for goal setting in sustainability initiatives. As part of other research, life cycle assessment has been proposed as a tool for developing new methods of evaluating progress toward sustainability. For example, the Sustainable Materials Management framework uses LCA to shape policy decisions. The next section details what has been learned about primary recycling materials using LCA, and it is followed by a review of the sustainable materials management framework.

7.3 LIFE CYCLE ASSESSMENT

Decision-makers and policymakers frequently use LCA to evaluate the impacts of systems and policies on the environment, economy, and society. As a computer-based tool, LCA quantifies the environmental benefits or burdens associated with a material throughout its entire life cycle (Khandelwal et al., 2019). The life cycle stages typically start with the extraction of raw materials and extend through processing those materials, manufacturing, sale, use, and end-of-life management (Blikra Vea et al., 2018; Kirkeby et al., 2006; Laurent et al., 2014a; Reap et al., 2008a).

The International Organization for Standardization (ISO) has developed guidelines—ISO 14040 and ISO 14044—that outline the requirements for conducting an LCA (Guinée et al., 2011; Khandelwal et al., 2019; Pryshlakivsky and Searcy, 2013; Reap et al., 2008a; Yadav and Samadder, 2018). These guidelines describe four key phases: (1) goal and scope definition, (2) inventory analysis, (3) impact assessment, and (4) interpretation.

The results of an LCA are not exact measurements but potentials. For example, the greenhouse gas emissions estimated using LCA is the potential emissions footprint associated with the systems, evaluated as part of the goal and scope definition.

7.3.1 Waste Versus Product Life Cycle Assessment

When applying LCA to a waste management system, users can choose between a generic product LCA model or a specialized waste LCA model (Gentil et al., 2010). *Product LCA models* handle a single product from extraction to end of life, whereas *waste* LCA models assess heterogeneous materials comprising various waste fractions (see Table 7-1; see also Clavreul et al., 2014).

Product LCA models typically do not include the flexibility needed to model a functional unit with multiple waste fractions directly. However, some product LCA models offer supplementary add-on modules for landfilling or incineration end-of-life treatments. Practitioners may prefer waste LCA models over product LCA models, because waste LCA models provide an environment that includes all the necessary life cycle inventory and impact assessment methods to model multiple waste fractions under various treatment processes (Clavreul et al., 2014; Gentil et al., 2010). However, not all waste LCA models can handle complex systems with different waste treatment technologies. Waste management environmental modeling usually reflects the environmental footprint associated with each material for various common end-of-life management options:

- *Recycling* includes transportation to the MRF and remanufacture facilities, sorting and processing at the MRF, and use of sorted recyclables as a secondary feedstock (to avoid emissions by not extracting virgin materials).
- *Biological treatment* covers transport, equipment uses, fugitive emissions, virgin fertilizer avoidance, and soil carbon storage.
- *Thermal treatment* involves the combustion process of waste fractions, use of ancillary materials for facility operation (e.g., ammonia), the transport of ashes, management of ash at a landfill, recovery of ferrous and nonferrous metal fractions from the ash, and the avoided emissions from electricity production and from not extracting virgin metals.
- *Landfilling* accounts for the use of materials for landfill construction, operation, closure, and postclosure care; emissions from landfill gas (e.g., methane generation) and from landfill leachate; and the offsets due to carbon storage and avoided emissions from natural gas extraction and electricity production from landfill gas recovery.
- *Production* encompasses all upstream stages before a product is consumed, including raw materials acquisition, processing, manufacturing, and associated transportation.
- *Source reduction* accounts for avoiding the production, consumption, or reuse of a product. Results are reported as negative values relative to production impacts.

TABLE 7-1 Produce and Waste Life Cycle Assessment (LCA) Models

	Description	**Systems Covered**	**Example Programs**	**Example Uses**
Product LCA Model	Handles a single product from its extraction to end of life. Does not directly model a functional unit with multiple waste fractions. Requires users to select the desired life cycle inventory (LCI) databases and life cycle impact assessment (LCIA) methods, which typically are not preloaded into the user working environment.	Extraction of raw materials, processing of raw materials to a desired form, manufacturing form into a product, distributing to market, consumer use, and end-of-life management.	SimaPro, OpenLCA, LCA for Experts (formerly GaBi), Waste Resources Assessment Toolkit for the Environment (WRATE)	Measure the environmental impacts associated with a single product or material throughout its life cycle or at a certain life stage (e.g., manufacturing).
Waste LCA Model	Assesses a heterogeneous material containing various waste fractions and typically provides an environment that contains all the necessary LCI analysis and LCIA methods to model many waste fractions under various treatment processes.	Follows a "zero-burden" assumption, where the system starts at the collection of the product from a waste generator (e.g., home, business) then is transported to a waste management facility and treated. In cases where the material treatment involves recycling or remanufacturing, the emissions associated with a material's extraction, processing, and manufacturing (i.e., upstream stages) are accounted for in the system.	Municipal Solid Waste Decision Support Tool (MSW-DST), Solid Waste Optimization Lifecycle Framework (SWOLF), Waste Reduction Model (WARM), Environmental Assessment of Environmental Technologies (EASETECH)	Compare the environmental impacts associated with various prospective solid waste management approaches.

SOURCE: Anshassi and Townsend, 2020.

7.3.2 Economic Input-Output Life Cycle Assessment

The LCA models described in Table 7-1 predict environmental impacts at the scale of a single MSW management system, or for a unit of material (mass or dollar value) processed and managed with an assumed set of technologies. Characterization factors are then used, expressed per unit mass (e.g., emissions or environmental health effect). More detailed and accurate calculations can be supported for a single site, a single MSW system, or a given technology, as permitted by data, expertise, models, and time. Furthermore, their results can be summed and aggregated to a state or national level, but only with knowledge of the inventory of sites, systems, and technologies across the area of aggregation. The effort needed to gather and evaluate this information is high, even for a single state, and might be prohibitive.

The economic input-output (EIO) LCA framework provides a rapid, although highly aggregated alternative. It starts with an estimate of material or dollar flows between sectors of the economy and then assigns a quantity of environmental impact to each unit of activity within each sector. Sectorial emissions or impacts are then calculated and summed to determine the total economy-wide impact, accounting for current or projected levels of direct sectorial activity and the indirect upstream activities that support them through the supply chain.

In its simplest form, the EIO-LCA framework allows users to evaluate environmental impacts associated with economic activities, drawing from environmental data that converts cost data to environmental impacts, plus financial data (e.g., import and export prices, purchase prices, value-added, price index data). The method has been applied in a number of national and regional studies of alternative product and process options (Castellani et al., 2019; San Miguel et al., 2024). In some cases, it is used in a hybrid approach, in which direct product or process models are first evaluated and then linked to an EIO model (Ercan and Tatari, 2015; Zhao et al., 2016). Databases and tutorial support for EIO-LCA models are available (Hauschild et al., 2018; Hendrickson et al., 2010; Nakamura, 2023).

The U.S. Environmentally Extended Input-Output model, developed by the U.S. Environmental Protection Agency (EPA), is based on EIO-LCA and tailored to the structure of the U.S. economy. It provides environmental impacts for various sectors and relies not only on EPA databases but also on financial data from the U.S. Census and the Bureau of Economic Analysis.

A few studies using the EIO-LCA approach are described here. First, Huang and Matthews (2008) conducted an economy-wide assessment of U.S. goods and services consumption using the EIO-LCA model, reporting:

> Power generation contributes to nearly one fifth of the total embodied energy and greenhouse gas equivalent emissions in manufactured goods; and for the services and other institutions sectors, its contributions are more than one third. . . . Consumer purchases of waste management services are found to contribute to nearly a quarter of all cancer and non-cancer impacts in the entire economy, signaling the need for producer responsibility policy aimed to reduce toxic materials that eventually enter the waste stream. Subtotal supply chain analysis of packaging materials found that on an energy basis, there exist opportunities to expand the existing applications of deposit-refund programs on beverage containers to other goods. Agencies, companies, and industry groups can use sectoral and supplier contribution analyses to identify opportunities for reducing the life cycle impacts of their products.

Kumar and colleagues (2016) adapted and tailored an EIO-LCA model to estimate Indiana's statewide greenhouse gas emissions from wind turbine electricity generation over its life cycle, from manufacturing through operations and decommissioning. They demonstrate that wind energy production is not entirely free of greenhouse gas emissions when considering all costs and life cycle stages.

DiStefano and Belenky (2009) explore the U.S. nationwide impact of converting (MSW) to methane in anaerobic digesters to generate renewable energy, reduce greenhouse gas emissions, and save landfill space. The authors used the EIO-LCA model from Hendrickson et al. (2010) to calculate carbon dioxide–equivalent emissions from landfill activity and the projected reduction achievable from implementation of nationwide anaerobic digesters. They project that these systems would result in greenhouse gas emissions savings equivalent to a nationwide emissions reduction of 1.9 percent, compared with U.S. greenhouse gas emissions in 2006 (DiStefano and Belenky, 2009). A significant portion of this projected reduction is achieved within the waste management and remediation sector.

Deniz and colleagues (2023) predicted environmental emissions and energy and material consumption from alternative packaging waste collection rates in Avcilar Municipality, Istanbul City, Türkiye. The EIO-LCA model results showed that, as the amount of packaging waste collection increased, the top two sectors responsible for the most greenhouse gas emissions were the oil and gasoline production and the electricity production and supply sectors (Deniz et al., 2023).

In summary, compared with alternative, process-detailed methods using life cycle inventory and impact assessment steps, the EIO-LCA approach is at a coarser scale and has difficulty assessing effects of specific technologies in the MSW management and recycling system. Use of a hybrid LCA model may be able to address this problem. The EIO-LCA model may also be appropriate for evaluating national and state components of MSW impacts, in parallel with more detailed system studies or as a first step of analysis.

7.3.3 General Formula for Calculating Environmental Impacts

The simplified formula for calculating the potential environmental impact from recycling requires taking the difference between two estimated values: (1) calculated impacts of producing the same quantity of product from virgin materials; and (2) calculated impacts of producing the same quantity of product from recycled (or secondary) materials. When the net difference is taken, a net negative value means that the product made from recycled materials results in environmental damage avoidance. However, if the net difference is a positive value, then production using recycled materials results in no environmental avoidance and instead an additional environmental damage. For example, the greenhouse gas emissions for recycling aluminum cans (i.e., used beverage containers) typically results in net negative greenhouse gas emissions because producing one can using recycled material will generate fewer emissions than producing one can using virgin material. Using aluminum cans as a feedstock to produce a new aluminum can avoids the need to mine virgin aluminum ore (i.e., bauxite) and all the related emissions, energy, waste, and resources.

7.3.4 Key Recycling Modeling Assumptions in Life Cycle Assessments

Recycling primarily generates emissions or offsets related to MRF sorting, transporting to remanufacturing facilities, and the remanufacturing process itself (Anshassi and Townsend, 2021).[3] While emissions from MRF sorting can vary by material and model, their contributions are generally negligible. For instance, Anshassi and Townsend (2021) reported that the average contribution of MRF sorting to the net greenhouse gas emissions potential environmental impact factor is approximately 15 percent across several models.

Table 7-2 presents key input assumptions and their defaults for each model along with other key assumptions related to recycling modeling. These models use technological separation efficiency to allocate the life cycle impact associated with MRF sorting on a material-specific basis. The Municipal Solid Waste Decision Support Tool assumes the most aggressive separation efficiency is 99 percent for all materials, while the other models also assume high efficiencies (although specific to each material). Note the MRF separation efficiencies reported in Table 7-2 may differ from those cited earlier in the report (e.g., 87 percent), as the earlier values reflect more recent data, whereas the values in Table 7-2 represent default assumptions from the time the models were originally developed (or in the case of any updates).

---

[3] Waste collection (i.e., transportation from household to MRFs) was not included in the models cited in Anshassi and Townsend (2021).

TABLE 7-2 Key Input Assumptions and Their Defaults for Each Assessment Model

Parameter	WARM	MSW-DST	SWOLF	EASETECH	WRATE
Facility Management					
Facility Type	Not reported	Commingled recyclables with both manual and mechanical sorting with no advanced sorting technology	Commingled recyclables with both manual and mechanical sorting with some advanced sorting technology (i.e., optical glass, PET, HDPE sorter)	Commingled recyclables with both manual and mechanical sorting with no advanced sorting technology	Commingled recyclables with both manual and mechanical sorting with some advanced sorting technology (i.e., optical glass, PET, HDPE sorter)
Facility Lifetime	Not reported	Not reported	30 Yrs	Not reported	25 Yrs
Emissions from facility construction, operation, and decommission	Not Included	Not Included	Not Included	Not Included	Included
MRF Process					
MRF separation Efficiency					
Newspaper	95%	99%	93%	100%	99%
Cardboard	100%	99%	93%	100%	99%
Office Paper	91%	99%	93%	100%	99%
HDPE	92%	99%	93%	90%	49%
PET	95%	99%	93%	90%	49%
Glass	90%	99%	96%	95%	100%
Aluminum Cans	100%	99%	93%	100%	100%
Steel Cans	100%	99%	93%	100%	78%
Remanufacturing Process					
Distance to remanufacture (km), market type					
Newspaper	414, U.S.	322, U.S.	100, U.S.	250, European	250, European
Cardboard	1,086, U.S.	322, U.S.	100, U.S.	250, European	250, European
Office Paper	414, U.S.	322, U.S.	100, U.S.	250, European	250, European
HDPE	800, U.S.	483, U.S.	100, U.S.	800, European	144, European
PET	800, U.S.	483, U.S.	100, U.S.	800, European	250, European
Glass	573, U.S.	322, U.S.	100, U.S.	100, European	250, European
Aluminum Cans	533, U.S.	644, U.S.	100, U.S.	250, European	250, European
Steel Cans	533, U.S.	805, U.S.	100, U.S.	250, European	250, European
Remanufacture type					
Newspaper, Cardboard, Office Paper, HDPE, PET, Glass, Aluminum Cans, Steel Cans	All Closed-loop				
Substitution ratio (Mg of recycled material substituted per Mg of virgin material)					
Newspaper	1	1	1	1	1
Cardboard	1	0.85	1	1	1
Office Paper	1	1	1	1	1
HDPE	1	1	1	1	1

continued

TABLE 7-2 continued

Parameter	WARM	MSW-DST	SWOLF	EASETECH	WRATE
Substitution ratio (Mg of recycled material substituted per Mg of virgin material)					
PET	1	1	1	1	1
Glass	1	1	1	1	1
Aluminum Cans	1	1	1	1	1
Steel Cans	1	1	1	1	1
Recycled Input Ratio (Mg of product made per 1 Mg of recycled material)					
Newspaper	0.94	1	0.94	0.86	1.32
Cardboard	0.93	1	0.93	0.92	0.89
Office Paper	0.66	1	0.65	0.84	0.99
HDPE	0.93	1	0.86	0.93	0.85
PET	0.94	1	0.86	0.80	0.76
Glass	0.98	1	0.97	1	1
Aluminum Cans	0.93	1	0.93	0.93	1
Steel Cans	0.98	1	0.84	0.84	1
Forest Carbon Offset	Included	Not Included	Not Included	Not Included	Not Included

NOTES: The data presented here reflect the time at which the study was published. Since then, updates to these models have been made, including the release of MSW-DST v2, which now shares more similarities with the assumptions used in SWOLF. EASETECH = Environmental Assessment System for Environmental Technologies; HDPE = high-density polyethylene; MRF = materials recovery facility; MSW-DST = Municipal Solid Waste Decision Support Tool; PET = polyethylene terephthalate; SWOLF = Solid Waste Optimization Lifecycle Framework; WARM = Waste Reduction Model; WRATE = Waste and Resources Assessment Tool for the Environment.
SOURCE: Anshassi and Townsend, 2021.

Depending on the material, transportation contributions to greenhouse gas emission footprints typically range from less than 1 percent to 3 percent (Anshassi and Townsend, 2021). These models account for the transportation of recovered materials to remanufacturing facilities that are often hundreds of miles away. Some LCA practitioners have observed that transportation-related emissions are relatively minor compared with the emissions associated with sorting and remanufacturing materials (Rigamonti et al., 2017).

According to Anshassi and Townsend (2021), remanufacturing contributes the largest share to the net recycling potential environmental footprint across many environmental indicators and waste LCA models, and it exhibits the greatest variability in magnitude. For instance, aluminum cans generally show the largest environmental avoidance for most indicators and models, while glass tends to show the least avoidance. Most models indicate that cardboard may result in net added emissions (though the Waste Reduction Model associates cardboard with an avoidance [Anshassi and Townsend, 2021]). These materials are modeled using a closed-loop recycling approach, where the material is remanufactured back into the same product without significant quality degradation (Schrijvers et al., 2016). However, this modeling approach does not match reality for certain materials, such as plastics, where most recycling is open. That is, recycled plastic bottles are generally used for the manufacture of different products, such as textile fibers.

The models estimate the environmental impacts of remanufacturing based on the assumption that each ton of recycled material replaces a particular amount of virgin material in producing the same product (called the substitution ratio). Most models use a one-to-one substitution ratio. Although a closed-loop approach is used, model developers acknowledge that complete quality preservation is idealistic and therefore use a recycling–input ratio. This ratio is specific to each material and represents the amount of new product per metric ton of recycled material input (e.g., material-specific ratios ranging from 0.65 to 1.32).

7.4 REPORTED ENVIRONMENTAL FOOTPRINTS FOR RECYCLING MATERIALS

The overall environmental impacts of recycling material categories vary widely, depending on the environmental indicator considered, because of the different processes involved in creating these products, spanning raw material extraction, manufacturing, usage, and end-of-life management. As noted in the previous section, assumptions and methodologies used to measure the impacts of recycling must be referenced against the production of products using virgin materials. Table 7-3 reports the environmental footprint—listed as emissions, pollutants, and resource usage avoided—for the most commonly recycled materials. The table is followed by a narrative explanation of the units used for each area.

TABLE 7-3 Environmental Footprints for Recycling Materials

Material	Carbon Dioxide (kgCO$_2$eq)	Energy (MJ)	Water (gallons)	Human Health Toxicity (CTUH)	Ecotoxicity (CTUe)	Eutrophication (kgNeq)	Acidification (kgSO$_2$eq)
Aluminum cans	−8.3 to −16.3	−135 to −194	−8.8	-3.4×10^{-9} to -3.0×10^{-6}	−0.4 to −53	-1.2×10^{-3} to -5.5×10^{-3}	−0.044 to −4.7
Steel cans	−0.8 to −2.5	−10 to −19	−0.6	3.9×10^{-6} to -2.6×10^{-7}	0.6 to −7.1	-4.1×10^{-5} to -5.6×10^{-4}	−0.0012 to −0.4
Plastic (PET and HDPE)	−0.4 to −2	−19 to −55	−0.4	2.8×10^{-8} to -4.0×10^{-7}	0.01 to −15	-1.7×10^{-3} to 8.9×10^{-4}	−0.0019 to −0.7
Glass (mixed by color)	−0.2 to −0.4	−1.8 to −2.3	−0.05	4.1×10^{-8} to -4.5×10^{-8}	0.06 to −1.8	-1.3×10^{-5} to -5.2×10^{-5}	−0.0010 to −0.03
Paper (newspaper and office paper)	0.5 to −2.6	3.3 to −19	0.03 to −0.6	5.7×10^{-8} to -2.9×10^{-7}	1.7 to −8.9	3.9×10^{-4} to -9.9×10^{-4}	0.027 to −0.3
Cardboard	0.2 to −2.8	0.8 to −14	0.10	2.4×10^{-9} to -1.3×10^{-6}	0.6 to −38	-3.1×10^{-4} to -9.2×10^{-3}	−0.0003 to −0.14

NOTES: All negative units are avoidances per kilogram of recycled material; all positives are emissions per kilogram of recycled material. Data based on modeling emissions using WARM, MSW-DST v2, SWOLF, EASETECH, and WRATE.

Carbon Dioxide

Greenhouse gases absorb energy and slow it from escaping into space, which causes the Earth to warm. Greenhouse gas quantities are expressed as kilograms of carbon dioxide equivalents (kgCO$_2$eq) per kg of material, to allow for comparison of global warming impacts of different gases relative to CO$_2$; kgCO$_2$eq is a measure of how much energy the emission of 1 kg of gas will absorb over a given period, relative to the emissions of 1 kg of CO$_2$. Emissions levels, and their avoidance, depend on several factors, including the local recycling infrastructure, the type of energy used (renewable versus fossil fuels), and the specific recycling processes.

Energy

The amount of direct and indirect energy used throughout the life cycle of product from both nonrenewable and renewable energy sources is included in the energy use indicator. The energy savings achieved through recycling depend on several factors, including the efficiency of local recycling processes and the type of energy used in manufacturing (renewable versus nonrenewable). The justification for energy avoidance or additional energy needed for each material are the same as provided in the above section on carbon dioxide, because the largest sources of emissions or avoidances come from the use or offset of fossil-based energy sources in the remanufacturing process. Often the energy levels required to manufacture a new product using virgin sources are greater than when those required to recycle feedstocks.

Water

The water use indicator considers the amount of the water evaporated, incorporated into products, transferred to other watersheds, or disposed into the sea. Water savings from recycling translates to a reduction in the demand of water resources that are used heavily in production of new materials from raw sources.

Human Health Toxicity and Ecotoxicity

The release of toxic materials and exposure to humans via inhalation or ingestion are considered for various human health effects during LCA. The units are expressed as comparative toxic units (CTUh), interpreted as disease cases per kg of substance emitted. This measure indicates adverse impacts and includes cancer and other noncancer diseases (or total human toxicity potential). In contrast, ecotoxicity considers the release of toxic materials into an aquatic ecosystem. The units are expressed as comparative toxic units (CTUe), interpreted as the potentially affected fraction of species over time and volume per kg of substance emitted (or total ecotoxicity potential). These metrics provide a comparative measure of the potential health and ecological risks associated with different materials and processes.

Eutrophication

Eutrophication considers the enrichment of aquatic ecosystems from nutrients that cause undesirable algal growth (e.g., nitrates, phosphates). The units are expressed as kilograms nitrogen equivalence (kgNeq) to allow for comparison of nutrients in the water relative to nitrogen. For most materials, potential eutrophication impacts arise from the use of fossil fuels in the extraction of the virgin material and the electricity used in its manufacture. Other potential causes may be nitrogen or phosphate-based chemicals in the extraction and manufacturing processes.

Acidification

Acidification considers the increasing concentration of hydrogen ions within the environment due to the addition of acids. The units are expressed as kilograms sulfur dioxide equivalence (kgSO$_2$eq) to allow for comparison of acids in the air relative to sulfur dioxide. A main concern from acidification is the impacts of acid rain on ecosystems, infrastructure, and human health. Like eutrophication potential, acidification sources come from the use of fossil fuels in the extraction of the virgin material and the electricity used in its manufacture, with other sources from the acids used in the beneficiation process of metals (aluminum and steel), processing of petroleum fractions for those used in plastics, processing of silica sand for glass, and pulping process needed to dissolve the wood fibers to produce paper pulp.

Not covered in this section are the environmental impacts of a curbside recycling system. For example, Anshassi and Townsend (2023) estimate that the average greenhouse gas emissions from a typical residential curbside recycling program amount to 0.046 metric tons of carbon dioxide equivalent per household per year. The greatest sources of emissions are waste collection and landfilling waste, while the most significant offsets are achieved through recycling instead of landfilling materials such as metals, paper, and plastics.

7.5 SUSTAINABILITY MATERIALS MANAGEMENT

The concept of sustainable materials management first emerged from the EPA (2002) publication *Beyond RCRA: Waste and Materials Management in the Year 2020*. It was further developed in the EPA (2009) report *Sustainable Materials Management: The Road Ahead*. These and other documents describe the framework as a set of resource-efficient strategies implemented throughout the entire life cycle of materials and products, encompassing extraction, refinement, manufacturing, assembly, distribution, use, and end-of-life management (see Figure 7-1).

FIGURE 7-1 Life stages included in the sustainable materials management framework.
SOURCE: EPA, 2009.

Unlike traditional waste management approaches, which often focus solely on disposal, sustainable materials management aims to optimize the use of resources and minimize waste and pollutants at every stage. Local governments adopting this framework seek to create policies that promote the most efficient use of resources while mitigating environmental impacts. From a policy perspective, sustainable materials management represents a long-term, systemic approach to waste management that considers the interests of all stakeholders, both public and private.

As policymakers integrate sustainable materials management principles into regulatory frameworks, their goal is to foster sustainable production and consumption practices while transforming end-of-life management into a driver of enhanced sustainability and productivity. To guide their decisions, they rely on LCA models that quantify material flows across all life-cycle stages in terms of environmental, economic, and social impacts. LCA models not only track the pathways of material flows but also identify which economic sectors generate the most waste. Additionally, they evaluate the environmental and economic effectiveness of various waste management strategies, providing valuable insights into how different approaches impact sustainability.

7.6 REDUCING WASTE ENVIRONMENT FOOTPRINTS

Although specific needs and strategies apply to each material, this discussion focuses on two key categories that show exceptional promise for reducing waste: mixed paper—the largest component found in recycling bins—and food waste, which presents some of the most complex environmental challenges and has one of the lowest recovery rates. Addressing these two streams is critical for maximizing diversion and improving system-wide sustainability outcomes. Following a description of current policies and considerations for each of these materials, Table 7-4 describes the political feasibility of each scenario.

TABLE 7-4 Political Feasibility of Each Scenario, Defined by Evaluating Similar Programs and Policy Challenges

Scenario	Has Local Government Adopted a Similar Scenario?	If So, How Is It Applied?	Policy Challenges
Junk Mail Ban	Yes	• **United States:** The U.S. national "Do Not Call Registry": subscribers input their contact information; for a small fee, the Federal Trade Commission removes their phone number and email address from unwarranted telemarketing calls (Cain, 2005; Wambuguh, 2011). • **France, Germany, Denmark, United Kingdom:** "Stop Pub," "No Junk Mail," or "No Junk Mail, Please!" stickers that residents can place onto their mailbox to indicate they are opting out of receiving junk mail (Liebig and Rommel, 2014; Resse, 2005; Simon, 2016).	• Low retention rate for adherence to stickers and registry, because of consumer attachment to store flyers, monetary savings, and pastimes.
Food Donation Mandate	Yes	• **United States:** Food Waste Challenge and the U.S. Environmental Protection Agency Food Recovery Challenge are federal programs designed to prevent food waste. Nine states and several cities enacted state-level food donation tax incentives and organic disposal bans (Chen and Chen, 2018). • **France:** Legislation bans certain grocery stores of a certain size from tossing unsold food and ensures the food is circulated within the French food bank network (Busetti, 2019; Mourad, 2015; Redlingshöfer et al., 2020). • **Denmark, Belgium:** Advancement of food waste reduction through collaborations among multiple stakeholders along the food supply chain (De Boeck et al., 2017; Halloran et al., 2014). • **Italy, Romania:** Regulations permit a more streamlined process for food donations (e.g., reduced the number of forms required for recording purposes, removed sanctions associated with donating food beyond its "sell-by" date, and reduced disposal waste taxes for food donors).	• Unprofessionalism in food safety management, consumer misconception that donated food is unsafe for consumption, fragmented business structure of charity organizations, lack of flexibility and clarity in the regulations, incurred expenses by donors and donation organizations due to new personnel needed to manage donations, storage needs, and distribution costs, etc. (Busetti, 2019; De Boeck et al., 2017; Diaz-Ruiz et al., 2019; Sakaguchi et al., 2018; Schneider, 2013).

7.6.1 Reducing the Use of Mixed Paper

Several strategies can be employed to reduce the environmental footprint of mixed paper effectively. Encouraging residents to receive utility bills and bank statements electronically, rather than by mail, can significantly cut down on paper usage. Providing incentives for this transition can further boost participation. Additionally, instituting policies to minimize or prohibit unsolicited junk mail can also contribute to paper reduction. Since much of the junk mail comes from credit card, mortgage, and insurance companies (Wambuguh, 2011), targeting these industries with specific policies could be particularly effective. In office, university, and business settings, reducing paper use can be achieved by limiting the number of prints allowed per employee and displaying individual print volumes to promote more mindful printing. Another simple yet impactful measure is configuring printers to default to double-sided printing, which can be implemented easily across various work environments. These measures collectively support a reduction in mixed paper consumption and its associated environmental impacts.

Over the past 2 decades, numerous countries have implemented national programs to help residents reduce the amount of junk mail they receive. In the United States, the "Do Not Mail Registry" allows residents to opt out of unsolicited advertising, such as preapproved credit card offers and insurance advertisements (Cain, 2005; Wambuguh, 2011). Similarly, France passed regulations requiring both producers and distributors of junk mail to participate in recycling and providing residents with "Stop Pub" stickers to place onto their mailbox indicating they opted out of receiving junk mail (Resse, 2005). Germany, Denmark, and the United Kingdom have also enacted legislation providing their residents with stickers such as "No Junk Mail" or "No Junk Mail, Please!" (Liebig and Rommel, 2014; Resse, 2005; Simon, 2016). While these regulations have contributed to source reduction of paper waste, the retention rates for adherence to these stickers were relatively low, only 9 percent in France and 25 percent in Germany (Simon, 2016), further indicating ongoing challenges in consumer adoption.

7.6.2 Reducing Food Waste

Global recognition is growing about the environmental, social, and economic issues concerning the copious mass of food waste (Hannibal and Vedlitz, 2018; Papargyropoulou et al., 2014), and many countries have established food donation regulations and programs to combat the generation of waste (Busetti, 2019; Chen and Chen, 2018; De Boeck et al., 2017; Diaz-Ruiz et al., 2019; Halloran et al., 2014; Nomura, 2020; Redlingshöfer et al., 2020; Thyberg and Tonjes, 2016). Several U.S. states (e.g., California, Oregon) and cities (e.g., San Francisco, California; Portland, Oregon) passed state-level food donation tax incentives (Chen and Chen, 2018). In Europe, France implemented a national policy mandating supermarkets to donate edible fractions to charitable organizations and placed a disposal ban on edible food (Busetti, 2019; Mourad, 2015; Redlingshöfer et al., 2020). Denmark, Belgium, Italy, and Romania advanced food donations through various similar regulations (Busetti, 2019; Redlingshöfer et al., 2020).

Several researchers have documented potential issues arising from food donation programs, including consumer misconception that donated food is unsafe for consumption, fragmented business structure of donation organizations, lack of flexibility and clarity in the regulations, and additional expenses incurred by donors and donation organizations (Busetti, 2019; De Boeck et al., 2017; Diaz-Ruiz et al., 2019; Sakaguchi et al., 2018; Schneider, 2013). The political viability of these programs is improved when local governments follow a multifaceted approach that includes consumer education and proactive planning with regulators and food supply stakeholders (Hamilton et al., 2015; Thyberg and Tonjes, 2016). Tevapitak and colleagues (2019) identified the criticality of collaboration between local governments and stakeholders to the beneficial management of water, an approach that can be extended to food waste. For example, both the United States and Italy passed a "Good Samaritan Act" to prevent liability disputes for food donors and donation organizations (De Boeck et al., 2017; Sakaguchi et al., 2018).

REFERENCES

Anshassi, M., S. Laux, and T.G. Townsend. 2018. Replacing recycling rates with life-cycle metrics as government materials management targets. *Environmental Science & Technology* 52(11):6544–6554. https://doi.org/10.1021/acs.est.7b06007.

Anshassi, M., and T.G. Townsend. 2020. Looking Beyond Florida's 75% Recycling Goal: Development of a Methodology and Tool for Assessing Sustainable Materials Management Recycling Rates in Florida. Hinkley Center for Solid and Hazardous Waste Management, University of Florida.

Anshassi, M., and T.G. Townsend. 2021. Reviewing the underlying assumptions in waste LCA models to identify impacts on waste management decision making. *Journal of Cleaner Production* 313:127913. https://doi.org/10.1016/j.jclepro.2021.127913.

Anshassi, M., and T.G. Townsend. 2023. The hidden economic and environmental costs of eliminating kerbside recycling. *Nature Sustainability* 6:919–928. https://doi.org/10.1038/s41893-023-01122-8.

Anshassi, M. and T.G. Townsend. 2024. Residential recycling in Florida: A case study on costs, environmental impacts, and improvement strategies. *Resources, Conservation and Recycling* 206:107627. https://doi.org/10.1016/j.resconrec.2024.107627.

Bian, Z., J.-X. Lu, Y. Huang, D. Xuan, G. Ou, and C. Sun Poon. 2023. Recycling of waste glass in lightweight geopolymer using incineration bottom ash as a foaming agent: Towards energy conservation. *Construction and Building Materials* 400:132632. https://doi.org/10.1016/j.conbuildmat.2023.132632.

Blikra, V.E., V. Martinez-Sanchez, and M. Thomsen. 2018. A review of waste management decision support tools and their ability to assess circular biowaste management systems. *Sustainability* 10(10):3720. https://doi.org/10.3390/su10103720.

Busetti, S. 2019. A theory-based evaluation of food waste policy: Evidence from Italy. *Food Policy* 88:101749. https://doi.org/10.1016/j.foodpol.2019.101749.

Cain, R.M. 2005. Federal do not call registry is here to stay: What's next for direct marketing regulation? *Journal of Interactive Marketing* 19:54–62. https://doi.org/10.1002/dir.20031.

Chen, C., and R. Chen. 2018. Using two government food waste recognition programs to understand current reducing food loss and waste activities in the U.S. *Sustainability* 10:2760. https://doi.org/10.3390/su10082760.

Christensen, T.H., A. Damgaard, J. Levis, Y. Zhao, A. Björklund, U. Arena, M.A. Barlaz, V. Starostina, A. Boldrin, T.F. Astrup, and V. Bisinella. 2020. Application of LCA modelling in integrated waste management. *Waste Management* 118:313–322. https://doi.org/10.1016/j.wasman.2020.08.034.

Clavreul, J., H. Baumeister, T.H. Christensen, and A. Damgaard. 2014. An environmental assessment system for environmental technologies. *Environmental Modelling & Software* 60:18–30. https://doi.org/10.1016/j.envsoft.2014.06.007.

De Boeck, E., L. Jacxsens, H. Goubert, and M. Uyttendaele. 2017. Ensuring food safety in food donations: Case study of the Belgian donation/acceptation chain. *Food Research International* 100:137–149. https://doi.org/10.1016/j.foodres.2017.08.046.

de Oliveira, F.D.G., N.M. Robey, T.J. Smallwood, C.J. Spreadbury, and T.G. Townsend. 2022. Landfill gas as a source of anthropogenic antimony and arsenic release. *Chemosphere* 307:135739. https://doi.org/10.1016/j.chemosphere.2022.135739.

Deniz, R.F., C. Xiaoju, and N.H. Orak. 2023. Investigation of greenhouse gas emissions from municipal solid waste management practices using an economic input-output life cycle analysis approach. *Journal of Cleaner Production* 420:138450. https://doi.org/10.1016/j.jclepro.2023.138450.

Diaz-Ruiz, R., M. Costa-Font, F. López-i-Gelats, and J.M. Gil. 2019. Food waste prevention along the food supply chain: A multi-actor approach to identify effective solutions. *Resources, Conservation and Recycling* 149:249–260. https://doi.org/10.1016/j.resconrec.2019.05.031.

DiStefano, T.D., and L.G. Belenky. 2009. Life-cycle analysis of energy and greenhouse gas emissions from anaerobic digestion of municipal solid waste. *Journal of Environmental Engineering* 135(11):1097–1105.

EPA (U.S. Environmental Protection Agency). 2009. Sustainable Materials Management: The Road Ahead. https://www.epa.gov/smm/sustainable-materials-management-road-ahead.

Ercan, T., and O. Tatari. 2015. A hybrid life cycle assessment of public transportation buses with alternative fuel options. *International Journal of Life Cycle Assessment* 20:1213–1231. https://doi.org/10.1007/s11367-015-0927-2.

Erkisi-Arici, S., J. Hagen, F. Cerdas, and C. Herrmann. 2021. Comparative LCA of municipal solid waste collection and sorting schemes considering regional variability. *Procedia CIRP* 98:235–240. https://doi.org/10.1016/j.procir.2021.01.036.

Fullerton, D. 2025. The circular economy. In T. Lundgren, M. Bostian, and S. Managi (eds.), Encyclopedia of Energy. *Natural Resource, and Environmental Economics* 2e(3):254–265. Elsevier. https://doi.org/10.1016/B978-0-323-91013-2.00050-2.

Galavote, T., G. de Lorena Diniz Chaves, L.H. Yamane, and R.R. Siman. 2024a. The effects of municipal waste reduction and recycling policies on the economic feasibility of landfill gas generation. *Energy for Sustainable Development* 81:101493. https://doi.org/10.1016/j.esd.2024.101493.

Galavote, T., L.H. Yamane, N.S. de S.L. Cano, G. de L.D. Chaves, and R.R. Siman. 2024b. Waste management policies and impact of landfill diversion goals in gas-to-energy recovery. *Waste and Resource Management* 177:122–138. https://doi.org/10.1680/jwarm.22.00011.

Gentil, E.C., A. Damgaard, M. Hauschild, G. Finnveden, O. Eriksson, S. Thorneloe, P.O. Kaplan, M. Barlaz, O. Muller, Y. Matsui, R. Ii, and T.H. Christensen. 2010. Models for waste life cycle assessment: Review of technical assumptions. *Waste Management* 30:2636–2648. https://doi.org/10.1016/j.wasman.2010.06.004.

Geyer, R., J.R. Jambeck, and K.L. Law. 2017. Production, use, and fate of all plastics ever made. *Science Advances* 3:e1700782. https://doi.org/10.1126/sciadv.1700782.

Guinée, J.B., R. Heijungs, G. Huppes, A. Zamagni, P. Masoni, R. Buonamici, T. Ekvall, and T. Rydberg. 2011. Life cycle assessment: past, present, and future. *Environmental Science & Technology* 45(1):90–96. https://doi.org/10.1021/es101316v.

Halloran, A., J. Clement, N. Kornum, C. Bucatariu, and J. Magid. 2014. Addressing food waste reduction in Denmark. *Food Policy* 49:294–301. https://doi.org/10.1016/j.foodpol.2014.09.005.

Hamilton, H.A., M.S. Peverill, D.B. Müller, and H. Brattebø. 2015. Assessment of food waste prevention and recycling strategies using a multilayer systems approach. *Environmental Science & Technology* 49(24):13937–13945. https://doi.org/10.1021/acs.est.5b03781.

Hannibal, B., and A. Vedlitz. 2018. Throwing it out: Introducing a nexus perspective in examining citizen perceptions of organizational food waste in the U.S. *Environmental Science & Policy* 88:63–71. https://doi.org/10.1016/j.envsci.2018.06.012.

Huang, L. 2024. Fintech inclusion in natural resource utilization, trade openness, resource productivity, recycling and minimizing waste generation: Does technology really drive economies toward green growth? *Resources Policy* 90:104855. https://doi.org/10.1016/j.resourpol.2024.104855.

ICF International. 2016. Documentation for Greenhouse Gas Emission and Energy Factors Used in the Waste Reduction Model (WARM): Management Practices Chapters.

Jain, M., A. Ashwani, and A. Kumar. 2023. Landfill mining: A review on material recovery and its utilization challenges. *Process Safety and Environmental Protection* 169:948–958. https://doi.org/10.1016/j.psep.2022.11.049.

Jambeck, J.R., R. Geyer, C. Wilcox, T.R. Siegler, M. Perryman, A. Andrady, R. Narayan, and K.L. Law. 2015. Plastic waste inputs from land into the ocean. *Science* 347:768–771. https://doi.org/10.1126/science.1260352.

Khandelwal, H., H. Dhar, A.K. Thalla, and S. Kumar. 2019. Application of life cycle assessment in municipal solid waste management: A worldwide critical review. *Journal of Cleaner Production* 209:630–654. https://doi.org/10.1016/j.jclepro.2018.10.233.

Kirkeby, J.T., H. Birgisdottir, T.L. Hansen, T.H. Christensen, G.S. Bhander, and M. Hauschild. 2006. Environmental assessment of solid waste systems and technologies: EASEWASTE. *Waste Management & Research* 24(1):3–15. https://doi.org/10.1177/0734242X06062580.

Klemeš, J.J., P. Stehlík, and E. Worrell. 2010. Waste treatment to improve recycling and minimise environmental impact, Selected papers from 11th Conference Process Integration, Modelling and Optimisation for Energy Saving and Pollution Reduction. *Resources, Conservation and Recycling* 54(5):267–270. https://doi.org/10.1016/j.resconrec.2009.11.005.

Laurent, A., I. Bakas, J. Clavreul, A. Bernstad, M. Niero, E. Gentil, M.Z. Hauschild, and T.H. Christensen. 2014a. Review of LCA studies of solid waste management systems—Part I: Lessons learned and perspectives. *Waste Management* 34:573–588. https://doi.org/10.1016/j.wasman.2013.10.045.

Laurent, A., J. Clavreul, A. Bernstad, I. Bakas, M. Niero, E. Gentil, T.H. Christensen, and M.Z. Hauschild. 2014b. Review of LCA studies of solid waste management systems—Part II: Methodological guidance for a better practice. *Waste Management* 34:589–606. https://doi.org/10.1016/j.wasman.2013.12.004.

Lebreton, L., and A. Andrady. 2019. Future scenarios of global plastic waste generation and disposal. *Palgrave Communications* 5. https://doi.org/10.1057/s41599-018-0212-7.

Liebig, G., and J. Rommel. 2014. Active and forced choice for overcoming status quo bias: A field experiment on the adoption of "no junk mail" stickers in Berlin, Germany. *Journal of Consumer Policy* 37:423–435. https://doi.org/10.1007/s10603-014-9264-2.

Lott, D.J., S.J. Laux, and T.G. Townsend 2024. Analysis of ammonia-nitrogen removal kinetics by stage in pilot scale vertical flow wetlands treating landfill leachate in series. *Chemosphere* 360:142409. https://doi.org/10.1016/j.chemosphere.2024.142409.

Ma, J., Y. Gu, L. Liu, Y. Zhang, M. Wei, A. Jiang, X. Liu, and C. He 2023. Study on the effect of landfill gas on aerobic municipal solid waste degradation: Lab-scale model and tests. *Science of the Total Environnent* 869:161875. https://doi.org/10.1016/j.scitotenv.2023.161875.

Meijer, L.J.J., T. van Emmerik, R. van der Ent, C. Schmidt, and L. Lebreton. 2021. More than 1000 rivers account for 80% of global riverine plastic emissions into the ocean. *Science Advances* 7:eaaz5803. https://doi.org/10.1126/sciadv.aaz5803.

Mourad, M. 2015. France moves toward a national policy against food waste (No. R-15-08-b). Natural Resources Defense Council.

Mueller, W. 2013. The effectiveness of recycling policy options: Waste diversion or just diversions? *Waste Management, Special Thematic Issue: Urban Mining* 33:508–518. https://doi.org/10.1016/j.wasman.2012.12.007.

Nakamura, S. 2023. A practical guide to industrial ecology by input-output analysis: Matrix-based calculus of sustainability. Springer.

Nomura, A., 2020. The shift of food value through food banks: a case study in Kyoto, Japan. *Evolutionary and Institutional Economics Review* 17:243–264. https://doi.org/10.1007/s40844-019-00154-0.

Papargyropoulou, E., R. Lozano, J.K. Steinberger, N. Wright, and Z.-bin Ujang. 2014. The food waste hierarchy as a framework for the management of food surplus and food waste. *Journal of Cleaner Production* 76:106–115. https://doi.org/10.1016/j.jclepro.2014.04.020.

Plastic Pollution Coalition. 2019. 157,000 Shipping Containers of U.S. Plastic Waste Exported to Countries with Poor Waste Management in 2018. Plastic Pollution Coalition.

Pryshlakivsky, J., and C. Searcy. 2013. Fifteen years of ISO 14040: A review. *Journal of Cleaner Production* 57:115–123. https://doi.org/10.1016/j.jclepro.2013.05.038.

Psyrri, G., M.Z. Hauschild, T.F. Astrup, and A.T.M Lima. 2024. Recycling for a sustainable future: Advancing resource efficiency through life cycle assessment resource indicators. *Resources, Conservation and Recycling* 209:107759. https://doi.org/10.1016/j.resconrec.2024.107759.

Reap, J., F. Roman, S. Duncan, and B. Bras. 2008a. A survey of unresolved problems in life cycle assessment Part 1. *International Journal of Life Cycle Assessment* 13:290. https://doi.org/10.1007/s11367-008-0008-x.

Reap, J., F. Roman, S. Duncan, and B. Bras. 2008b. A survey of unresolved problems in life cycle assessment Part 2. *International Journal of Life Cycle Assessment* 13:374. https://doi.org/10.1007/s11367-008-0009-9.

Redlingshöfer, B., S. Barles, and H. Weisz. 2020. Are waste hierarchies effective in reducing environmental impacts from food waste? A systematic review for OECD countries. *Resources, Conservation and Recycling* 156:104723. https://doi.org/10.1016/j.resconrec.2020.104723.

Reinhart, D., R. Joslyn, and C.T. Emrich. 2020. Characterization of Florida, U.S. landfills with elevated temperatures. *Waste Management* 118:55–61. https://doi.org/10.1016/j.wasman.2020.08.031.

Reinhart, D.R., N.D. Berge, S. Santra, and S.C. Bolyard. 2010. Emerging contaminants: Nanomaterial fate in landfills. *Waste Management, Special Thematic Section: Sanitary Landfilling* 30:2020–2021. https://doi.org/10.1016/j.wasman.2010.08.004.

Resse, A. 2005. "Stop Pub": Can banning of junk mail reduce waste production? *Waste Management & Research* 23(1):87–91. https://doi.org/10.1177/0734242X05051523.

Rigamonti, L., A. Falbo, L. Zampori, and S. Sala. 2017. Supporting a transition towards sustainable circular economy: sensitivity analysis for the interpretation of LCA for the recovery of electric and electronic waste. *International Journal of Life Cycle Assessment* 22:1278–1287. https://doi.org/10.1007/s11367-016-1231-5.

Robey, N.M., Y. Liu, M. Crespo-Medina, J.A. Bowden, H.M. Solo-Gabriele, T.G. Townsend, and T.M Tolaymat. 2024. Characterization of per- and polyfluoroalkyl substances (PFAS) and other constituents in MSW landfill leachate from Puerto Rico. *Chemosphere* 358:142141. https://doi.org/10.1016/j.chemosphere.2024.142141.

Ruan, W., and B. Zou. 2024. Optimal strategies of critical mineral depletion and recycling. *Resources, Conservation and Recycling* 209:107793. https://doi.org/10.1016/j.resconrec.2024.107793.

Sahoo, K., R. Bergman, S. Alanya-Rosenbaum, H. Gu, and S. Liang. 2019. Life cycle assessment of forest-based products: A review. *Sustainability* 11:4722. https://doi.org/10.3390/su11174722.

Sakaguchi, L., N. Pak, and M.D. Potts. 2018. Tackling the issue of food waste in restaurants: Options for measurement method, reduction and behavioral change. *Journal of Cleaner Production* 180:430–436. https://doi.org/10.1016/j.jclepro.2017.12.136.

Schmidt, C., T. Krauth, and S. Wagner. 2017. Export of plastic debris by rivers into the sea. *Environnemental Science & Technology* 51:12246–12253. https://doi.org/10.1021/acs.est.7b02368.

Schneider, F. 2013. The evolution of food donation with respect to waste prevention. *Waste Management, Special Thematic Issue: Urban Mining* 33:755–763. https://doi.org/10.1016/j.wasman.2012.10.025.

Schrijvers, D.L., P. Loubet, and G. Sonnemann. 2016. Critical review of guidelines against a systematic framework with regard to consistency on allocation procedures for recycling in LCA. *International Journal of Life Cycle Assessment* 21:994–1008. https://doi.org/10.1007/s11367-016-1069-x.

Simon, F. 2016. Consumer adoption of No Junk Mail stickers: An extended planned behavior model assessing the respective role of store flyer attachment and perceived intrusiveness. *Journal of Retailing and Consumer Services* 29:12–21. https://doi.org/10.1016/j.jretconser.2015.11.003.

Suknark, P., S. Buddhawong, and K. Wangyao. 2023. Investigating the effect of waste age and soil covering on waste characteristics prior to landfill mining using an electrical resistivity tomography technique. *Journal of Environmental Management* 339:117898. https://doi.org/10.1016/j.jenvman.2023.117898.

Tevapitak, K., and A.H.J (Bert) Helmsing. 2019. The interaction between local governments and stakeholders in environmental management: The case of water pollution by SMEs in Thailand. *Journal of Environmental Management* 247:840–848. https://doi.org/10.1016/j.jenvman.2019.06.097.

Thyberg, K.L., and D.J. Tonjes. 2016. Drivers of food waste and their implications for sustainable policy development. *Resources, Conservation and Recycling* 106:110–123. https://doi.org/10.1016/j.resconrec.2015.11.016.

Townsend, T.G., and M. Anshassi. 2023. Sustainable construction materials management. In T.G. Townsend and M. Anshassi (eds.), *Construction and demolition debris*. Cham: Springer International Publishing, pp. 389–421. https://doi.org/10.1007/978-3-031-25013-2_11.

Wambuguh, O. 2011. Junk mail in residential homes in the United States: Insights from a sub-urban home in California. *Resources, Conservation and Recycling* 55:782–784. https://doi.org/10.1016/j.resconrec.2011.03.011.

Yadav, P., and S.R. Samadder. 2018. A critical review of the life cycle assessment studies on solid waste management in Asian countries. *Journal of Cleaner Production* 185:492–515. https://doi.org/10.1016/j.jclepro.2018.02.298.

Yang, Y., H. Zhang, L. Wu, and M. Wang. 2024. Supply potential, carbon emission reduction, energy conservation, and sustainable pathways for aluminum recycling in China. *Sustainable Production and Consumption* 50:239–252. https://doi.org/10.1016/j.spc.2024.07.034.

Zhao, Y., M. Chen, S. Wu, Y. Zhang, D. Chen, and J. Zhang. 2024. Recycling of waste edible oil derivatives as phase change materials for building energy conservation. *Journal of Cleaner Production* 472:143451. https://doi.org/10.1016/j.jclepro.2024.143451.

Zhi, Y., S. Ma, J. Qin, Z. Zhao, and C. Zhou. 2023. Assessing the city-level material stocks in landfills and the landfill mining potential of China. *Environmental Research* 236:116737. https://doi.org/10.1016/j.envres.2023.116737.

Appendix A
Committee Member Biographical Sketches

Don Fullerton is visiting professor at the University of California (UC), Santa Barbara (2023–present). From 2008 to 2023, he was Gutgsell Professor of Finance at the University of Illinois Urbana-Champaign, having previously taught at Princeton, Virginia, Carnegie Mellon, and Texas. Dr. Fullerton's research analyzes distributional and efficiency effects of environmental and tax policies. After completing research on taxation, he switched his focus to environmental economics and policy, where he analyzed household behavior in response to a price per bag of garbage by measuring the reduction of curbside garbage and the increase in curbside recycling. Dr. Fullerton's current research is on the circular economy, including policies to reduce negative effects of extraction, production, and disposal. He studies economic effects of climate policy, energy efficiency mandates, renewable portfolio standards, cap-and-trade permits for emissions, and carbon taxes, including the distribution of the burdens of policy as well as effects on the environment, on cost efficiency, and on overall social welfare. From 1985 to 1987, he was Deputy Assistant Secretary for Tax Analysis at the Treasury Department, and in 2014, he was a lead author of the United Nations. Intergovernmental Panel on Climate Change (IPCC Fifth Assessment Report). Dr. Fullerton received a B.A. from Cornell, and a Ph.D. in economics from UC Berkeley.

Debra Reinhart is a Pegasus Professor Emerita at the University of Central Florida. Prior to her retirement in June 2021, she was Associate Vice President for Research and Scholarship at the University of Central Florida and a member of the Civil, Environmental and Construction Engineering Department. Dr. Reinhart's research area is solid waste management, with a focus on optimized waste collection, recycling, and processing as well as sustainable operation of landfills. From 2011 to 2013, she was a program manager at the National Science Foundation. Dr. Reinhart is an associate editor for the journal Waste Management and is a registered professional engineer in Florida and Georgia, a Board-Certified Environmental Engineer, and a Fellow of the American Society of Civil Engineers, the American Association for the Advancement of Science, and the Association of Environmental Engineering and Science Professors. She holds a B.S. in engineering from Florida Technological University and an M.S. in sanitary engineering and a Ph.D. in environmental engineering from the Georgia Institute of Technology.

Malak Anshassi is an assistant professor at Florida Polytechnic University teaching solid waste management, sustainability, and life cycle assessment courses. Her research focuses on incorporating life cycle thinking into solid waste management. Dr. Anshassi previously conducted research using principles from sustainable materials management (SMM) to analyze the application of life cycle thinking into Florida's solid waste management system to achieve the 75 percent recycling rate target. In her current research she formulates SMM-based solid waste management and policy approaches that decision makers from any region of the world can use to measure their waste management system's environmental and economic impacts. She received her doctorates degree from the University of Florida in 2020 and is an active member of Solid Waste Association of North America, Air and Waste Management Association, and Recycle Florida Today.

Fatima Hafsa is a circular economy specialist and materials engineer. Her research focuses on sustainable supply chains, recycling economics, and voluntary governance. Dr. Hafsa's work experience spans private businesses (Mitsubishi Chemicals, Engro Foundation), non-profits (World Wide Fund for Nature-Pakistan), and government agencies in the United States, Pakistan, and West Africa. She is current working as a circular economy specialist at the World Bank, where she is assisting Global South governments in implementing circular economy at a national level, and she has extensive experience in recycling systems within the United States. As a postdoctoral scholar at Arizona State University, Dr. Hafsa assessed the efficacy of U.S. recycling systems. She mapped local recycling supply chains, assessed waste trade, identified recycling innovations, and identified ways in which U.S. local governments can use sustainable purchasing to improve recycling systems. Dr. Hafsa has a master's in materials engineering from Georgia Institute of Technology, and she completed her Ph. D. in sustainability at Arizona State University in 2022. She is also a Fulbright scholar from Pakistan.

Tracy Horst is the Environmental Compliance and Recycling Director for the Choctaw Nation of Oklahoma and has been with the tribe since 2007. In her current position, she has assisted with creating and maintaining a compliance program to ensure the tribe is following environmental regulations. Horst has helped to establish the tribe's recycling program and set up and manages two recycling centers. She assists rural communities in southeastern Oklahoma in learning and understanding more about what recycling is, how to start recycling, and options for recycling. Horst currently serves on the Tribal Waste and Response Steering Committee for the Institute for Tribal Environmental Professionals and is active with the Oklahoma Recycling Association including serving on the Board of Directors, as vice president, and president. She also serves on the Board of Directors for the Durant Sustainability Coalition and the planning committee for the Texoma Earth Day Festival. Horst has degrees in biology and chemistry from Southwestern College in Winfield, Kansas.

Derek Kellenberg is a professor in the Economics Department at the University of Montana. He specializes in the fields of international and environmental economics, with particular interests in international environmental policy, international pollution haven effects, and the determinants of hazardous waste trade across countries. Dr. Kellenberg has published articles in the *Journal of the Association of Environmental and Resource Economics*, *Journal of Urban Economics*, and *Journal of Environmental Economics and Management*, among others, and he currently serves on the editorial board for the *Journal of the Association of Environmental and Resource Economists* and the *International Journal of Sustainable Society*. He was the keynote speaker at the Symposium on International Waste Trade: Economic Research and Policy Implications in Brussels, Belgium, in 2015 and authored a chapter in the *Annual Review of Resource Economics* on the economics of international waste trade. Dr. Kellenberg received his Ph.D. in economics from the University of Colorado.

Jeremy O'Brien is a solid waste management researcher and consultant who has more than 45 years of experience in the industry. From 1979 to 1989, O'Brien served as a program manager for Public Technology, Inc., a local government research organization with 160 member governments, which served as the technical arm of the National League of Cities and the International City Management Association. From 1989 to 1998, he was a senior professional associate and project manager for HDR Engineering, Inc., a national civil/environmental engineering firm with 3,300 employees in 90 offices throughout the United States. O'Brien was HDR's Recycling Technical Director in 1991–1992. Since 1999, he served as the director of applied research for the Solid Waste Association of North America (SWANA), is a professional association of over 10,000 solid waste and recycling practitioners in the United States and Canada. During his tenure, O'Brien established the SWANA Applied Research Foundation that has developed over 60 reports in the areas of solid waste collection, sustainable materials management, waste conversion and energy recovery and landfill disposal. While he retired from SWANA in January 2025, he continues to work for the association on a part time basis. O'Brien received two degrees from Duke University: a B.A. in religion in 1974 and an M.S. in urban and environmental engineering in 1978.

Susan Robinson retired from Waste Management (WM) in 2021 after 23 years of service. She has 40 years of experience in the waste/recycling (W/R) industry and has worked in the public sector, as well as for a private recycling company. Robinson currently consults for trade organizations, consultancies, and corporate clients.

Appendix A

At WM, she managed municipal contracts and implemented several large W/R programs before moving to a corporate role leading WM's policy team and ultimately WM's sustainability team. Robinson served as a subject matter expert on recycling markets and operations, and was a frequent speaker on waste, recycling, and sustainability issues. She worked on an internal team that developed an analysis of the cost and emissions impact for the entire U.S. solid waste stream, presenting the results publicly upon completion. Robinson published a monthly sustainability column for Waste 360, as well as articles for Air and Waste Management Magazine.

She is currently an appointed member of the U.S. Environmental Protection Agency's National Advisory Council. She has served on multiple industry-related boards and on the Washington State Governor's Beyond Waste Task Force. She was a five-time recipient of the WM's Circle of Excellence Award. She attended Stanford University before receiving her B.S. from the University of Washington.

Hilary Sigman is a professor in the Department of Economics at Rutgers University and a research associate of the National Bureau of Economic Research. She conducts empirical research on the effects of environmental policy and has focused particularly on the economics of waste and contaminated land. Dr. Sigman's research has been published in journals such as the *American Economic Review*, *RAND Journal of Economics*, *Journal of the Association of Environmental and Resource Economists*, and *Journal of Environmental Economics and Management*, and she has been funded by grants from the National Science Foundation, the U.S. Environmental Protection Agency (EPA), and the World Bank. She has served on EPA Science Advisory Board committees and on the Board of Directors and other committees of the Association of Environmental and Resource Economists. Dr. Sigman holds a B.A. in economics and environmental studies from Yale, an M.Phil. in economics from Cambridge University, and a Ph.D. in economics from the Massachusetts Institute of Technology.

Mitchell Small is a professor emeritus in civil and environmental engineering and engineering and public policy at Carnegie Mellon University. His research involves mathematical modeling of environmental systems, environmental statistics, and risk assessment, and his current projects include models for leak detection at geologic CO_2 sequestration sites; the value of scientific information for conflict resolution among polarized stakeholders; and multi-objective design of solid waste management systems. Dr. Small has published over 200 manuscripts in peer reviewed journals, books and conference proceedings.

He has participated in a number of Environmental Protection Agency Science Advisory Board (SAB) panels and U.S. National Research Council (NRC) committees, serving as chair of the 2012–2014 NRC Committee on Risk Management and Governance Issues in Shale Gas Extraction. Dr. Small was elected fellow of the Society for Risk Analysis (SRA) in 2003, received Distinguished Educator Awards from the SRA in 2013 and the Association of Environmental Engineering and Science Professors (AEESP) in 2017, and served as an associate editor for the journal Environmental Science & Technology between 1995 and 2011, where he helped to introduce new areas of focus for the journal, including life-cycle analysis, industrial ecology, and environmental policy analysis.

Rebecca Taylor is an assistant professor at the University of Illinois Urbana-Champaign in the Department of Agricultural & Consumer Economics. She is a leading expert on the economics of regulating waste from consumer items, and particularly the regulation of plastic carryout bags. Dr. Taylor analyzes the effectiveness of environmental and food policies in changing consumer behavior, especially when there is debate over optimal policy design. She also studies how these policies interact with issues of equity and how these policies displace consumption in unintended ways, such as consumers buying more garbage bags when plastic carryout bags are banned. Dr. Taylor's research has gained international attention and has been covered by news outlets in the United States, Australia, New Zealand, Canada, Mexico, and Indonesia. Dr. Taylor's papers have also influenced policy recommendations, such as those given by the Unites Nation's Environmental Programme "Legislative Guide for the Regulation of Single-Use Plastic Products," and have won distinguished award, such as the 2020 Best Paper Prize for the Journal of Environmental Economics & Management. Dr. Taylor received her Ph.D. in agricultural and resource economics at the University of California, Berkeley, and her B.A. in economics at Washington & Lee University.

Sofia Berto Villas-Boas is a professor in the Department of Agricultural and Resource Economics at University of California (UC), Berkeley. She holds the Class of 1934 Robert Gordon Sproul Chair in Agricultural Economics. She has published in top economics and field journals such as *Review of Economic Studies*, *RAND Journal of Economics*, *American Journal of Agricultural Economics*, *Journal of Environmental Economics and Management*, *Marketing Science*, *Management Science*, and *Review of Economics and Statistics*. Dr. Berto Villas-Boa's research interests include industrial organization, consumer behavior, food policy, and environmental regulation. Her recent empirical work estimates the effects of policies on consumer behavior, such as a bottled water tax, a plastic bag ban, and a soda tax campaign and its implementation. Other published work has focused on the economics behind wholesale price discrimination banning legislation, contractual relationships along a vertical supply chain, and identifying the role of those contracts in explaining the pass- through of cost shocks along the supply chain into retail prices that consumers face. Dr. Berto Villas-Boas received her Ph.D. in Economics from UC Berkeley in May 2002, and she has an undergraduate degree in economics from the Universidade Católica Portuguesa.

Appendix B
Disclosure of Unavoidable Conflict of Interest

The conflict-of-interest policy of the National Academies of Sciences, Engineering, and Medicine (https://www.nationalacademies.org/about/institutional-policies-and-procedures/conflict-of-interest-policies-and-procedures) prohibits the appointment of an individual to a committee like the one that authored this Consensus Study Report if the individual has a conflict of interest that is relevant to the task to be performed. An exception to this prohibition is permitted only if the National Academies determine that the conflict is unavoidable and the conflict is promptly and publicly disclosed.

When the committee that authored this report was established a determination of whether there was a conflict of interest was made for each committee member given the individual's circumstances and the task being undertaken by the committee. A determination that an individual has a conflict of interest is not an assessment of that individual's actual behavior or character or ability to act objectively despite the conflicting interest.

Susan Robinson was determined to have an unavoidable financial conflict of interest because holds stock in Waste Management (WM) from her prior employment at WM, and she performs consulting work for recycling industry actors related to tracking the implementation of recycling policy and the setup of recycling program operations.

The National Academies determined that the experience and expertise of the individual was needed for the committee to accomplish the task for which it was established. The National Academies could not find another available individual with the equivalent experience and expertise who did not have a conflict of interest. Therefore, the National Academies concluded that the conflict was unavoidable and publicly disclosed it on its website (www.nationalacademies.org).

Appendix C
Public Meeting Agendas

**COMMITTEE ON COSTS AND APPROACHES FOR
MUNICAPAL SOLID WASTE RECYCLING PROGRAMS**

MARCH 13, 2024
Virtual
OPEN SESSION

4:00	**Welcome and Session Objectives** Don Fullerton and Debra Reinhart

NOTE: After remarks from the co-chairs, members introduce themselves: name, affiliation, main area of expertise.
If congressional staff are available, they will be asked to make brief remarks.

4:10   **Environmental Protection Agency's (EPA's) Perspectives on the Committee's Task**
EPA representatives

NOTE: EPA representatives will provide a presentation of about 30 minutes. They will be asked to discuss relevant work within the agency, how EPA plans to use the committee's report, and important sources of information for the committee.
After EPA's presentation, there will be a Q&A session with the committee.

5:00   End of open session

**COMMITTEE ON COSTS AND APPROACHES FOR
MUNICAPAL SOLID WASTE RECYCLING PROGRAMS**
Keck Center 500 Fifth Street, NW Washington, DC 20001

APRIL 16, 2024
Room 101
The committee held a 2-day hybrid meeting.

Open Session: Information Gathering Panels

9:30–10:25   **Panel 3: Data and Local Decision Support Tools for Assessing Internal Costs and Externalities of MSW Recycling**

Moderator: Debbie Reinhart, *University of Central Florida, study committee co-chair*
Panelist:
Susan Thorneloe, *U.S. EPA*

After brief remarks, the committee will engage in Q&A and discussion

Appendix C

10:25–10:40	Break
10:40–11:40	**Panel 4: Drivers and Challenges in Developing and Implementing MSW Recycling Programs**

    Moderator: Fatima Hafsa, *World Bank, study committee member*
    Panelists:
    Erin Murphy, *Ocean Conservancy & University of Toronto*
    David Allaway, *Oregon Department of Environmental Quality*
    Susan Fife-Ferris, *Seattle Public Utilities*

    After brief remarks from each, the committee will engage with the panelists in Q&A and discussion

5:00   End of Open Session

**COMMITTEE ON COSTS AND APPROACHES FOR
MUNICAPAL SOLID WASTE RECYCLING PROGRAMS**
Keck Center 500 Fifth Street, NW Washington, DC 20001

APRIL 17, 2024
Room 201

<u>*Open Session: Information Gathering Panels*</u>

9:30–10:25   **Panel 3: Data and Local Decision Support Tools for Assessing Internal Costs and Externalities of MSW Recycling**

    Moderator: Debbie Reinhart, *University of Central Florida, study committee co-chair*
    Panelist:
    Susan Thorneloe, *U.S. EPA*

    After brief remarks, the committee will engage in Q&A and discussion

10:25–10:40   Break

10:40–11:40   **Panel 4: Drivers and Challenges in Developing and Implementing MSW Recycling Programs**

    Moderator: Fatima Hafsa, *World Bank, study committee member*
    Panelists:
    Erin Murphy, *Ocean Conservancy & University of Toronto*
    David Allaway, *Oregon Department of Environmental Quality*
    Susan Fife-Ferris, *Seattle Public Utilities*

    After brief remarks from each, the committee will engage with the panelists in Q&A and discussion

5:00   End of Open Session

COMMITTEE ON COSTS AND APPROACHES FOR MUNICAPAL SOLID WASTE RECYCLING PROGRAMS

JUNE 11, 2024
Virtually

Open Session: Livestreamed for Public Viewing

1:10–2:15 **Financing and End Markets of MSW Recycling**
Moderator: Susan Robinson
Panelists:
Judy Sheahan, U.S. Conference of Mayors
Joe Pickard, Recycled Materials Association (ReMA)
Bryan Staley, Environmental Research and Education Foundation

NOTE: Each panelist will provide a 10-minute presentation followed by a 5-min individual Q&A, then followed by a roundtable panel discussion and Q&A.

2:15–2:30 Break

2:30–3:20 **Technologies and Design for the MRF: Cost, Efficiency, and Other Implications**
Moderator: Debbie Reinhart
Panelists:
Nathiel (Nat) Egosi, RRT
Jim Frey, Resource Recycling Systems (RRS)

NOTE: Each panelist will provide a 10-minute presentation followed by a 5-min individual Q&A, then followed by a roundtable panel discussion and Q&A.

3:20–3:30 Break

End of Open Session

COMMITTEE ON COSTS AND APPROACHES FOR MUNICAPAL SOLID WASTE RECYCLING PROGRAM

JULY 23, 2024
Virtually

Open Session: Livestreamed for Public Viewing

4:10–4:55 **Social and Behavioral Considerations for Recycling Participation**
Moderator: Malak Anshassi

Kip Viscusi, Vanderbilt University
Bevin Ashenmiller, Occidental College

NOTE: Each panelist will provide a 10-minute presentation followed by a 5-min individual Q&A, then followed by a roundtable panel discussion and Q&A.

4:55–5:05 End of Open Session